Biological Insulating Liquids

Ernst Peter Pagger · Norasage Pattanadech ·
Frank Uhlig · Michael Muhr

Biological Insulating Liquids

New Insulating Liquids for High Voltage Engineering

Springer

Ernst Peter Pagger
EPP Consulting GmbH
Vienna, Austria

Frank Uhlig
Graz University of Technology
Graz, Austria

Norasage Pattanadech
Electrical Engineering Department
School of Engineering
King Mongkut's Institute of Technology
Ladkrabang
Bangkok, Thailand

Michael Muhr
Graz University of Technology
Graz, Austria

ISBN 978-3-031-22459-1 ISBN 978-3-031-22460-7 (eBook)
https://doi.org/10.1007/978-3-031-22460-7

This Springer imprint is published by the registered company Springer Nature Switzerland AG
The registered company address is: Gewerbestrasse 11, 6330 Cham, Switzerland

Preface

Mineral oil, silicon oil, synthetic ester and natural ester—what do they have in common? All these liquids can be and are used as insulating liquids in high-voltage engineering. In some points, their behavior is very different. The book should provide the reader above all with information about *Biological Insulating Liquids* used in transformer cooling compared to other insulating liquids. Transformers are critical and expensive pieces of equipment in the power generation and distribution network. Any failure could have drastic consequences to an entire electrical network.

Environmental friendliness, risk reduction and, above all, sustainability are key words that are currently shaping the transformation in the energy industry. Of particular importance are the used materials regarding to the electrical equipment. Driven by climate change and global warming, the industry has to look in all fields of their activities to find sustainable solutions and reducing the use of fossil fuels to a minimum. Biological, renewable insulating liquids have a number of benefits combined with a very small carbon footprint. Some of them already used in the high-voltage engineering and other great candidates for replacing mineral oil are in the pipeline.

The first liquid filled transformers were not filled with mineral; they were filled with vegetable oil. But because of the bad oxidation stability of the pure vegetable oils, the missing of sealed devices and antioxidants, manufactures switched to mineral oil. Mineral oil is already used for approximately 100 years as insulating liquid and is still the liquid with the widest utilization. In the 1970s, mixtures of polychlorinated biphenyls (PCBs) were because of their non-flammable nature and their chemical stability the highlight on the market of transformer liquids. Polychlorinated biphenyls were widely used for about fifty years under a variety of trade names, the most common of which were Askarel®. As chemically stable, polychlorinated biphenyls would only slowly biodegrade. That is that they tended to persist in nature as opposed to decomposing into basic elements. In numerous health studies, their negative effects on both humans and wildlife were documented. During the 1980s and 1990s, a great deal of time, money and effort was necessary to retro-fill transformers with more acceptable liquids.

In the late 1970s, General Electric went to market with a transformer design called 'Vaportran'. Vaportran used a dielectric coolant called R-113, also known as 'Freon'. Freon was a very effective cooling insulating liquid. In the transformer tank, it was in liquid state, and then, the heat from the windings vaporizes the liquid, and finally, the vapor is condensed in the heat exchanger again. After getting knowledge of the problems with polychlorinated biphenyls, this was an effective replacement for PCB due to its performance characteristics and non-flammable nature. With global concerns about damage to the ozone layer, Freon is being phased out worldwide. Nearly, the same happened to hexafluoride (SF_6). Nowadays, widely used in switchgears, circuit-breakers and transformers, it will be panned by the European Commission as its global warming potential is approximately 23,000 times higher compared to carbon dioxide. These are examples where enthusiasm had to give way to reality once again.

Silicone oil was for several decades the fluid of choice when less flammable dielectric liquid was desired. It has a relatively high fire point and is generally considered to self-extinguish when the source of a fire is removed. But the liquid is not miscible with conventional mineral oil, not biodegradable and very expensive.

These examples show that you must be very careful when switching to new insulating liquids and to have consider all side effects; otherwise it is easy to come out of the frying pan into the fire.

Because of sometimes large amounts of insulating oil and direct contact with high-voltage elements, power transformers filled with mineral oil are one of the most dangerous devices. Fires in transformers filled with mineral oil, in some cases with many dead people, are evidence of this. Synthetic—and natural ester filled transformers are due to the higher fire point of the liquid much saver. Synthetic—and natural esters are also environmentally friendly; hence, they are biodegradable. The basis of the natural esters is of vegetable origin (wheat, soy, sunflower, castor, palm oil, peanut oil, rapeseed, etc.). This also opens the discussion about the use of food for industrial purposes. We must take this into account in all our deliberations very carefully. One way is to use non-edible vegetable oils. An assessment and estimation of the effects on the ecology and how this can be minimized is necessary. Otherwise, the same thing happens that has been shown with some of the insulating liquids listed above; they will disappear from the market for a short or long time. Regarding the entire route from the cradle to the grave and weighing up all the pros and cons, the *Biological Insulating Liquids* show a sensible alternative to the other insulating liquids—above all to mineral oil liquids.

The different properties of the different insulating liquids also influence the design of the devices and must be taken into account accordingly.

Vienna, Austria — Ernst Peter Pagger
Bangkok, Thailand — Norasage Pattanadech
Graz, Austria — Frank Uhlig
Graz, Austria — Michael Muhr

Contents

Chapter 1
Introduction

As infrastructure comes due for replacement and nearly everything, we own becomes electrified, the aging of equipment and additional electrical load demand means the need for more sustainable transformers in the future. The change in global energy politics, motivated by multiple reasons (above all because of climate change), has driven the electrical power industry not only to devising more efficient solutions but also to an increased use of energy resources. The Global Status Report shows that, above all, renewable forms of energy have a significant increase in recent years (Figs. 1.1, 1.2, [1]).

Energy production today has become more decentralized, with renewable energy being generated regionally, away from the main consumers. The decentralized power generation meant that electrical networks needed to be reconfigured.

Electricity is supposed to be a fundamental need of modern society and is usually the backbone for economic development and the welfare of society. The electrical power system grid connects the power plants through transmission- and distribution lines to the end user (Fig. 1.3). The main goal is providing electrical power to the consumers in ready-to-use form. The network is mainly responsible for integrating cities and appreciates social activities combining with economic, public, and environmental systems with population and urban development. In addition, the power system also contributes to socioeconomic growth and improves the standard of living through developing inter-to-intra city grids during expansion. The demand for development of a prospective low carbon grid has increased enormous stresses on the stability and efficiency of dielectric materials used in power systems to meet impulsive and vivid working conditions. The future electrical power system must possess the ability to cope with the prompt progression of power load and its asymmetric distribution demand for huge capacity, extended distance, and low-loss power transmission. This kind of network interconnects the grids in various countries and even different continents to confirm a secure and consistent supply of energy write Rafiq et al. [2]. Transformers are essential parts in the power system for voltage level conversion and maintaining the power flow.

E. P. Pagger et al., *Biological Insulating Liquids*,
https://doi.org/10.1007/978-3-031-22460-7_1

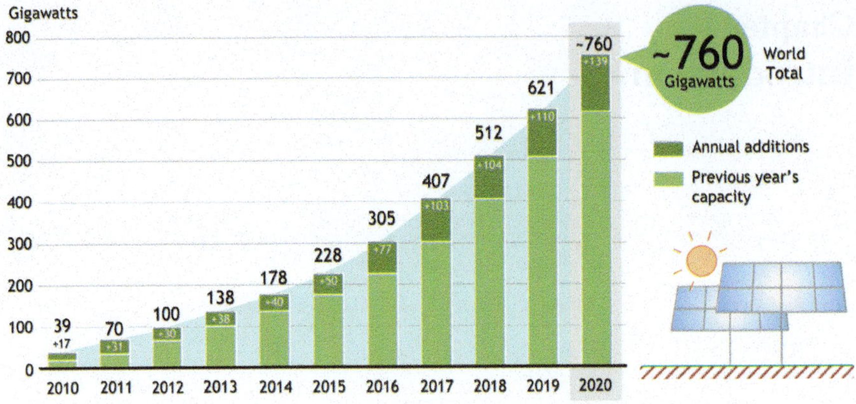

Fig. 1.1 Solar PV global capacity and annual additions, 2010–2020 [1]

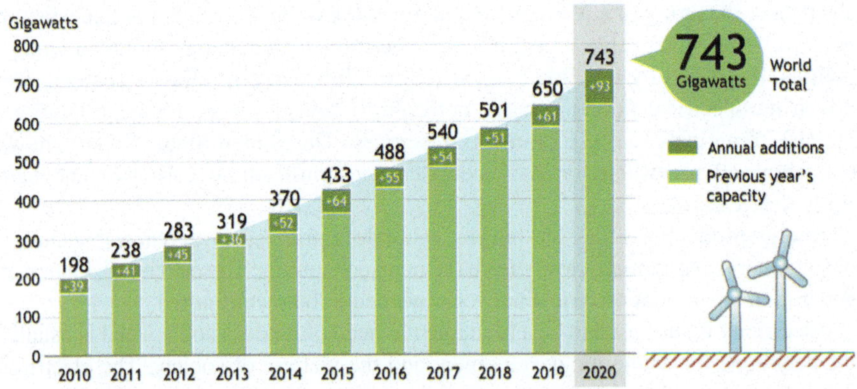

Fig. 1.2 Wind power global capacity and annual additions, 2010–2020 [1]

Transformers convert electrical energy from one voltage level to another. They are an important factor of the electricity network and therefore special attention regarding condition and risk assessment based on available requirements is necessary. "If the transformer is the heart of the electric power system, the insulating liquid is the blood, and it can tell you an incredible amount of information about the transformer's condition", writes Jason Dennison [3]. Transformers are applied in five major regions.

- At power plants, where power is generated and raised to transmission. After generating in power stations or at renewable sources such as windmills and solar panels, electricity energy needs find is way to the consumer. This transport is more efficient at higher voltage, which is why power industrially generated at 10 kV to 30 kV is converted by transformers into typical voltages of 220 kV up to 400 kV, or in some countries with long distances even higher.
- At switching stations, where the transmission voltage is changed.

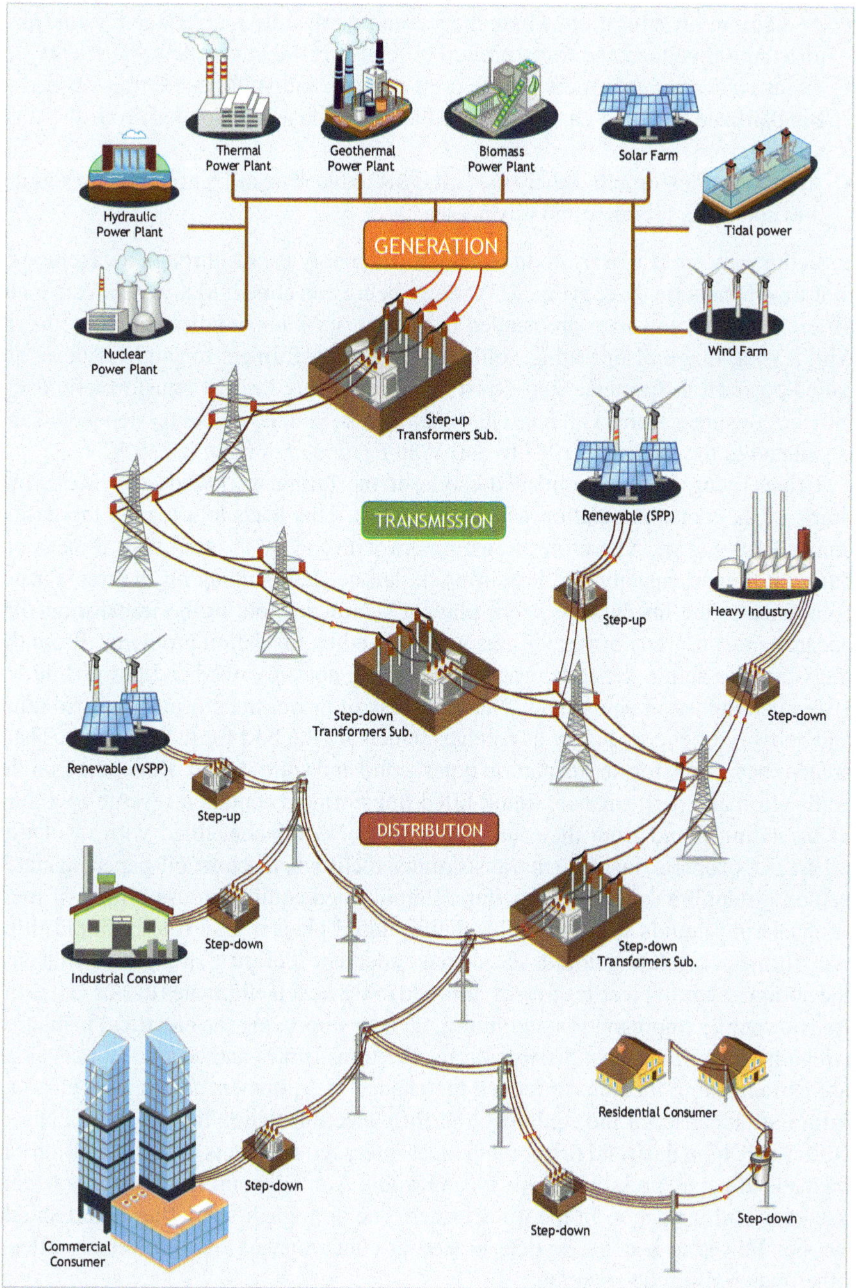

Fig. 1.3 The power system grid

- At distribution substation, where the incoming transmission voltage is reduced to distribution voltage and transform up excessive residential renewable energy.
- At instrument transformers, where high currents and voltages are transformed to standardized low and easily measurable values that are isolated from the high voltage.
- At service transformers, where the voltage is reduced to utilization level for routing into consumers' homes and businesses.

In this way, industrially produced electrical energy passes through an average of four transformation stages (Fig. 1.3) before being consumed. Many transformers of different classes and sizes are needed in the transmission and distribution network, with a wide range of operating voltages. Large transformers for high voltages are called power transformers. Step-down transformers are the last transformation step into the consumer mains voltage. This single-phase voltage is country dependent and is situated in the range of 110 V to 240 V. In Europe, this value is 230 V.

Urban living is the dominant lifestyle of the future worldwide. By 2050, two thirds of the world population will live in cities. This leads to a locally increasing demand for energy. Therefore, transformers with the same or smaller dimension, but with higher capacity and less fire risk, are needed. Among all the transformer components, the insulation system plays a significant role in the transformer life because most of transformer failures were caused by insulation problems. From the transformer's action depends proper functioning not only overhead power line but also much industrial equipment. The main task of power transformer is distribution and delivery of energy power at voltages which are the best for the economy. There are two basic transformer insulation types, solid and solid–liquid. Depending on the application and performance, liquid filled transformers contain a seventh to a third of insulating liquid from the total weight. Other equipment filled with insulating oil are, for example, instrument transformers, rectifiers, reactors, oil-paper-insulated cables, circuit breakers and capacitors. Liquid filled equipment use billion of liters of insulating liquids worldwide. Insulating liquid plays a vital role in liquid filled transformers. Insulating liquids should have adequate dielectric strength to withstand the normal range of electric stresses imposed in service. It eliminates the air gap in the transformer by impregnation and filling, thereby improving the electrical insulation strength and helping heat dissipation for the transformer and second, it acts as an arc extinguishing medium to reduce the chance of equipment failure as well as an information carrier of the equipment performance. Insulating liquids should have a sufficiently high flash and fire point to meet safety requirements. Solid insulation can be made of paper, pressboard, epoxy, and wood. Among them, Kraft paper is widely used as solid insulation in the transformer. The biological insulating liquid should not be allowed to become so deteriorated or contaminated that it adversely affects other materials in the apparatus.

When it comes to selecting an insulating liquid that best meets the needs of a particular installation or retro-fill an existing equipment, there are several factors that need to be considered. Today, most transformers are still filled with mineral oil, which will not help the construction of green power grid and environmental-friendly

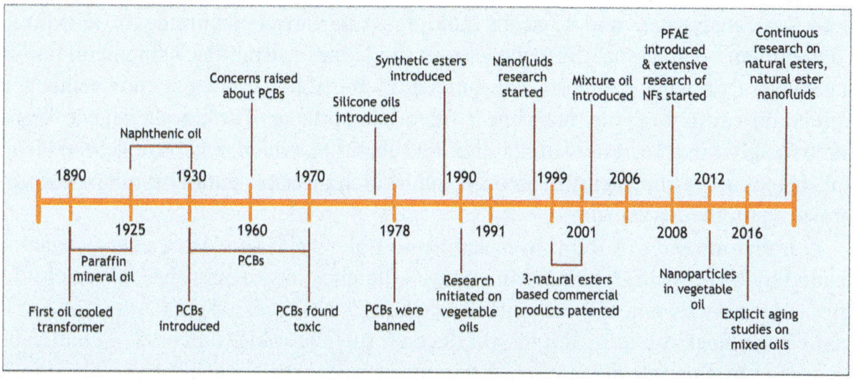

Fig. 1.4 Timeline of research and development on insulating liquids for high voltage devices [2]

society. This has been the case since the late nineteenth century when chemist Elihu Thomson patented the use of mineral oil in transformers to help disperse heat from the core of a transformer to prolong its life. In the meantime, a wide variety of insulating liquids have come onto the market, and some have disappeared again (see timeline Fig. 1.4). Synthetic ester insulating liquids were developed to answer a difficult problem in the high voltage engineering, when PCB (polychlorinated biphenyl) based liquids were effectively banned in the late 1970s [4]. In the early days, synthetic esters were used for retrofilling PCB units and to produce replacement transformers for locations of high fire risk [5]. Another development in the late 1990s was the introduction of natural esters to the transformer market. It is worth noting that despite both liquids being named ester, there are fundamental differences between synthetic and natural types. The key difference between the two is that natural esters can only be used in sealed transformers, while synthetic esters are suitable for either sealed or breathing configurations. The development of synthetic ester and natural ester power transformers has been conducted almost in parallel (Fig. 1.4), with the majority of the natural ester projects being carried out in the USA and Brazil, while the synthetic ester transformers are mainly installed in Europe.

Presently, alternative insulating liquids are becoming more and more popular as we look for options that further prolong transformer life, lower environmental risks, and save money. Not the lease of these would be first cost. As far as is known, the biological isolation fluids are just not the cheapest. In determining first cost, there is more to consider than just the initial equipment cost. There are installation requirements specific to different liquid types that can add a significant burden to project costs or reduce them significantly. Fire safety is one of the main concerns of today's research community due to the applications of insulating liquids in some sensitive areas for instance, subway, channels, ships, offices, shops, workplaces, etc. With a fire point of > 300 °C, most of the biological insulating liquids are classified as K-class insulating liquid according to IEC 61039 [6]. This allows smaller distances to buildings and adjacent transformers. International Electrotechnical Commission [7] and Factory Mutual [8] have published data sheets that define separation distances between transformers and adjacent buildings, fire barrier requirements, and liquid containment systems specific to the various fluid types and ratings. Mineral oil liquids have several drawbacks such as, non-biodegradability, low flash point, non-renewable and could cause a serious problem if there is a spillage. These issues have sought huge focus from researchers to search for a substitute, which may be developed from substances accessible plentiful naturally and which probably could submit paramount prospects in the longer run.

A transformation of the power supply to renewable sources of energy, accompanied by a substantial increase in energy efficiency, is the appropriate strategy for this and the development of new insulating liquids has received great attention, especially biological insulating liquids, produced from renewable sources. The challenges presented by this transformation of the power system are considerable. The aim in the final result is an emission-free power supply. The sustainability and the reduction of fossil materials does not stop at the insulating liquids either. A variety of insulating liquids—above all mineral oil liquids—are cost effective and have good electric properties, but they are not considered being biodegradable in a reasonable time frame.

But a biodegradable insulating liquid would be desirable for electrical apparatus such as transformers, no matter if used in popular or remote areas. Most of the ester liquids are fully biodegradable and natural esters have additionally a very low carbon footprint. Therefore, transformers manufacturers must face new specification related to these new requirements.

There are three factors that influence the chemical stability of the insulating liquid: temperature, oxygen availability and catalyst presence. The oil degradation process might be caused by decomposition of the hydrocarbon molecules at high temperature. The oxygen content in insulating liquid might lead to a rise of the acidity number and to sludge formation. Catalysts such as copper and iron are dissolved in the insulating liquid during operation and might accelerate the aging process. Yuliastuti writes [9] that insulating liquid should fulfill the following health and requirements: non-toxic, biodegradable, thermostable, recyclable, recondition able, readily disposable, and not listed as a hazardous material.

Biological insulating liquids used in transformers are not as widely known as more traditional transformer insulating liquids such as mineral oil, silicon oil or dry type transformers. Hence, therefor there is still widespread confusion about the nature and application of biological insulating liquids. The biological insulating liquids are naturally obtained from seeds as well as from flowers. Often, they are by-products, where the crop is grown for another purpose other than seed oil: cotton (fabric), corn (grain), soybean (protein-rich meal).

In addition to the environmental aspects mentioned above, the use of biological insulating liquids also has technical advantages. Biological insulating liquids absorb more moisture compared to mineral oils and keep the solid insulation dry. However, high concentration of unsaturated fatty acids in the ester molecule makes them unstable and prone to oxidation. Biological insulating liquids have higher acidity content than mineral oil due to hydrolysis reaction which forms above mentioned acids (high molecular acids) like stearic and oleic acids, whereas mineral oil produces

Table 1.1 Characteristic comparison between mineral oil and biological insulating liquids

Criteria	Mineral oil characteristic	Biological liquid characteristic
Key properties	Produced from increasingly scarce and non-renewable special petroleum crude	Refined vegetable oil, produced from domestically grown, renewable sources
Environmental properties	Contains compounds that do not readily biodegrade	Highly biodegradable, non-toxic
Leaks and spills	Spill clean-ups are required by regulation and typically necessitate special equipment and material to help capture contaminated runoff	Relatively rapid biodegradation may eliminate the need for environmentally related clean-up operations
Fire risk	Catches fire more easily, leading to higher probability of transformer fires	Higher fire point reduces the frequency and impact of transformer fires
Transformer performance	Does not slow down the standard insulation aging rate, requires special and expensive processing to dry out the paper insulation	Proven to slow down the aging rate of the insulation system, resulting in an increase in the expected life of a transformer by decades, also promotes automatic dry-out of paper insulation
Utility cost	Smaller investment leads to shortened life of transformer and diminished economic returns, increases liability	Upfront investment promotes transformer life and leads to longer-term economic benefits

low molecular acids, like formic and acetic acids during degradation. The high molecular acids are hardly dissociated and less aggressive. One particular problem is that over time and with substantial exposure to high voltage electricity, the beneficial characteristics of these liquids, such as insulating and/or heat dissipation properties, may degrade.

Biological insulating liquids are deliberated as the suitable substitute of mineral oil in the applications where fire protection, ecological vulnerabilities or better insulation qualities are necessary. A roughly general side to side characteristic comparison between mineral oil and biological insulating liquids is described in Table 1.1 [9].

The first vegetable oil was used for a capacitor insulation in 1962 in electrical engineering and gave a good match with cellulose due to its higher dielectric constants [9]. Biological insulating liquids developed in the early 1990s in the USA as a green and environmentally friendly alternative due to enhanced environmental issues associated with traditional mineral oil and silicon oil. The first transformer prototype with biological insulating liquids as an insulating medium was developed in 1996. However, standard fabrication of transformer immersed with biological insulating liquids initiated in 1999. 2013 first commissioning of a 300 MVA transformer in the 420 kV grid filled with biological insulating liquids in Germany. In Austria, the first distribution transformer filled with biological insulating liquid was put into operation

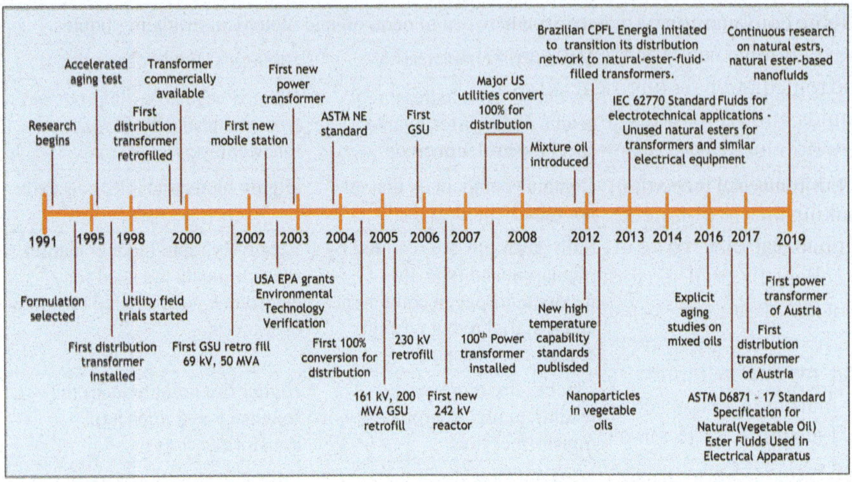

Fig. 1.5 Timeline of proceedings in the development and application of biological insulating liquids [2]

in 2017 and the first power transformer in 2019. Additional information about the time-based progress of biological insulating liquids is depicted in Fig. 1.5.

The market of biological insulating liquids is very volatile. The book represents data, properties, and research results of biological insulating liquids about approximately the last fifteen years mostly compared with mineral oil Nynas 4000X. During this period, ABB has withdrawn the sunflower seed-based vegetable oil Biotemp®. Since then, biological insulating liquids like rapeseed oil named Midel® eN 1204 from Midel and soyabean oil named Envirotemp® FR3™ from Cargill are still on the market and new insulating fluids based on esters—without claim to completeness—such as Midel® eN 1215 from Midel, Paryol Electra 7426® from Fratelli Parodi or NeuGenplus and MATROL-BI® FDV01A from IMCD—have come onto the market. With the deepening of the thought of sustainable development and people's increasing awareness of environmental protection, in the future we can be sure there will be more players in the field of biological insulating liquids. For all who see the conversion of food materials into biological insulating liquid as a substitute of mineral oil and synthetic oils with great scepsis, it is necessary to look for other alternative materials which are non-edible oils. Possibilities, for example, are using *Jatropha curcas* oil extracted from the fruits of *Jatropha curcas* tree or oil from pongamia pinnata tree or castor oil. These vegetable oils are not used in food. *Jatropha curcas* plants can grow well on barren lands with low precipitation in tropical and sub-tropical areas and do not require a lot of fertilizer. Pongamia pinnata tree is a nitrogen fixing plant producing their own fertilizer from the air.

Driven by ecological considerations, Nynas, a big player in mineral oil for transformers, has launched a new dielectric insulating liquid on the market. It is a biodegradable synthetic liquid whose hydrocarbons come from biologically renewable sources.

The combination of fire safety and biodegradability typically eliminates very often the traditional need for firewalls and deluge systems and can reduce spill containment and remediation requirements. It can be expected that vegetable oil-based transformer liquids will increasingly replace mineral oil-based products in the marketplace. The power industry is a conservative, slow changing industry. But changing market forces are demanding new capabilities. New standards are being adopted around the world, making the environmental profile of biological insulating liquids even more desirable. Mohan Rao et al. write in [10]: "Ester fluids (biodegradable resources), which exhibit excellent thermal performance, enhanced fire safety and increased environmental protection, are foreseen as the workhorse for the decades to come".

To enhance the dielectric, thermal and chemical stability of insulation with biological insulating liquids, research and evaluating on nanofluids for application as insulating liquids have been done.

Feasibility study on the use of vegetable oil showed that in countries with climatic and weather condition with high humidity, frequent rain, and a lot of lightening, like in Malaysian biological insulating liquids can be introduced in power transformers either as new insulation liquid or as retro-fill [11].

Renewable, sustainable and environmentally friendly insulating materials will continuously substitute conventual insulating items on the market places more and more.

Gas-to-liquid (GTL) insulation liquids, produced by converting synthetic gas into liquid by using the Fischer–Tropsch process, are not described in this book.

References

1. Renewables 2021 Global Status Report. 2022. https://www.ren21.net/reports/global-status-report/.
2. Rafiq, M., et al. 2020. Sustainable, renewable and environmental-friendly insulation systems for high voltage applications, review, MDPI. *Molecules*. https://doi.org/10.3390/molecules25173901.
3. Dennison, J. 2022. Insulating Liquids: The Lifeblood of Transformers Reliability. *Transformer Technology* 17.
4. Stockton, D. P., et al. 2009, January | February. Seed-Oil-Based Coolants for Transformers, Safety, Reliability, and Environmental Performance of Natural Ester Fluids. *IEEE Industry Application Magazine* 15(1). https://doi.org/10.1109/MIAS.2008.930887.
5. Lashbrook, M., et al. 2016, October. The Development of 400 kV Transformers with Ester-Based Dielectric Liquids. *ARWtr2016*, La Toja Island–Spain.
6. IEC 61039. 2008. Classification of Insulating Liquids.
7. IEC 61936-1. 2010. Power Installations Exceeding 1 kV a.c.—Part 1: Common Rules.
8. FM Global. 2019, July. Property Loss Prevention Data Sheets 5-4.
9. Yuliastuti, E. 2010, June. Analysis of Dielectric Properties, Comparison Between Mineral Oil and Synthetic Ester Oil, Thesis, Faculty of Electrical Engineering, Mathematics, and Computer Science, Delft University of Technology.
10. Mohan Rao, U., et al. 2022, January, Decay particles and regeneration of ester dielectric liquids, A challenge, Transformer Technology, Issue 17.
11. Muhamad, N., and A. Suleiman. 2014, June 25. *Transactions on Electrical and Electronic Materials* 15(3): 113–116.

Chapter 2
Dielectric Insulating Liquids

In this chapter, the origin of the various insulating liquids and their basic chemical structure, which are very different, are described. Also, additives that positively influence the properties for use as an insulating liquid are presented. The idea is to replace mineral oil, which is currently the most commonly insulation liquid, by other insulating liquids especially by biological liquids. In addition to technical and economic points, the focus is on the combustibility of the insulating liquids and ecological aspects. As reference for all these considerations and investigations, mineral oil serves as the current dominant insulating oil. The main area of the application of these oils is the use as insulating and cooling material in transformers. In general, a distinction between polar and non-polar insulating liquids can be made. Polar liquids have a constant dipole moment, which results in a displacement of positive and negative charge centers, even when no external influences (electrical field, temperature, pressure,...) are present. A typical characteristic of polar liquids is their higher (relative) permittivity.

Liquid filled transformers are often not considered as an option for indoor installations due to historical issues of fire safety, environmental concerns, and special containment. Because of these perceptions, vacuum pressure impregnated dry-type and cast-resin transformers have often replaced liquid-filled transformers for indoor installations. With this trend, significant liquid filled advantages have been lost like superior life, efficiency, sound level, overload capacity, contamination resistance, and online diagnostics [1].

The new generation of dielectric insulating liquids now overcomes the fire safety and environmental issues so that the benefits of liquid-filled transformers can be retained for indoor installations. Additionally, biological insulating liquids are renewable and have a very low carbon footprint. It is generally agreed that any new insulating liquid must meet a number of operational requirements, such as: good electric

E. P. Pagger et al., *Biological Insulating Liquids*,
https://doi.org/10.1007/978-3-031-22460-7_2

and thermal properties; low viscosity; chemical and thermal stability; low flammability; compatibility with other transformer materials; miscibility with other liquids; environmental acceptability and low cost.

2.1 Mineral Oil

Mineral oil liquids are still the most widely used insulating liquids. Due to the large amounts used in transformer they are also referred to as "transformer oil". Mineral oil stems from crude oil and is obtained by refining a fraction of the hydrocarbons collected during the distillation of a petroleum crude stock. The boiling range of the collected fraction and the kind and degree of the refining process are selected so that the resulting oil reaches requirements specified for use in transformers. Crude oil is a lipophilic mixture of substances stored in the earth's crust, consisting mainly of hydrocarbons. It is used to generate electricity and as fuel for almost all modes of traffic and transport. In addition, crude oil is widely used in the chemical industry to produce plastics and other chemical products. Unfortunately, the use (incineration) of petroleum products is one of the main drivers of climate change. This is linked to the task of replacing petroleum products with sustainable products wherever possible. Additionally, naphthenic crude oil reserves from which mineral oil is obtained are limited.

Mineral oil insulation liquid is essentially a mixture of hydrocarbon structures, being termed either paraffinic, iso-paraffinic, or naphthenic, depending on the dominant constituents. Mineral oil insulating liquids with a high aromatic content tend to be used for cable and capacitor applications where gas absorption is important. For transformer application, aromatic content is kept lower, as this gives a good balance of properties. Polycyclic aromatic content, where multiple aromatic rings are bonded together, is kept below 3% for health and safety reasons.

Within crude mineral oil, there will be some small quantities of polar molecules, such as acids and ketones. These are removed during the manufacturing process, but will increase as the mineral oil insulating liquid ages. The presence of these polar molecular structures has a direct influence on parameters such as water solubility, tan delta, and interfacial tension. Consequently, these parameters are monitored for mineral oil insulating liquid, but they are less relevant to esters, as these liquids contain polar structures in their molecular structure [2].

A great deal of information on the molecular structure of insulating liquids may be derived from their infrared spectra. The FTIR spectra of the investigated mineral oil (Nynas 4000X) shows absorption bands of stretching vibrations at wave number 3650 cm^{-1} because of hydroxyl groups of the inhibitor (dibenzyl-para-cresol). The rather strong bands between 3170 and 2400 cm^{-1} result from the stretching vibration of the $-CH_3$, $-CH_2$ and $-CH$ groups, while the bending vibrations of the same groups are responsible for the bands 1480 to 730 cm^{-1} in the region of the infrared spectrum. The intensity of the band at 1600 cm^{-1} reflects also the small amount

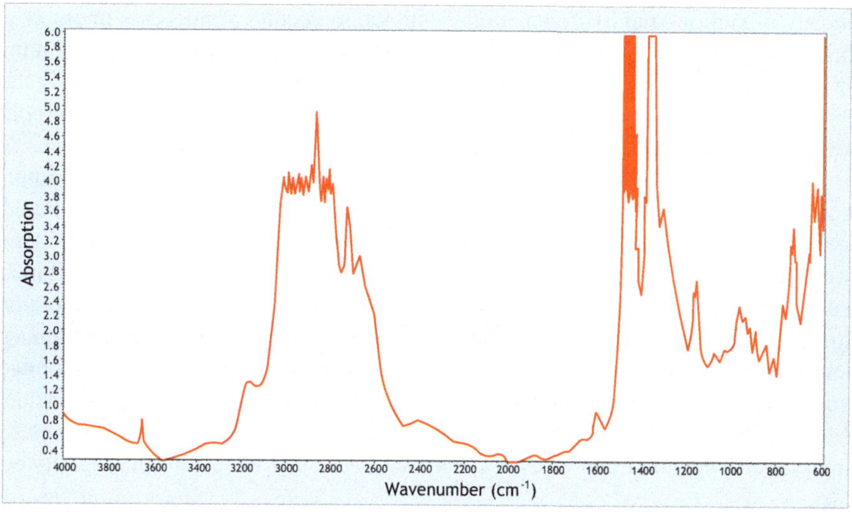

Fig. 2.1 FTIR spectra of mineral oil Nynas 4000X [3]

of aromatic parts present in the oil, which is characterized by the stretching vibrations of the C = C bonds within the benzene ring structures. The 720 cm^{-1} band results from the rocking vibrations of the chains containing the methylene groups ($-CH_2-$) consisting of four or more base units; consequently, its intensity represents an effective measure of the paraffinic content (Fig. 2.1).

2.2 Bio-based Hydrocarbon Insulating Liquid

Nynas has developed a new fully bio-based hydrocarbon dielectric insulating liquid called Nytro Bio 300X. Corps, vegetable oils and used cooking oils are used as starting materials.

2.3 Ester Liquids

In the last decade, esters are becoming a mainstream alternative to mineral oil insulating liquid not just for fire safety but for increased transformer performances and environmental characteristics. There are some similarities with mineral oil insulating liquid, as esters can be largely composed by hydrocarbon chains. However, there are fundamental differences, and it is important to focus on these areas to understand how to handle the different liquids and interpret laboratory data. The chemical structure of ester fluids molecules is different compared to mineral oil. Mineral oil consists mostly of carbon- and hydrogen atoms with the molecules arranged as linear, or as branched or cyclic. Ester liquids used in transformers have six to eight oxygen atoms in their molecule. Because of the high electronegativity of oxygen (Fig. 2.2) and because the charge balance points don't coincide (Fig. 2.3), ester fluids are polar versus the nonpolar mineral oil.

This is one of the main reasons for the different properties of ester liquids compared to mineral oil. The ester linkages are what give esters dielectric liquids some of their benefits, and also in the case of double carbon–carbon bonds, some weakness. There are some obvious chemical differences between mineral oil insulating liquid and ester based insulating liquids, and it is important for asset operators to understand how to monitor and maintain these liquids. The polar ester linkage leads to attributes such as much higher water solubility, higher tan delta and lower volume resistivity, when compared to mineral insulating liquids. It also means that some parameters routinely monitored for mineral insulating liquid are less useful in assessing the condition of esters. For example, there is no proven link between

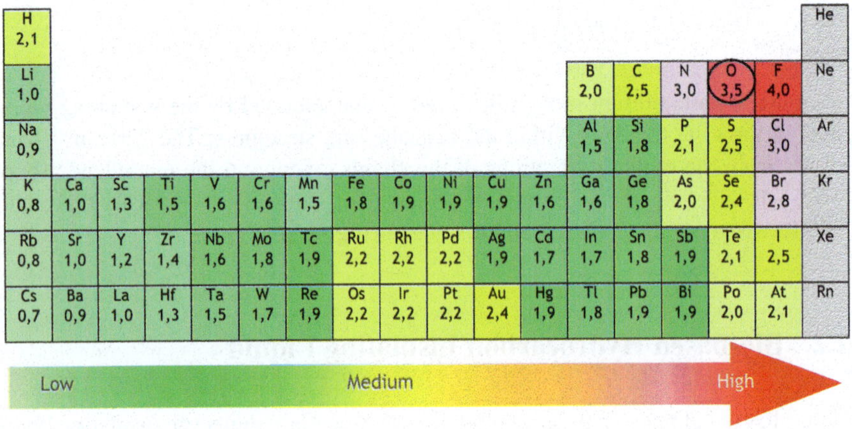

Fig. 2.2 Electronegativity of oxygen

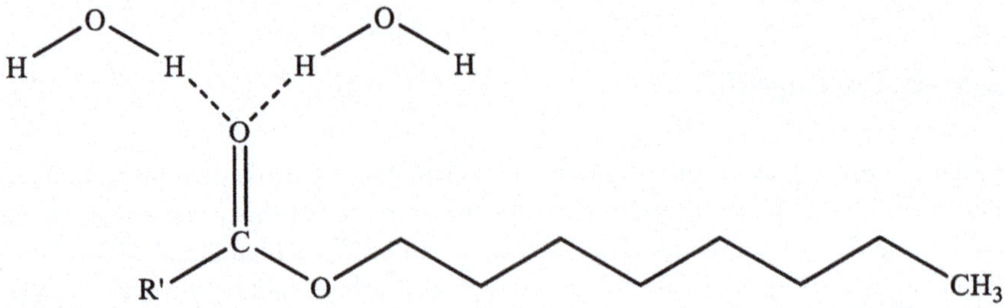

Fig. 2.3 Hydrogen bonds of water with organic acids

interfacial tension value and fluid condition in esters, despite this being a commonly used measurement for mineral insulating oil.

The polarity is an essential property that makes life on earth possible at all. In addition to the atomic forces involved in bonding of elements, there are molecular forces which influence the nature of matter. These forces are dipole–dipole interactions and van der Waals forces. They serve to hold neutral molecules to each other and appear to be electrostatic in nature.

Dipole–dipole interaction is the attraction of the positive end of one polar molecule to the negative end of another polar molecule. If a molecule is composed of a number of different elements, a dipole can result from the difference in electronegativity of two different atoms sharing a chemical bond. In a chemical bond between atoms of elements having different electronegativities, the electrons forming the bond will spend more time near the more electronegative element's atom. This has the effect of imparting a partial negative charge near that atom and a partial positive charge near the atom of the less electronegative element. The energy involved in this association is often much less than 4 kJ/mol. None the less, polar molecules are held together more strongly than nonpolar molecules because of this interaction.

A special type of dipole–dipole interaction is the hydrogen bond. In a hydrogen bond, a hydrogen atom serves as a bridge between two highly electronegative atoms. It is bonded covalently to one of these atoms and associated only through electrostatic forces with the second atom. When hydrogen is bonded to nitrogen, oxygen, or fluorine, the electron cloud bonding the two elements is distorted greatly toward the electronegative atom. This exposes the positively charged hydrogen nucleus. The hydrogen nucleus is attracted strongly to the negative charge of the electronegative atom of a second molecule. Compounds which display hydrogen bonding include organic acids (Fig. 2.3) alcohols, phenols, amines, amides, hydrogen fluoride and water. Why does HF have a lower boiling point (19.5 °C) than water even though fluorine has a higher molecular weight and is more electronegative than oxygen (Fig. 2.2)? Water, being a bent molecule with a tetrahedral electron geometry, can bond hydrogen in three dimensions (Fig. 2.4).

HF, being a linear molecule, has zigzag hydrogen-bonding interactions, in two dimensions, with angles of 116° (Fig. 2.5).

A more optimal angle would be at 180° since that would give more direct dipole interactions. This limits the hydrogen-bonding strength of HF. The hydrogen bond has a dissociation energy of about 21 kJ/mol, which is much greater than other dipole–dipole interactions. Hydrogen bonding strongly affects the boiling point, solubility properties and even the shape of large molecules. Figure 2.6 shows that without polarity, water would have a boiling point of approx. −130 °C. This is because of the low molecular weight of water. Compared to the linear alkanes with a polarity index close to zero, also toluene as an aromatic hydrocarbon has a higher boiling point. The hydrogen bond is much stronger than the pure dipole–dipole bond. This can be seen when comparing the boiling points of ethyl alcohol with dimethyl ether. Ethyl alcohol forms hydrogen bonds and has a boiling point of 78 °C, while dimethyl ether with the same molecular weight and pure dipole–dipole interaction has a boiling point of minus 25 °C.

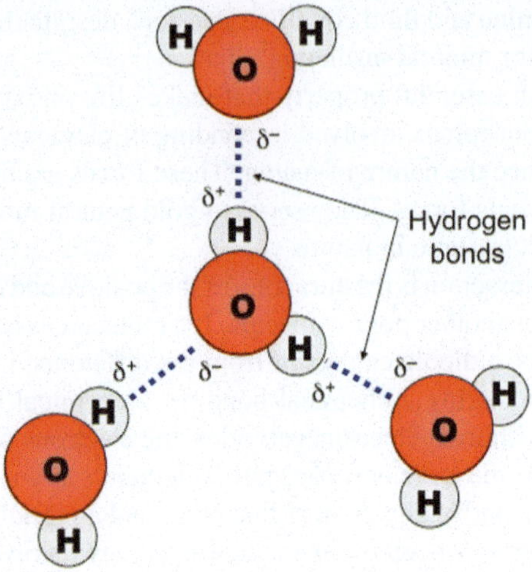

Fig. 2.4 Hydrogen bonds in water

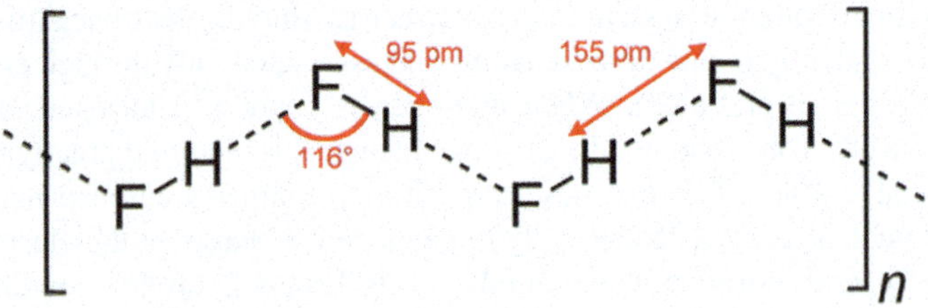

Fig. 2.5 Hydrogen bonds in hydrofluoric acid

An influence on the physical properties, albeit a very small one, has the van der Waals forces. Van der Waals forces are attractive intermolecular forces. They result from the polarizing effect molecules have on one another. The average distribution of charge in a nonpolar molecule is uniform. At any given instant, however, the electrons may not be distributed uniformly. Because of the momentary concentration of electrons in one portion of a molecule, a small dipole is created temporarily. The existence of this dipole induces a dipole of the opposite charge in an adjacent molecule. Because the dipoles have opposite charge, they attract each other. These dipoles rapidly disappear, only to be replaced by others. This interaction takes place at the surface of molecules; so, the more surface area a molecule has, the more it will interact with its neighbor through transient dipole formation. This molecular interaction is reflected in the boiling points of liquids, too. The boiling points of linear alkanes increases as their molecular weight (surface area) increases (Fig. 2.6). The boiling points of branched chain liquids are lower than that of the linear chain liquids. For example n-octane has a boiling point of 125.6 °C and isooctane 99.3 °C, respectively. This results from the lower surface area presented by the branched chain

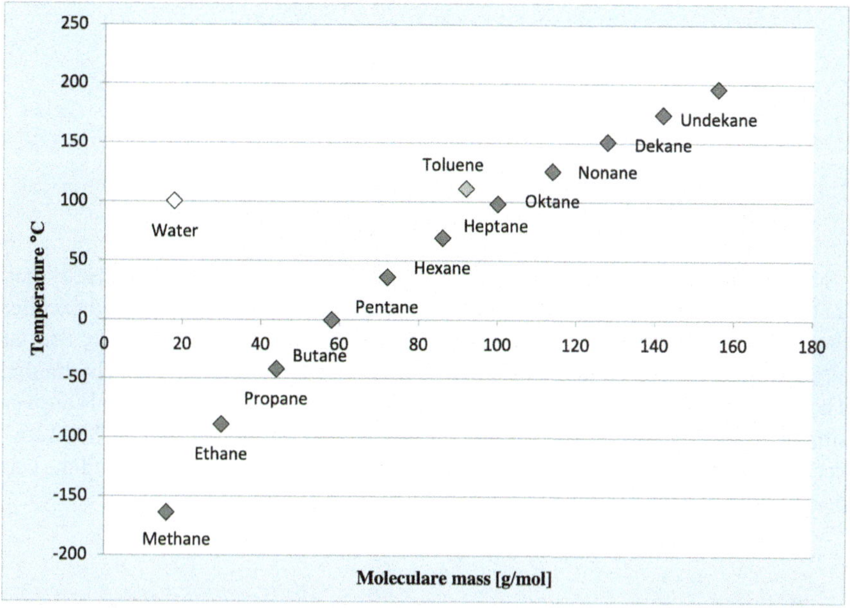

Fig. 2.6 The influence of the polarity on the boiling point

molecule. This effect is particularly true for small molecules which can approach a spherical shape. A sphere has the smallest possible surface area for a given volume.

Organic esters possess similar electric properties compared to mineral oil. Recent focus on health, safety, and the preservation of the environment has promoted the use of environmentally friendly and biodegradable liquids to avoid the effects of mineral oil liquids. Although electrical power transmission and distribution industry has showed its interest on these alternative liquids as alternative to mineral oil liquids, it is necessary to demonstrate with facts and data that these insulating liquids offer suitable performance, as well as they are safe and economic.

2.3.1 *Synthetic Ester*

The synthetic ester oils used as insulating liquids are mostly a tetra ester based on pentaerythritol synthesized from linear and branched organic acids (R, R', R'', R''') Eq. 2.1. The organic substituents R, R', R'' and R''' do not necessarily have to differ in their molecular structure.

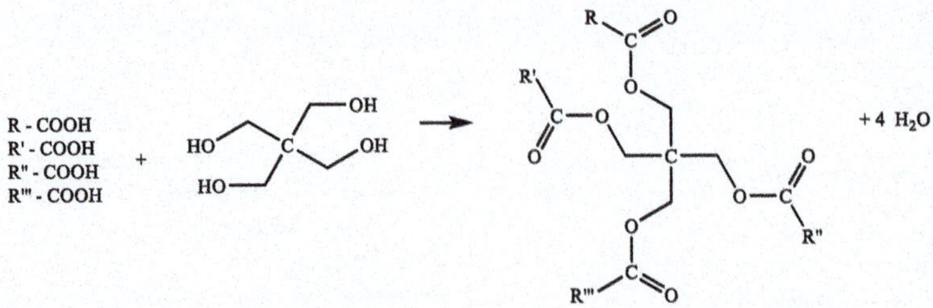

(2.1)

Synthetic ester dielectric liquids have suitable dielectric properties and are significantly more biodegradable than mineral insulating oil. They have proved excellent thermal stability, good low temperature properties and less toxic. Hence, they are already widely accepted in the fields of high-temperature lubrication and hydraulics. The FTIR spectra of the two studied ester liquids (Midel 7131, BecFluid) are very similar. There are adsorption bands in the range of wave number 3400 to 3600 cm^{-1} because of stretching vibrations of hydroxyl groups (alcohol) and above all the ester peak around the wave numbers 1700 to 1800 cm^{-1} (Figs. 2.7 and 2.8).

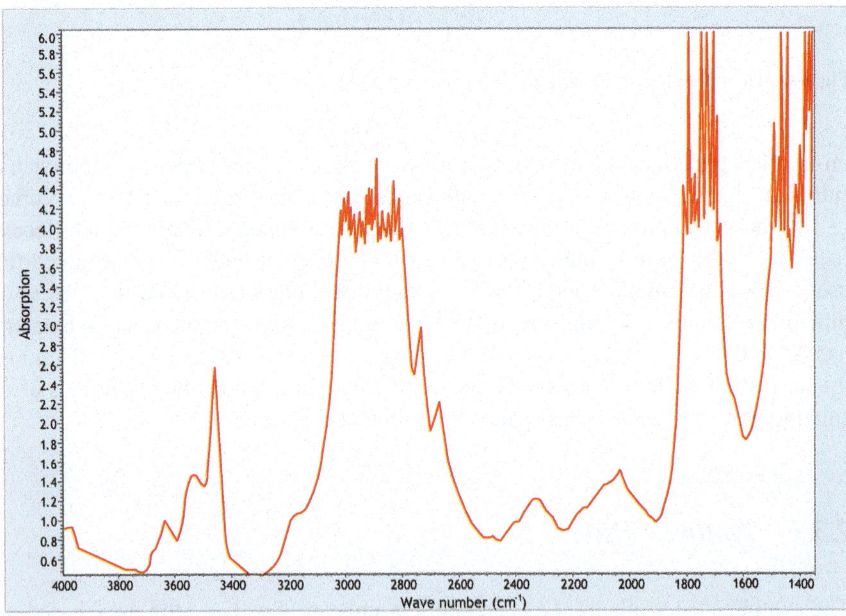

Fig. 2.7 FTIR spectra of synthetic ester Midel 7131 [3]

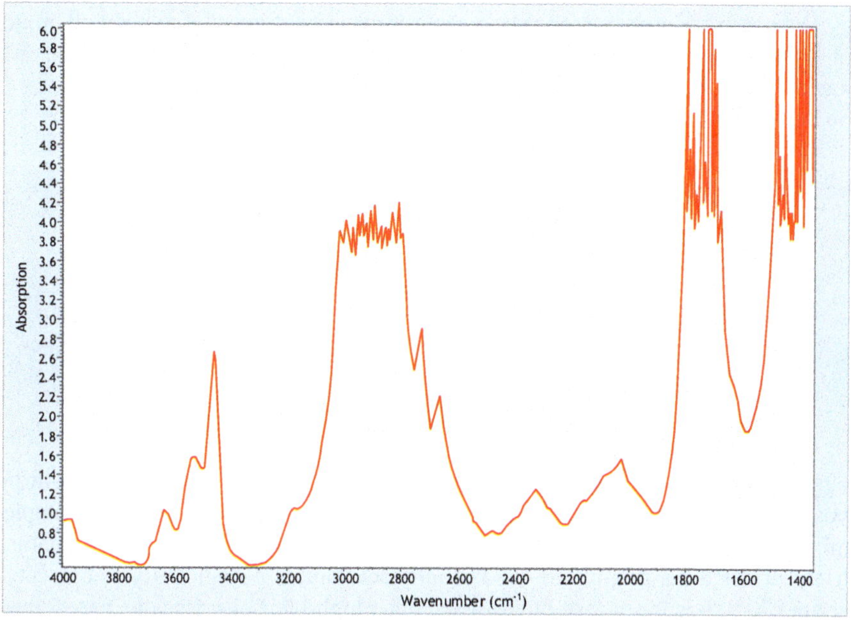

Fig. 2.8 FTIR spectra of synthetic ester Becfluid 9902 [3]

2.3.2 *Natural Ester (Vegetable Oil)*

Global production of vegetable oils is growing due to rising consumption in emerging countries and in biofuel application as shown in [4] (Fig. 2.9).

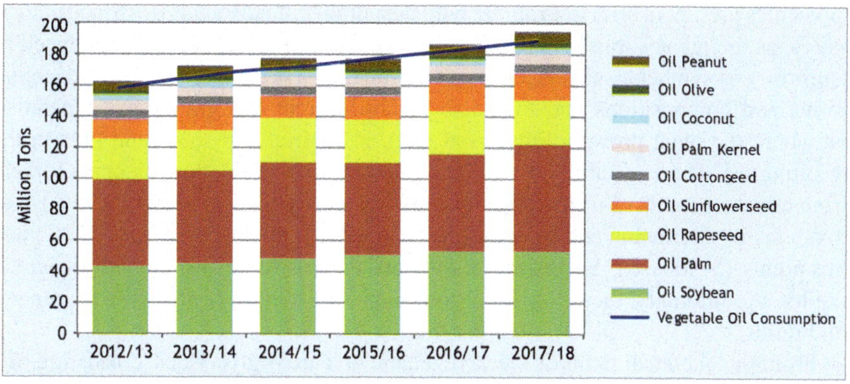

Fig. 2.9 Global vegetable oil production and consumption

Vegetable oils are assumed to be an appropriate replacement of mineral oil liquids in high voltage equipment. The base oil of natural ester is of plant origin. In a natural ester process, the alcohol (glycerol) is esterified with different fatty acids of predominately sixteen to eighteen carbon chains (Eq. 2.2).

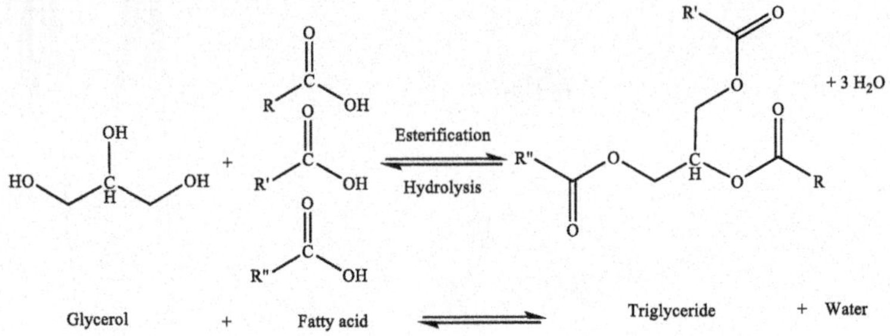

(2.2)

The presence of a high number of fatty acids in a vegetable oil, as well as multiple possibilities of their combination with glycerol, makes vegetable oils very complex mixtures with significantly different structures and properties. It is unusual for natural triglycerides to have only one kind of fatty acid unless a single fatty acid exceeds around 70%. It is the case of olive oil and other high-oleic oils. Usually, two or three different fatty acids are present in triglycerides and their chemical and thermophysical properties depend on fatty acid composition. For example, soybean oil contains eight triglycerides that account for over 70% of total [5]. These includes LLL (17.6%), LLO (15.3% Fig. 2.10), LLP (10.2%), PLO (6.9%), LnLO (4.8%), LLSt (4.2%), LLLn (7.9%) and OLO (6.3%) and their isomers (where L = linoleic, O = oleic, P = palmitic, Ln = linolenic, St = stearic acid).

However, two vegetable oils containing qualitatively and quantitively the same fatty acids will have different chemical or physical characteristics if the fatty acids are distributed in different ways in the triglycerides. In fact, each vegetable oil is characterized by its own fatty acid composition. The number of different triglyceride molecules present in oils rises rapidly with the number of fatty acids [6]. Since natural esters, as the name suggests, are natural substances, they can never be produced as uniformly as synthetic substances. Trace elements are present in the soil in different forms and compositions and thus also reach different concentrations, depending on where the plant grows (climate soil etc.), in its fruit and seeds and thus in the resulting oil. The vegetable oil will be extracted from seeds, flowers or less likely from other segments of fruits. Vegetable oils are highly biodegradable (> 95%), less toxic, lesser flammable and have an extraordinarily high flashpoint (> 300 °C) and fire points (> 300 °C). Furthermore, they are also more environmentally friendly liquids. Additionally, these liquids absorb more moisture in comparison to mineral oil liquids.

From the chemical point of view, vegetable oils are triglycerides, consisting of a glycerol molecule and three long-chain fatty acids. Over 1000 natural fatty acids have been identified. These vary in chain length, degree of unsaturation and the presence

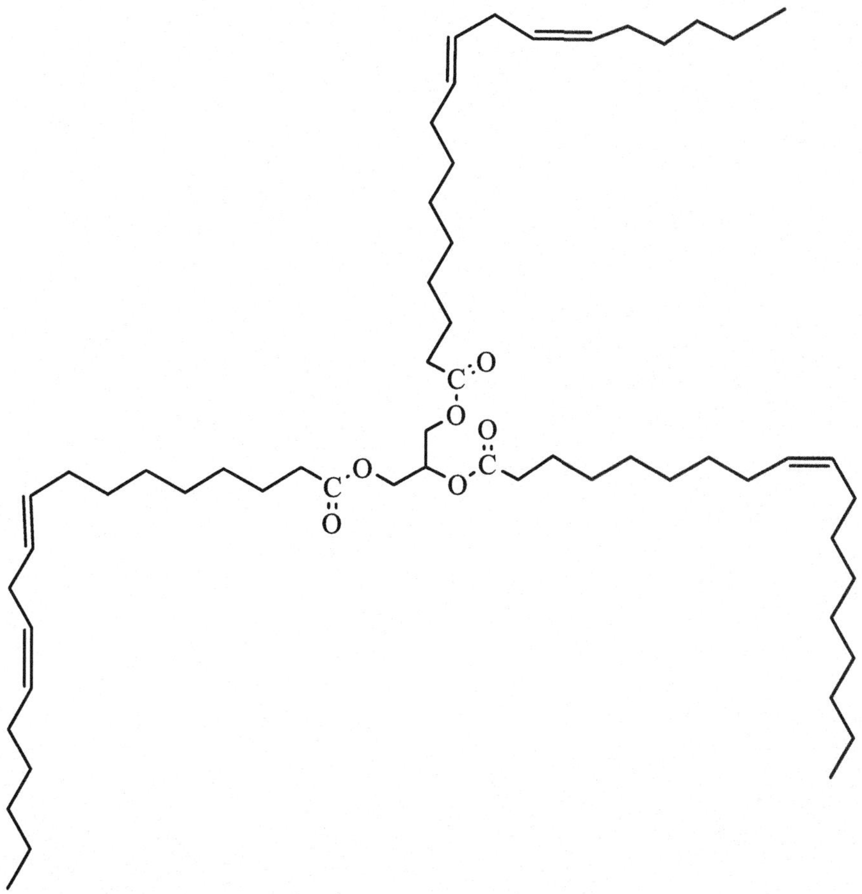

Fig. 2.10 Triacylglycerol LLO

or absence of other functional groups. Only a limited number of these—perhaps 25–50 at most—are likely to be important to most lipid scientists and technologists [5]. Glycerol is the backbone of the biological insulating liquids (natural ester) (Fig. 2.11).

The side chains R, R' and R" can consist of uniform or different fatty acid chains (Table 2.1) and they vary between the types of plant corps. In accordance with their biosynthetic origin, the fatty acids usually contain an integer carbon chain. There are four types of fatty acids—saturated, monounsaturated, polyunsaturated acids of the n-6 family and polyunsaturated acids of the n-3 family (the terms n-6 and n-3 relate to the first position of the double bound with respect to the methyl end of the chain). No oil in nature is composed entirely of any one of these three. The terms saturated, monounsaturated, and polyunsaturated refers to the degree of hydrogen saturation. A saturated fatty acid contains all the hydrogen atoms it possibly can have. In other words, it is fully saturated with hydrogen. A monounsaturated fatty acid

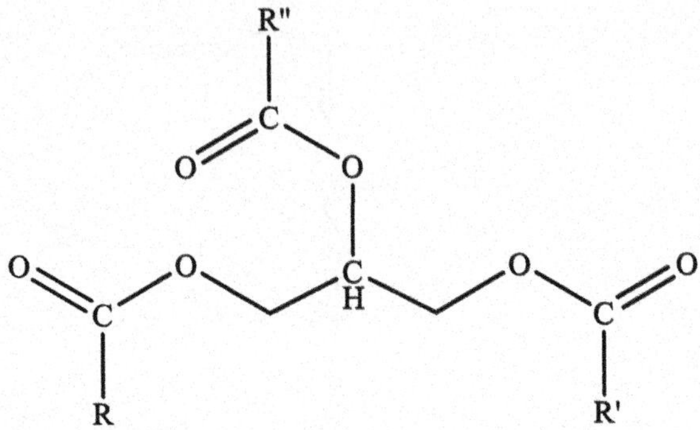

Fig. 2.11 Glycerol—backbone with fatty acids

contains all but one pair of hydrogen atoms it can hold. Polyunsaturated fatty acids are lacking two or more pairs of hydrogen atoms. Symbols such as *c* (Z), *t* (E), e and a are sometimes used to indicate *cis* olefinic, *trans* olefinic, ethylenic and acetylenic unsaturation. This molecular structure influences the properties of the natural ester. The melting point increases with increasing chain length but decreases with increasing number of double bonds. As the number of double bonds increases, the fatty acids become more reactive and facilitate oxidative attack and that lead to a lower level of oxidation resistance compared to other insulating liquids. These unsaturated links in the molecular structure are open to attack from oxygen and this will lead to increased viscosity and in the worst-case polymerization of the insulating liquid into a solid. For this reason, biological insulating liquids are only recommended for sealed equipment, without free access of oxygen from the air. Depending on the base seed oil the amount and type of unsaturated content will vary. The common fatty acids are easily recognized and separated from each other by gas chromatography and this technique is a standard analytical procedure. Qualitative information is also given by FTIR spectra. Figures 2.12, 2.13 and 2.14 shows the FTIR spectra of the natural ester, Biotemp, Envirotemp FR3 and Midel eN 1204. The double bounds are situated around wave number 3000 cm^{-1} and 1600 to 1700 cm^{-1}. From the biological insulating liquids, it seems that Envirotemp FR3 has the most double bounds.

Vegetable oils are available in plenty of variety across the world. That provides an economic (low transportation costs) and an ecological (low carbon emission) benefit as the source plant is not located far away from place of installation as it is with source material for mineral oil. Vegetable oils that are not suitable for human consumption can be used to produce insulating dielectric liquids (for example: jatropha curcas oil, pongamia pinnata oil, castor oil). Hammond shows in [7] the typical fatty acid composition from different seeds, nuts, cereals, and fruits (Table 2.2). The data fits very well with the results from Totzauer and Trnka (Table 2.3) [8] and Bartz [9]. The most common fatty acids in the natural ester are palmitic acid (Fig. 2.15), stearic acid (Fig. 2.16), oleic acid (Fig. 2.17), linoleic acid (Fig. 2.18) and linolenic acid (Fig. 2.19). Genetic modifications of the vegetable oils can achieve specific properties.

Table 2.1 Fatty acids in vegetable oils

ID	Number of C-atoms: number of double bonds	Trivial name	IUPAC[1] Nomenclature	Empirical chemical formula	Structure formula
A	8:0	Caprylic acid	n-Octanoic acid	$C_8H_{16}O_2$	Figure 2.20
B	10:0	Capric acid	n-Decanoic acid	$C_{10}H_{20}O_2$	Figure 2.21
C	12:0	Lauric acid	n-Dodecanoic acid	$C_{12}H_{24}O_2$	Figure 2.22
D	14:0	Myristic acid	n-Tetradecanoic acid	$C_{14}H_{28}O_2$	Figure 2.23
E	16:0	Palmitic acid	n-Hexadecanoic acid	$C_{16}H_{32}O_2$	Figure 2.15
F	16:1	Palmitoleic acid	(9Z)-Hexadec-9-enoic acid	$C_{16}H_{30}O_2$	Figure 2.24
G	18:0	Stearic acid	n-Octadecanoic acid	$C_{18}H_{36}O_2$	Figure 2.16
H	18:1	Oleic acid	(9Z)-Octadec-9-enoic acid	$C_{18}H_{34}O_2$	Figure 2.17
I	18:2	Linoleic acid	(9Z,12Z)-Octadeca-9,12-dienoic acid	$C_{18}H_{32}O_2$	Figure 2.18
J	18:3	Linolenic acid	(9Z,12Z,15Z)-octadeca-9,12,15-trienoic acid	$C_{18}H_{30}O_2$	Figure 2.19
K	20:0	Arachidic acid	Icosanoic acid	$C_{20}H_{40}O_2$	Figure 2.25
L	20:1	Gadoleic acid	(9Z)-9-Eicosensäure	$C_{20}H_{38}O_2$	Figure 2.26

(continued)

[1] International Union of Pure and Applied Chemistry.

Table 2.1 (continued)

ID	Number of C-atoms: number of double bonds	Trivial name	IUPAC Nomenclature	Empirical chemical formula	Structure formula
M	22:0	Behenic acid	n-Docosanoic acid	$C_{22}H_{44}O_2$	Figure 2.27
N	22:1	Erucic acid	(13Z)-Docos-13-enoic acid	$C_{22}H_{42}O_2$	Figure 2.28
O	24:0	Lignoceric acid	n-Tetracosanoic acid	$C_{24}H_{48}O_2$	Figure 2.29
P	18:1	Ricinoleic acid	(9Z,12*R*)-12-Hydroxyoctadec-9-enoic acid	$C_{18}H_{34}O_3$	Figure 2.30

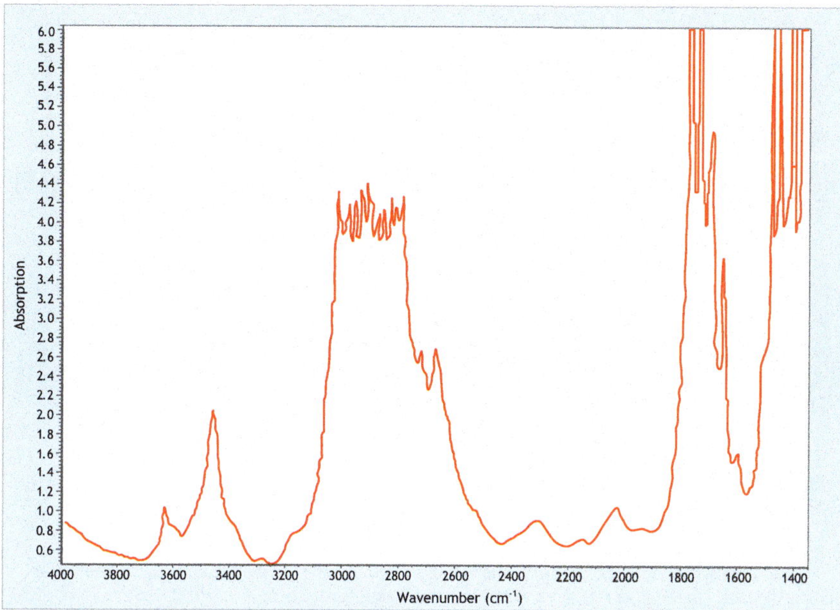

Fig. 2.12 FTIR spectra of natural ester Biotemp [3]

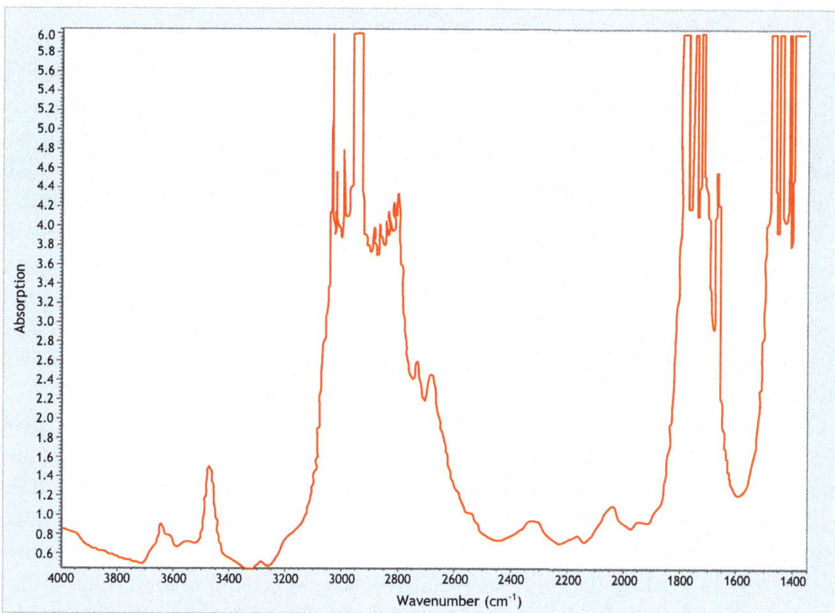

Fig. 2.13 FTIR spectra of natural ester Envirotemp FR3 [3]

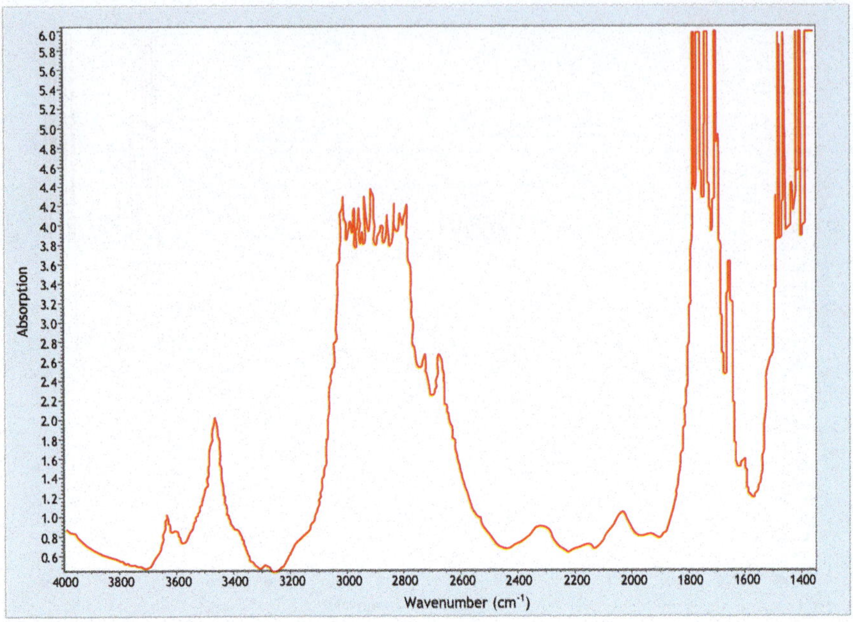

Fig. 2.14 FTIR spectra of natural ester Midel eN 1204 [3]

There are surprising differences between palm oil and palm kernel oil. Palm oil and palm kernel oil are both sustainable tropical oils which come from the fruit of the oil palm trees. Palm oil stems from the fruit and palm kernel oil from the seed. These versatile oils, however, have very different chemical compositions and physical characteristics. Palm oil is made by simply steaming and pressing the fleshy part of the fruit. Palm oil has a 50/50 balance of saturated and unsaturated fatty acids (Table 2.4). When grown sustainably and used regionally, palm oil can be an environmentally efficient oil crops that require a minimal area for maximum oil production compared to other crops. In terms of area, palm oil is the most productive vegetable oil (up to seven tons of oil per hectare; sunflowers and rapeseed up to a maximum of two tons of oil per hectare) and therefore—with clearly defined guidelines for cultivation—the more sustainable alternative in the Southeast Asian countries in the long term.

Vegetable oils are used as an ingredient or component in many manufactured products. Many vegetable oils are used to make soaps, skin products, candles, perfumes and other personal care and cosmetic products. Some oils are particularly suitable as drying oils and are used in making paints and other wood treatment products. Dammar oil (a mixture of linseed oil and dammar resin), for example, is used almost exclusively in treating the hulls of wooden boats. Vegetable oils are increasingly being used in the electrical industry as insulators. Vegetable oils are also used to make biodiesel, which can be used like conventional diesel. Some vegetable oil blends are used in unmodified vehicles but straight vegetable oil, also known as pure plant oil, needs specially prepared vehicles.

Table 2.2 Edible vegetable oils and their typically fatty acid composition [%]

Type of oil	A	B	C	D	E	F	G	H	I	J	K	L	M	N	O
	8:0	10:0	12:0	14:0	16:0	16:1	18:0	18:1	18:2	18:3	20:0	20:1	22:0	22:1	24:0
Babassu	5.4	6.0	44.3	15.8	8.6	–	2.9	15.2	1.7	–	0.1	–	–	–	–
Coconut	8.1	6.5	48.6	17.7	8.5	–	2.5	6.5	1.5	–	0.1	–	–	–	–
Palm kernel	4.0	4.1	49.7	16.0	8.0	–	2.4	13.7	2.0	–	0.1	–	–	–	–
Sunflower	–	–	0.1	0.2	6.8	0.1	4.7	18.6	68.6	0.5	0.4	–	–	–	–
Sunflower HOL[2]	–	–	0.1	0.2	3.0	0.4	5.9	82.4	7.4	–	0.3	0.3			
Rapeseed (leara)[3]	–	–	–	0.1	4.7	0.3	1.7	59.0	21.4	9.9	0.6	1.4	0.4	0.3	0.2
Soybean	–	–	–	–	10.0	0.2	3.5	21.0	55.3	9.2	0.5	–	0.3	–	–
Cottonseed	–	–	–	1.0	23.9	0.5	2.9	18.5	52.5	0.3	0.4	–	–	–	–
Peanut	–	–	–	–	10.1	0.2	3.5	51.4	27.3	0.1	1.6	1.3	3.1	–	1.4
Cocoa butter	–	–	–	0.1	26.2	0.3	34.4	34.8	2.9	1.1	0.2	–	–	–	–
Almond	–	–	–	0.1	8.5	1.1	1.0	57.0	31.4	0.6	0.1	0.1	0.1	–	–
Brazil nut	–	–	0.1	0.2	14.4	0.5	7.9	31.2	45.1	0.1	0.3	0.1	0.1	–	–
Corn	–	–	–	–	10.7	0.2	1.5	30.5	55.9	0.8	0.4	–	–	–	–
Rice bran	–	–	–	–	13.9	1.9	2.7	41.1	36.4	2.3	1.8	0.2	–	–	–
Wheat-germ	–	–	–	0.2	18.5	0.6	0.5	18.1	55.9	5.3	0.1	0.8	–	–	–

(continued)

[2] High oleic sunflower oil.
[3] Low erucic rapeseed oil.

Table 2.2 (continued)

Type of oil	A	B	C	D	E	F	G	H	I	J	K	L	M	N	O
	8:0	10:0	12:0	14:0	16:0	16:1	18:0	18:1	18:2	18:3	20:0	20:1	22:0	22:1	24:0
Palm Fruit	–	–	–	1.0	43.8	0.5	5.0	38.5	10.5	0.3	0.4	–	–	–	–
Olive	–	–	–	–	10.8	0.5	3.0	75.5	8.5	0.9	0.4	0.4	–	–	–
Sesame [11]	≤ 0.05	≤ 0.05	≤ 0.05	0.1	7.9–12.0	0.2	4.8–6.1	35.9–42.3	41.5–47.9	0.3–0.4	0.3–0.6	–	–	–	–
Camelina sativa [5]	–	–	–	–	–	–	–	10–20	16–24	30–40	–	15–23	–	–	–
Hazelnut [5]	–	–	–	–	–	–	–	65–75	16–22	–	–	–	–	–	–
Walnut [5]	–	–	–	–	–	–	–	–	50–60	13–15	–	–	–	–	–

Table 2.3 Fatty acid composition of vegetable oils in percent [8]

Fatty acid	Palmitic acid (E) 16:0	Stearic acid (G) 18:0	Oleic acid (H) 18:1	Linoleic acid (I) 18:2	Linolenic acid (J) 18:3
Palm oil	45	4	40	10	–
Olive oil	11	3	71	10	1
Rapeseed oil	4	2	62	22	10
Sunflower oil	7	5	19	68	1

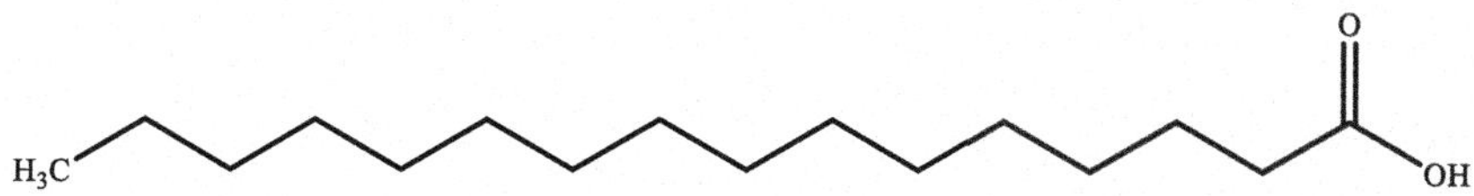

Fig. 2.15 Palmitic acid (**E**)

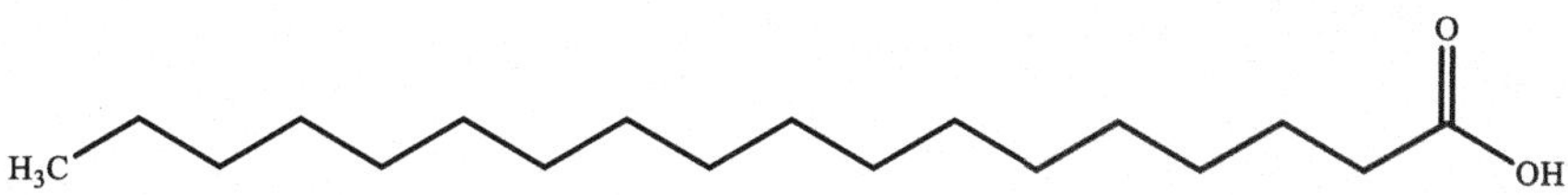

Fig. 2.16 Stearic acid (**G**)

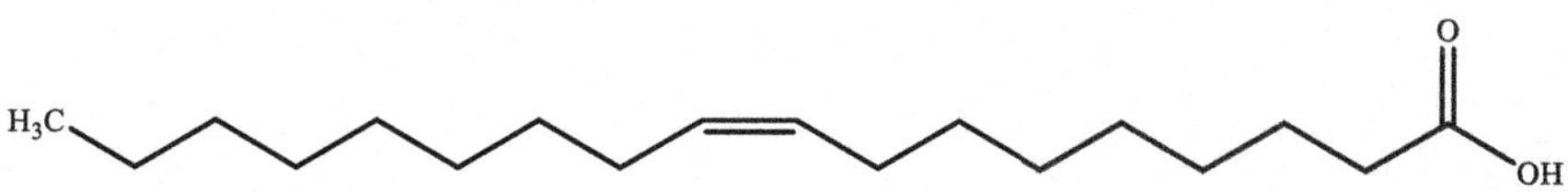

Fig. 2.17 Oleic acid (**H**)

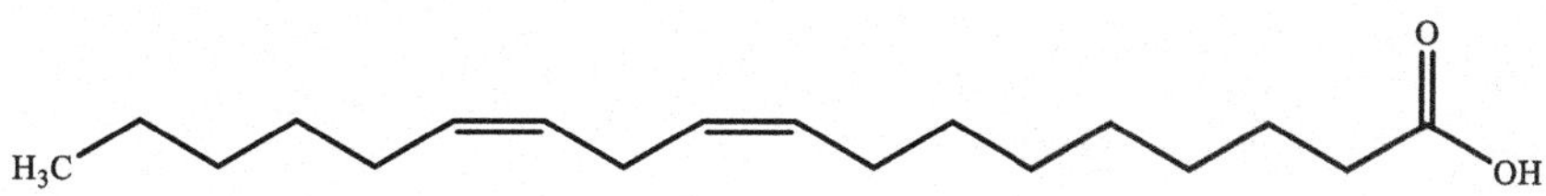

Fig. 2.18 Linoleic acid (**I**)

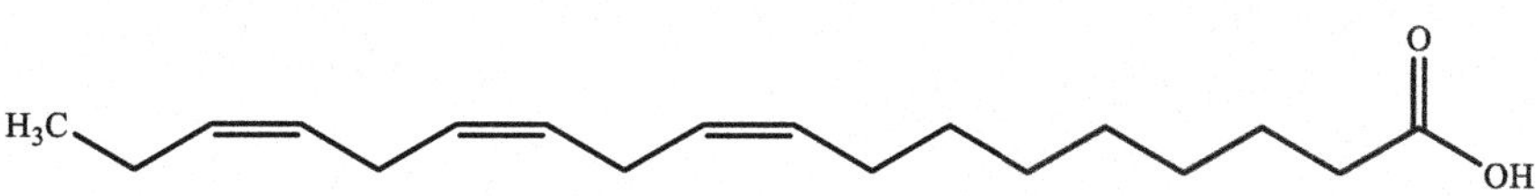

Fig. 2.19 Linolenic acid (**J**)

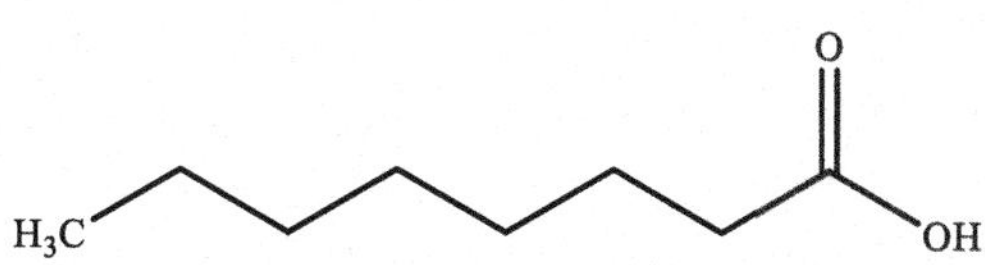

Fig. 2.20 Caprylic acid (**A**)

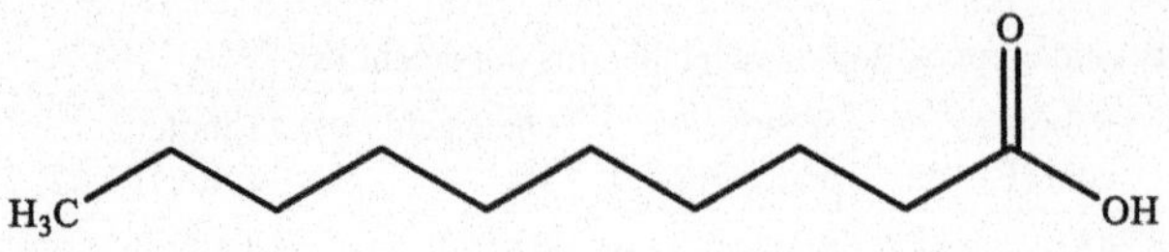

Fig. 2.21 Capric acid (**B**)

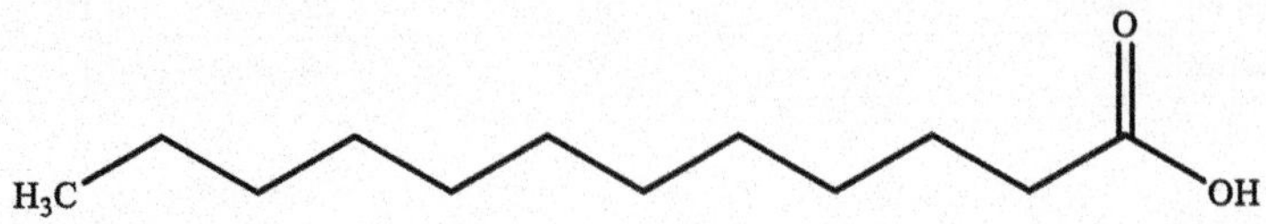

Fig. 2.22 Lauric acid (**C**)

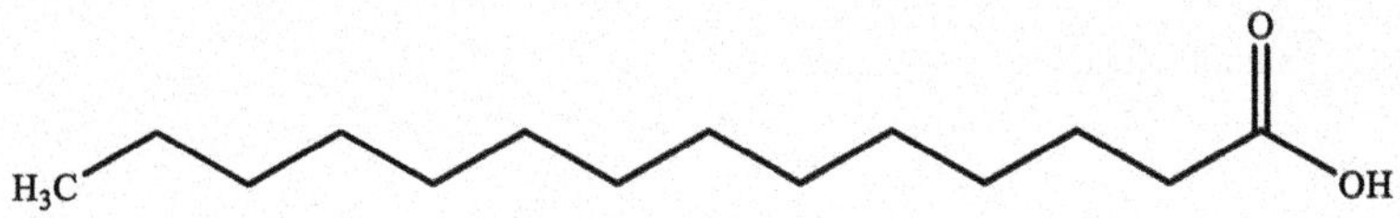

Fig. 2.23 Myristic acid (**D**)

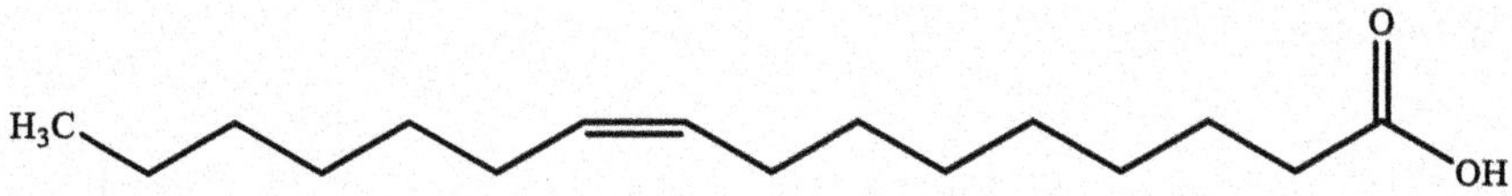

Fig. 2.24 Palmitoleic acid (**F**)

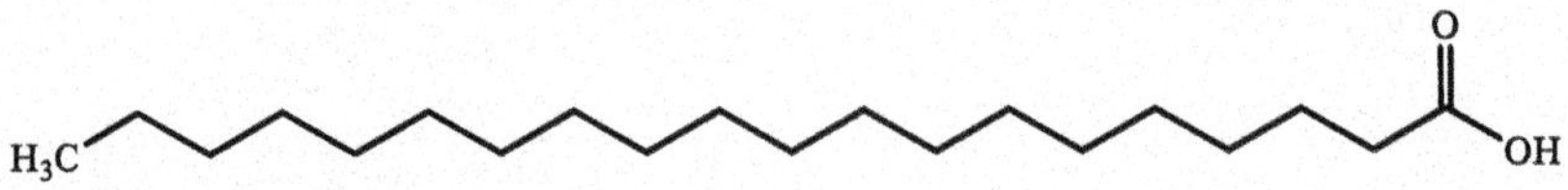

Fig. 2.25 Arachidic acid (**K**)

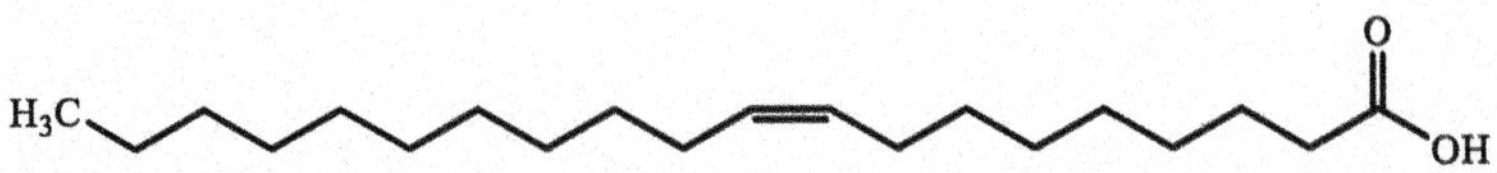

Fig. 2.26 Gadoleic acid (**L**)

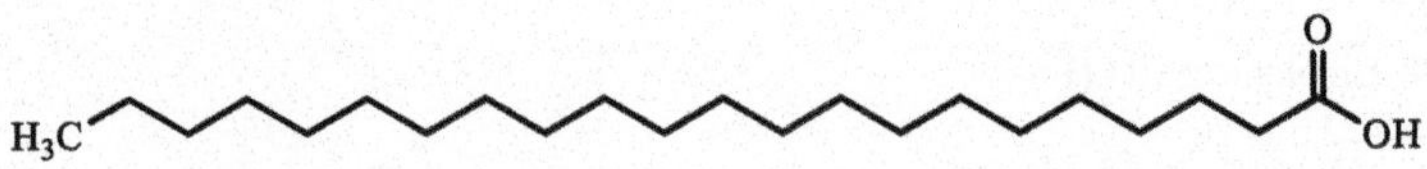

Fig. 2.27 Behenic acid (**M**)

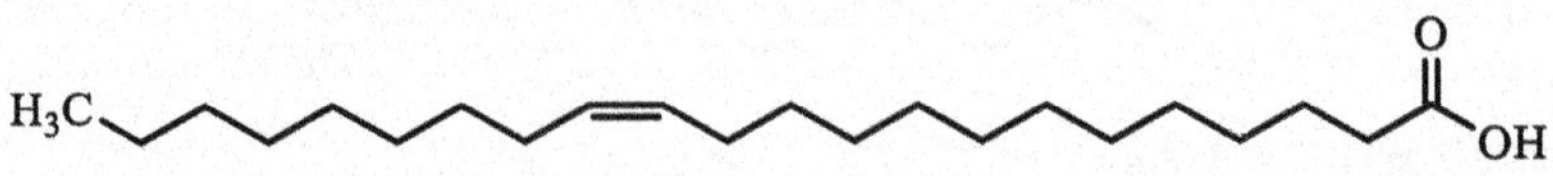

Fig. 2.28 Erucic acid (**N**)

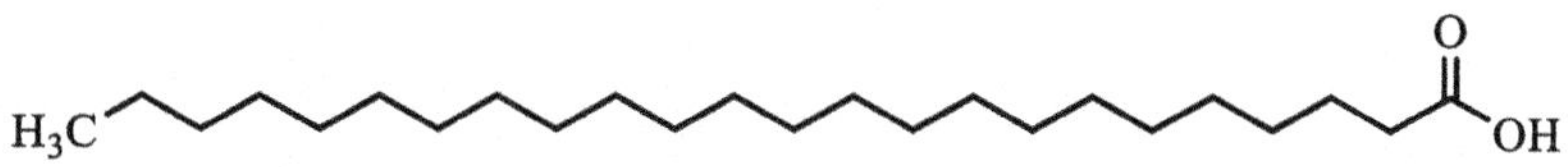

Fig. 2.29 Lignoceric acid (**O**)

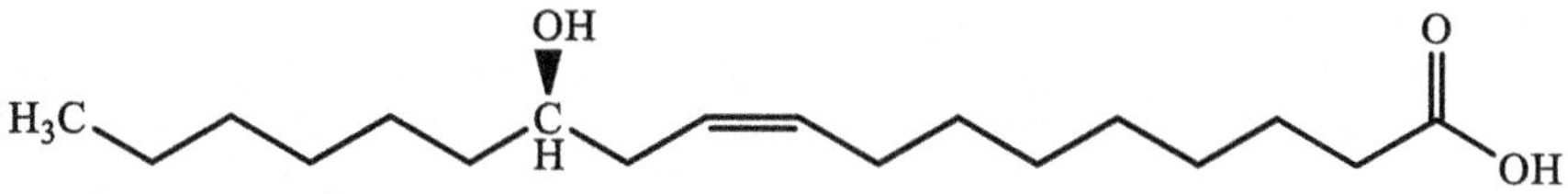

Fig. 2.30 Ricinoleic acid (**P**)

Table 2.4 Fatty acid composition of vegetable oils in percent [9]

Fatty acid	Palmitic acid (E) 16:0	Stearic acid (G) 18:0	Oleic acid (H) 18:1	Linoleic acid (I) 18:2	Linolenic acid (J) 18:3
Palm oil	40	4–6	38–41	8–12	–
Olive oil	7–16	1–3	64–86	4–15	0.5–1
Rapeseed oil	2–4	1–2	60	20	8
Sunflower oil	4–9	3–6	14–35	50–75	0.1
Corn oil	9–19	1–3	26–40	40–55	1
Pumpkin oil	7–13	6–7	24–41	46–57	–
Linseed oil	6–7	3–5	20–26	14–20	51–54
Soybean oil	7–10	3–5	22–31	49–55	6–11

Vegetable oils that are unsuitable for human consumption [10, 11] and have a similar chemical composition (Table 2.5) should also be increasingly considered for industrial purposes.

According to Table 2.2, the lauric acid (C), a saturated fatty acid is rich in babassu, coconut and palm kernel oil (about 50%). Most of the vegetable oils contains mainly oleic (H) and linoleic acids (I). Oleic acid (octadic-9-enoic acid, H) is a mono-unsaturated omega 9 fatty acid having C double bond at the 9th place (Fig. 2.17). Linoleic acid (9Z,12Z Octadecadienoic acid, I) is a polyunsaturated omega 6 fatty acid having two cis C bonds at the 9th and 12th places of the hydrocarbon chain (Fig. 2.18). Castor oil is mainly rich (about 90%) in ricinoleic acid (12-hydroxy-9-cis-octadecenoic acid, P), an unsaturated omega 9 fatty acid (Fig. 2.30). Ricinoleic acid having hydroxyl group in its 12th position of the hydrocarbon chain has unusual polar properties.

Figure 2.31 shows the major triglyceride component of castor oil. Castor oil is a vegetable oil obtained from the castor bean. The chemical structure of the vegetable oils influences the physical and chemical properties significantly.

Biological insulating liquids have the disadvantages that they have a higher viscosity and that they are prone to oxidation. A possibility is to reduce these disadvantages by a further procedural step (transesterification). Suwarno and Darma used

Table 2.5 Fatty acid composition of nonedible jatropha curcas oil [10] and castor oil [11]

Type of oil	8:0 A	10:0 B	12:0 C	14:0 D	16:0 E	16:1 F	18:0 G	18:1 H	18:2 I	18:3 J	20:0 K
Jatropha curcas	–	–	–	0.2	14.9	1.1	6.0	37.2	37.4	1.6	–
Castor	≤ 0.05	≤ 0.05	≤ 0.05	≤ 0.05	0.7–1.3	≤ 0.05	0.9–1.0	82–95	4.3–7.3	≤ 0.05	≤ 0.05

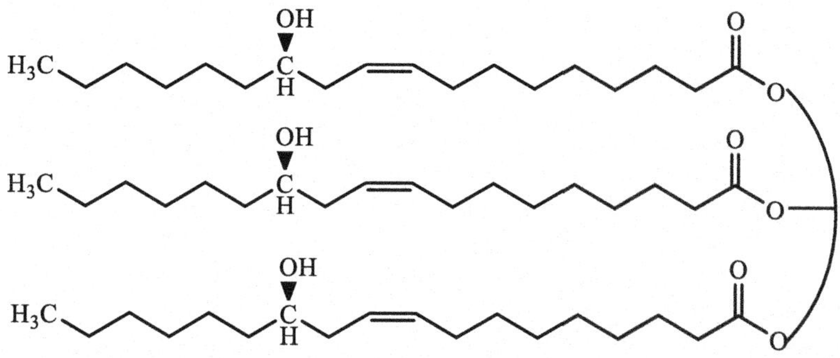

Fig. 2.31 Chemical structure of the major triglyceride of castor oil

methyl ester from palm oil [12], and Sitorus et al. from jatropha curcas oil [13] for their investigations. These methyl esters are made from palm oil and crude jatropha curcas oil by transesterification. Transesterification is a process in which simple alcohols react with ester (Eq. 2.3). Sitorus et al. neutralized the crude oil with caustic soda before transesterification process. The above mentioned benefits of the methyl ester come at the cost of a lower flash and fire point (see Sect. 3.2.3).

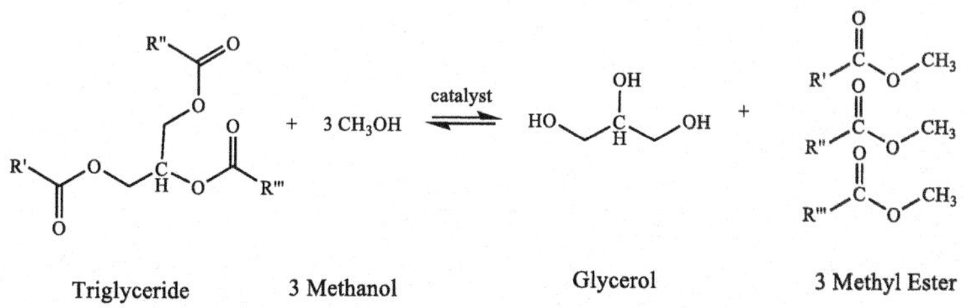

(2.3)

The chemical reaction is usually carried out with a basic catalyst (NaOH, KOH) in the complete absence of water. The bonding of alcohol and organic acid produces ester. In the transesterification process, alcohol combines with triglyceride molecule from acid to form glycerol and ester. The glycerol is then removed by density separation. Transesterification decreases the viscosity of oil, making it similar to diesel fuel in characteristics.

An excess of alcohol is needed to accelerate the reaction. With methyl alcohol, glycerol separation occurs readily. If water is present, the catalyst enter into the reaction and soap is the byproduct, which results in decreasing yield of ester (Eq. 2.4).

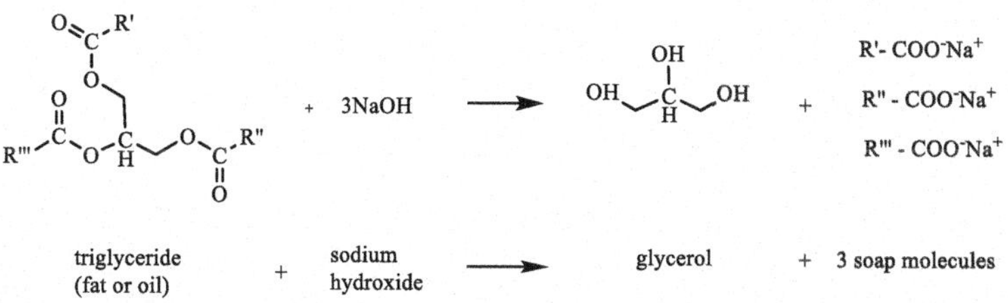

(2.4)

2.4 Silicon Oil

Silicone oils belong in many cases to the class of linear polydimethylsiloxanes (Me_3Si-(O-$SiMe_2$)$_X$-$OSiMe_3$). However, the methyl groups can be partially substituted by other organic groups such are phenyl, vinyl, respectively. These groups define the molecular weight of the silicone liquid, which in turn determines its viscosity.

$$RMe_2Si-(O-SiRMe)_X-OSiMe_2R$$

The FTIR spectra (Fig. 2.32) shows stretching vibration in the range of wave number 2800 to 3025 cm^{-1} and deformation vibration of the methyl groups in the range of 1425 to 1480 cm^{-1}, respectively. The $Si(CH_3)_2$ groups are responsible for the intense bands between 1300 to 1225 and 900 to 775 cm^{-1}.The silicone oil shows the clearest difference in the FTIR spectrum within the insulating liquids. Using the FTIR spectra, it is possible to differentiate between mineral oil, natural ester, synthetic ester, and silicone oil, but not within the respective ester liquids.

Silicone oils are odorless, colorless, hydrophobic, and thermally extremely stable. Their fire profile shows that they are difficult to ignite, and that the formation of silica crust inhibiting the combustion process. They are practically insoluble in water and in polar solvents, but very soluble in organic solvents. Because of the chemical stability, they are resistant to aging. Even in the presence of air, silicone fluids are practically indefinitely stable up to temperatures of 150 °C. Due to their excellent thermal properties (high flash- and fire point), these liquids were intended as a replacement for

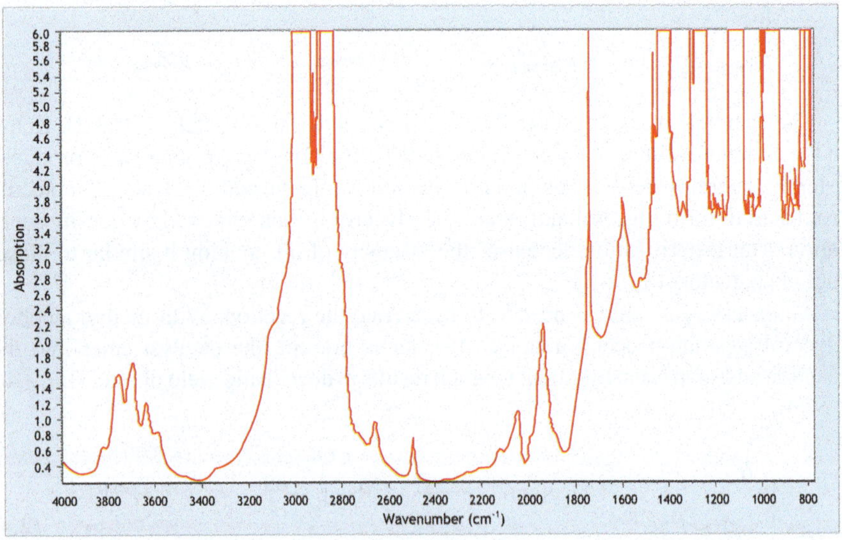

Fig. 2.32 FTIR spectra of silicone oil Powersil Fluid TR 50 [3]

the PCB-containing Askarel. In particular, the high price has hindered its widespread use up to now.

In addition to the silicone fluids, silicone rubber also consists of polydimethylsiloxane, which with molecular weights of 300,000 to 500,000 g/mol also has excellent electrical properties. The good insulating properties of silicones can also be transferred to silicone composite insulators. The high insulation strength paired with the mechanical strength and the advantages of the fracture properties compared to classic ceramics is the reason that this material is used in the field of over-voltage conductor and instrument transformers.

2.5 Iodine Number of Different Insulating Liquids

The degree of saturation/unsaturation is indicated by the iodine value (number) of the oil. The iodine value equals the number of grams of iodine required to saturate the fatty acids present in a sample of 100 g of oil or fat. Iodine values are often used to determine the amount of unsaturation in fatty acids. This unsaturation is in the form of double bonds, which react with iodine compounds. The higher the iodine number, the more $C = C$ bonds are present in the oil or fat.

Kurzweil et al. studied the properties—including the iodine number—of different insulating liquids (Table 2.6) [14]. The table shows that mineral oil based, and synthetic ester based insulating liquids have by a threshold value of 10 g iodine per 100 g insulating liquid nearly no unsaturated hydrocarbons. For this reason, they are robust against attacks of oxygen. The table depicts also that the high iodine number corelates with a higher pour point. This is because vegetable oils have long fatty acid with more or less double bounds in their molecule.

Table 2.6 Iodine number

Parameter	Nynas Nytro Taurus	Nynas Nytro Bio 300X	Midel 7131	Midel eN 1204	Envirotemp FR3
Iodine number [g/100 g]	1.9	1.6	1.1	118.8	133.0
Density (20 °C) [g/cm^3]	0.859	0.782	0.967	0.916	0.920
Kinematic viscosity (40 °C) [mm^2/s]	9.95	3.72	29.5	36.9	34.5
Pour point [°C]	−57	−60	−56	−31	−21

2.6 Content of Halogens of Different Insulating Liquids

The statement made by Shah and Tahir in [1] that natural esters do not contain halogens compared to mineral oil is incorrect. For biological insulating fluids, this depends on the soil and the environmental conditions where the plant is growing. Pagger's analyzes in [3] show that biological insulating liquids have a significantly higher halogen content compared to mineral oil. Surprisingly, the halogen content of the silicone oil is in the range of natural esters, which must be due to the production process (Table 2.7). The synthetic ester liquids are halogen free.

2.7 Additives

To adapt and improve the properties of the dielectric insulating liquids, various additives are added to them. The content of these substances is in the very low percentage range. Unfortunately, these are mostly not specified by the manufacturers of synthetic and natural esters, which makes it very difficult to control them during use.

2.7.1 Antioxidants

Insulating liquids will change over time when exposed to oxygen, radiation, excessive heat and/or corrosive environments. Oxygen accelerates the chemical breakdown of the insulating liquids and is therefore bad for all insulating liquids. Reaction between olefinic esters and triplet oxygen is a radical chain process involving three stages of initiation, propagation, and termination. In the initiation stage, an allylic hydrogen atom is removed and a resonance- stabilized radical is produced. For biological insulating liquids, oxidation is above all a process between molecules of unsaturated fatty acids and oxygen. The weak spot is the $C = C$ double bond in unsaturated fatty acids (that is why their high content means poor oxidation stability). During the oxidation process the highly reactive free radicals are created and released into the insulating liquid in which they further react. The oxidation is a slow process, but can be accelerated by the presence of metal (cooper), heat and light. Oxidation can be prevented or slowed down with the addition of antioxidants. These additives react with free radicals and other oxidative products and make them unable to further react, reducing the oxidation speed.

The ratio of monounsaturated to polyunsaturated is a measure of the tendency of vegetable oil to undergo autoxidation. The higher ratios indicate greater oxidative stability of vegetable oils. To prolong the storage life of vegetable oils, various natural and synthetic antioxidants such as α-tocopherol acetate (α-T, Fig. 2.33), citric acid (CA, Fig. 2.34), L-ascorbic acid (AA, Fig. 2.35), tertiary butyl hydroquinone (TBHQ,

Table 2.7 Content of halogens

Dielectric insulation liquid	Nynas Nytro 4000X	Envirotemp® FR3™ fluid	Biotemp®	Midel® eN 1204	Midel® 7131	BecFluid® 9902	Siliconöl Powersil® fluid TR 50
Sample 1 [ppm]	20	167	184	85	≤ 1	≤ 1	120
Sample 2 [ppm]	24	216	154	85	≤ 1	≤ 1	180
Sample 3 [ppm]	–	187	139	–	–	–	
Mean value [ppm]	**22**	**190**	**159**	**85**	**≤1**	**≤1**	**150**

Fig. 2.36), butylated hydroxytoluene (BHT, Fig. 2.37), butylated hydroxy anisole (BHA) a mixture of 2-tert-Butyl-4-hydroxyanisol and 3-tert-Butyl-4-hydroxyanisol (Fig. 2.38) and propyl gallate (PG, Fig. 2.39) are currently used. Except for non-inhibited mineral oil, almost all dielectric insulating fluids contain antioxidants.

The most common additive used to prevent oxidation is tertiary butyl hydroquinone. A substance that is readily soluble at use levels in fats and oils in a number of food-grade solvents and for this reason is also very often used in the food industry under the name E 319 as a preservative in the range up to 200 mg/kg fat or oil. In this case, tertiary butyl hydroquinone reacts with oxygen to water and tertiary butyl para benzoquinone (Eq. 2.5).

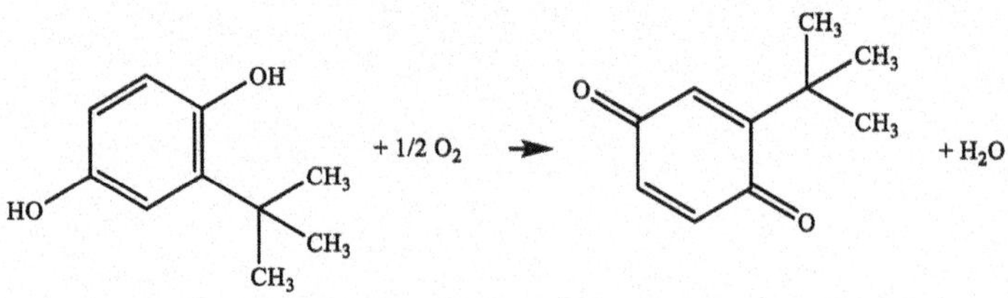

(2.5)

Studies demonstrate that phenolic antioxidants like TBHQ exhibit significant decomposition at elevated temperatures giving rise several breakdown products which may undergo further decomposition. At elevated temperatures both evaporation and decomposition result in a loss of antioxidative activity [15].

In the case of the "Biotemp" insulating liquid, this TBHQ could be clearly detected (Fig. 2.40). Most frequently used in insulation technology are the antioxidants dibutyl-para-kresol (2,6-di-tert-butyl-4-methylphenol) (Fig. 2.37) and 2,6-di-tert-butyl-phenol (Fig. 2.41). Pagger [3] detected dibutyl-para-kresol in mineral oil

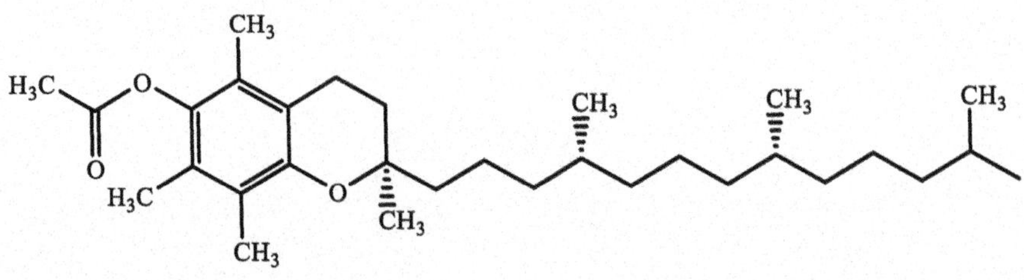

Fig. 2.33 α-Tocopherol acetate

Fig. 2.34 Citric acid

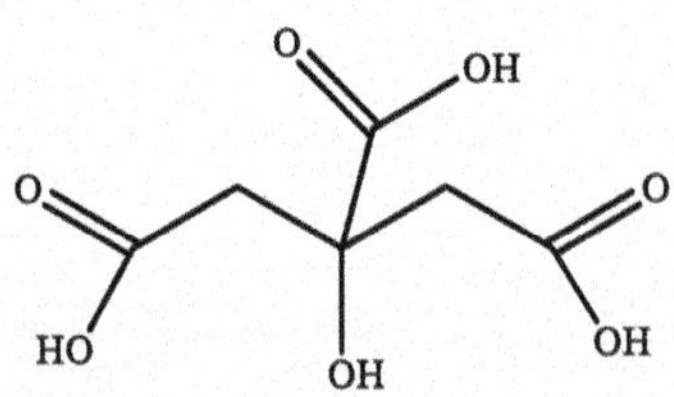

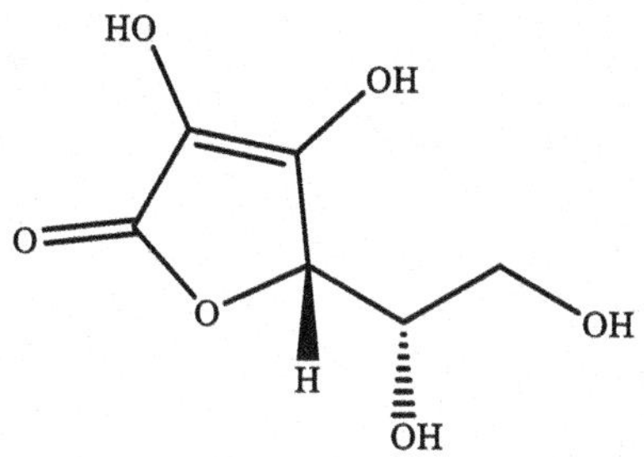

Fig. 2.35 L-Ascorbic acid

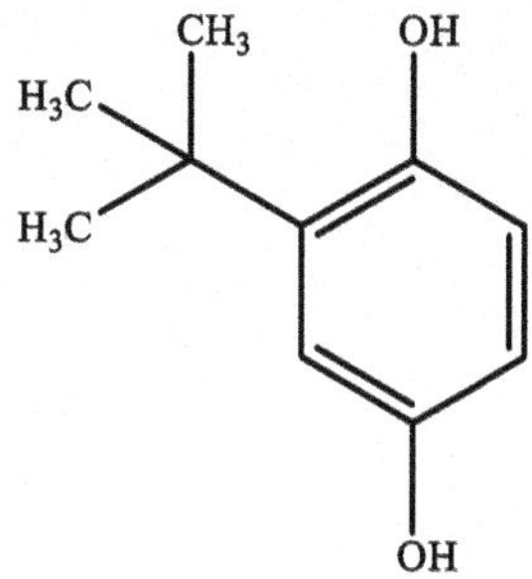

Fig. 2.36 Tert-Butylhydroquinone TBHQ

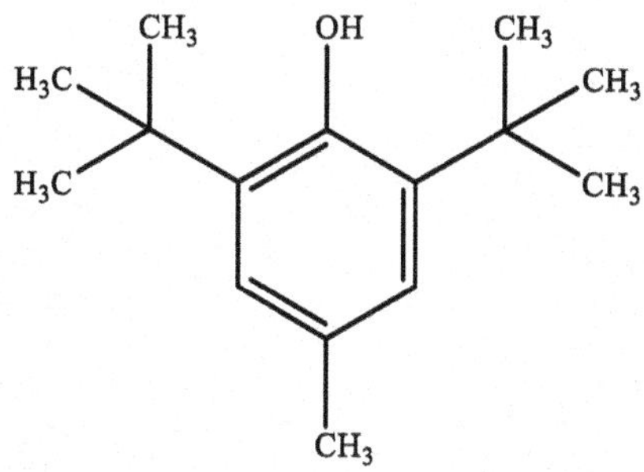

Fig. 2.37 2,6-Di-tert-butyl-p-kresol (BHT)

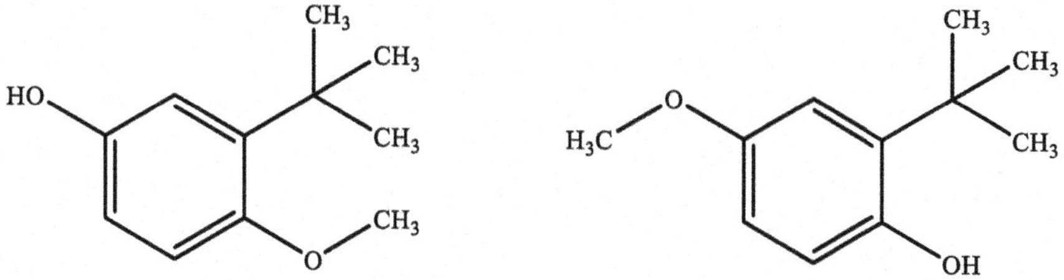

Fig. 2.38 2-tert-Butyl-4-hydroxyanisol (on the left) and 3-tert-Butyl-4-hydroxyanisol (on the right); BHA

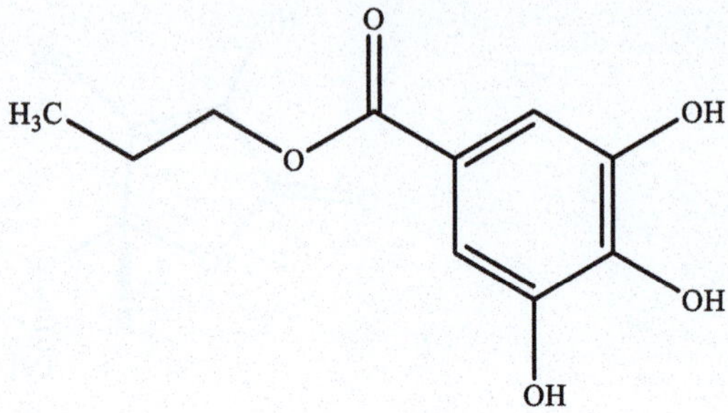

Fig. 2.39 Propyl gallate

Nynas 4000X (Fig. 2.42) and in the natural ester FR3 (Fig. 2.43), respectively. Some sulfur compounds present in mineral oils act as inhibitors, too.

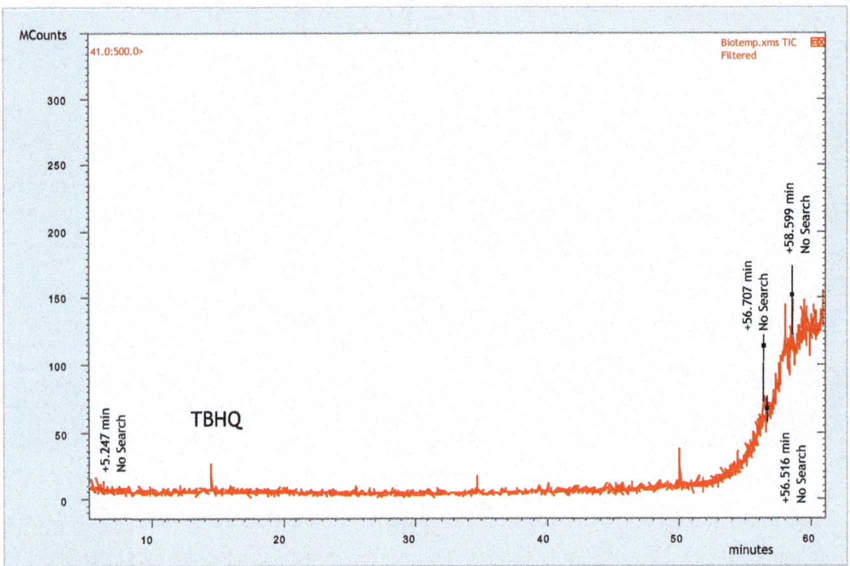

Fig. 2.40 Biotemp—MS chromatogram

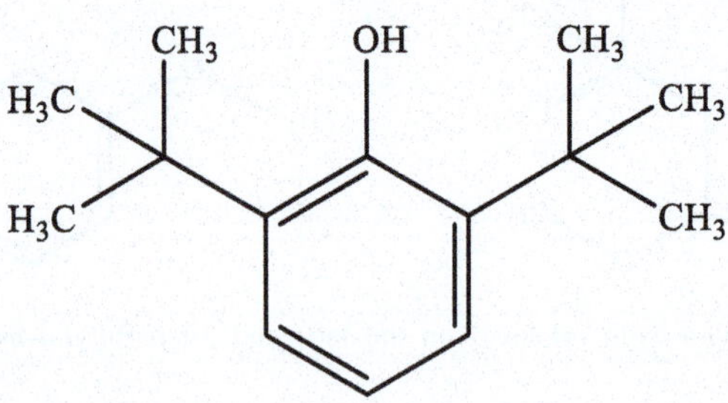

Fig. 2.41 2,6-Di-tert-butyl-phenol

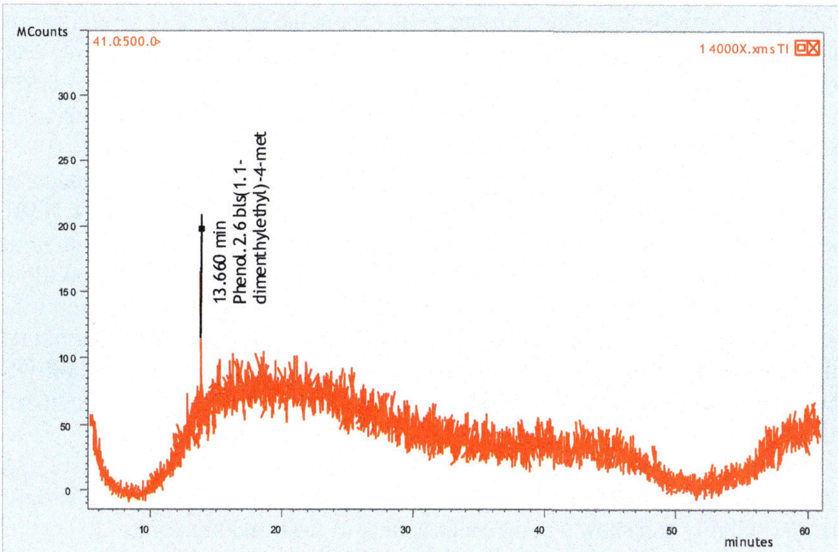

Fig. 2.42 Nynas Nitro 4000X—MS chromatogram

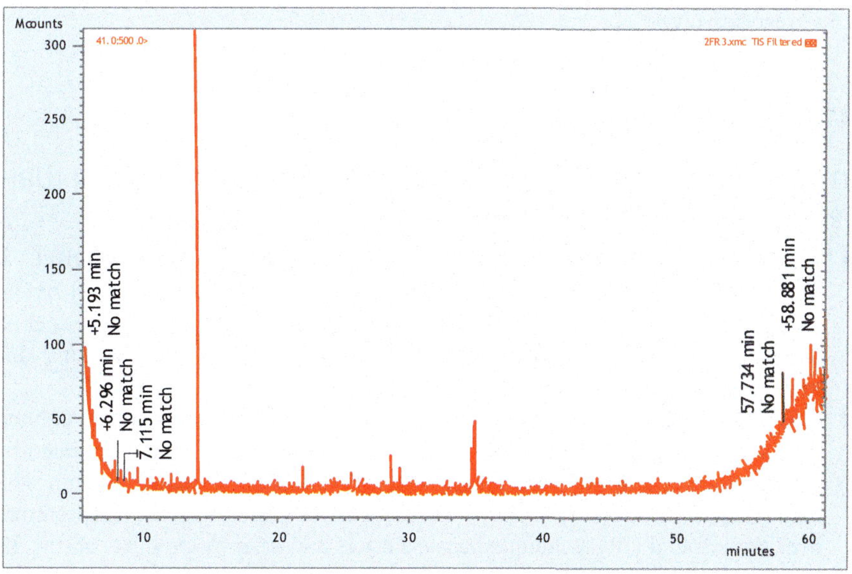

Fig. 2.43 Envirotemp FR3—MS chromatogram

In [16] Thanigaiselvan et al. present results about the influence of several antioxidants and mixture of them on properties of two insulating liquids (rapeseed oil and pongamia pinnata oil) like breakdown voltage, flash and fire point, and viscosity. In general, they found mostly positive effects on the properties compared to the insulating liquid without antioxidants.

A mixture of rapeseed oil (breakdown value: 32 kV) with BHA and BHT increased the breakdown voltage by 59% and with BHT and α-T by 63% respectively. Nearly the same effect could be observed with pongamia pinnata oil (breakdown value: 27 kV). In this case the mixture with BHA and BHT increased the break down voltage by 88% and with BHT and α-T by 59%, respectively. Though it seems that the percentage enhancement in breakdown voltage is more for pongamia pinnata oil, the mean value of breakdown voltage is less than that of the rapeseed oil. The combination of BHA and BHT, BHA and α-T enhances the dielectric strength to a satisfied level.

The main deviations in case of flash and fire points are tabulated in Table 2.8. Synthetic antioxidants like BHA, and BHT with higher concentration have more thermal stability and show a good enhancement of flash and fire points.

In [16] Thanigaiselvan et al. show that all antioxidants and their combination decrement viscosity at a temperature of 40 °C. For rapeseed oil with a starting value of 62 mm^2/s the reduction comprised up to 32% and for pongamia pinnata oil up to 25%, respectively.

2.7.1.1 Antioxidants—Test Methods

The 2,6-di-tert-butyl-4-methylphenol (DBPC) and 2,6-di-tert-butyl-phenol (DBP) content can be determined according to

- ASTM D2668: This method uses the infrared absorption band of DBPC at 3650 cm^{-1} and at 860 cm^{-1} for DBP, respectively. Interference by oil oxidation products may be difficult to eliminate with this method [17]. Pisareva et al. describe in [18] the characteristic absorption band for DBPC at 3631 cm^{-1} due to the stretching vibration of the hydroxyl group.
- ASTM D4768: After removing interfering substances on an extraction column with suitable solvent, the inhibitors are recovered in methanol and separated by gas chromatography (GC) on a nonpolar, silicone-bonded packed column with flam ionization detector (FID) [19]. The GC method is more specific and preferred over the infrared (IR) technique because esters and ester by-products absorb IR in the same region(s) as the inhibitor additives.

Table 2.8 Flash and fire point deviation because of antioxidants

Antioxidants	Concentration [%]	Rapeseed oil				Pongamia ponnata oil			
		Flash point		Fire point		Flash point		Fire point	
		[°C]	Deviation [%]	[°C]	Deviation [%]	[°C]	Deviation [%]	[°C]	Deviation [%]
α-T	0.5	338	+5.3	340	±0	232	−3.3	242	−5.1
AA	0.5	310	−3.1	315	−7.4	234	−2.5	246	−3.5
CA	0.5	305	−4.7	320	−5.9	230	−4.2	242	−5.1
BHA + BHT	0.5 + 0.5	360	+12.5	370	+8.8	258	+7.5	270	+5.9
BHT + α-T	0.5 + 0.5	359	+12.2	369	+8.5	265	+20.4	275	+7.8

2.7.2 Pour Point Depressant

The lowest temperature at which a liquid flow under prescribed conditions is known as the pour point. The ability of a dielectric insulating liquid to work under cold conditions is very important for liquid circulation and mechanical parts, like switchgears. Flow-improving additives enhance the viscosity of biological insulating liquids at low temperatures, without eliminating their Newtonian character. Without the proper selection and treat rate of a pour point depressant, nearly all biological insulating liquids will exhibit poor low temperature properties, leading, in the worst case to liquid hardening and equipment failure. These liquids contain triglycerides instead of paraffinic wax, but the structure of these molecules is wax like.

All biological insulating liquids are mixtures of different fatty acids with pour points in a wide range. As the temperature of the biological insulating liquid is decreased, saturated long chain acids come out of the liquid as tiny crystals, and the liquid begins to appear hazy to the naked eye. The temperature at which this occurs is called the cloud point.

Over the years a wide variety of synthetic materials has been introduced commercially as pour point depressant. Poly alkyl methacrylate (Fig. 2.44), one of the first of the polymeric pour point depressant, continue to be viewed as the best chemistry available today, with a worldwide market share that far outstrips alternatives. The primary reason for the widespread preference is the molecular structure of the polymers and the tremendous flexibility in chemical structure. Pour point depressing activity is only weakly dependent on molecular weight, and the degree of polymerization (n + m) of pour point depressants may vary from about 200 to 2000.

Mechanism of action: Pour point depressants do not in any way affect either the temperature at which fatty acids or wax crystalizes from solution or the amount of them that precipitates. Rather, when fatty acids or wax crystals form, pour point depressants co-crystalize along with the precipitation present in the liquid and modify the growing pattern of fatty acid and wax crystal structures. Additionally, the fatty

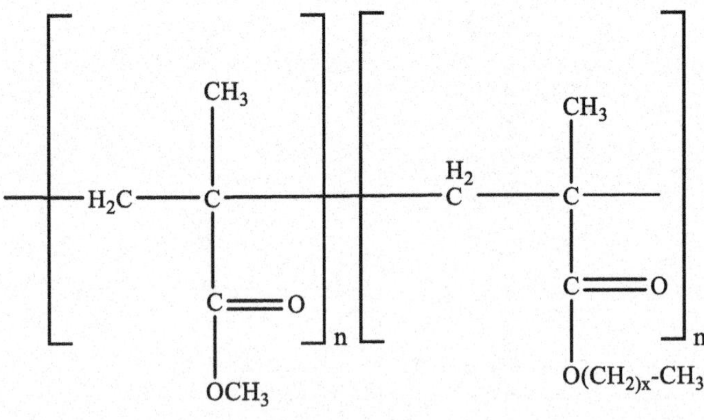

Fig. 2.44 Chemical structure of poly alkyl methacrylate

acid and wax crystals are kept apart from each other by the pour point depressant backbone, and as a result of this steric hindrance, the wax crystals are no longer able to form three-dimensional structures that inhibit flow. While prevention of gelation is aimed at ensuring the biological insulating liquid's pumpability, the liquid can be expected to still exhibit a wide range of behavior, from complete fluidity to borderline gelation. This means that mechanical parts within the transformer can still be moved.

An important fact to note is that as temperature is decreased all fluids and liquids will eventually solidify or, more accurately, become immobile, regardless of fatty acid or wax precipitation. This is simply a matter of the viscosity becoming so high that oil will not flow under the influence of gravity. This is normally called the viscous pour point. The viscosity at which the viscous pour point is reached is generally considered to be upwards of 100,000 mm^2/s.

When pour point depressants are used, it is important to recognize that these additives are themselves waxy materials. Therefore, in the process of adding a pour point depressant to a dielectric insulating liquid, one is in fact adding wax to the system. Hence, the dosage must be carefully selected to get an optimum response and pour point depressant should be avoided to prevent reversion of low-temperature properties. A properly selected pour point depressant will provide a dramatic enhancement of low-temperature performance even at low concentrations. Raising concentration may offer additional, minimal improvement. A too high pour point concentration can lead eventually to a reversal of performance gains. In the extreme, an over-treated dielectric insulating liquid may have poorer low-temperature properties than the untreated liquid (see Fig. 2.45 [20]).

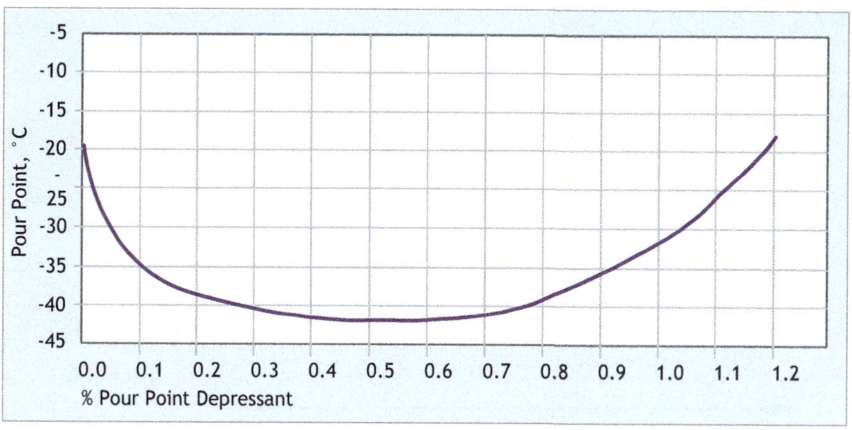

Fig. 2.45 Pour point reduction versus pour point depressant concentration

2.7.2.1 Pour Point Depressant—Test Method

The pour point of a liquid can be determined by

- ASTM D97: After preliminary heating, the sample is cooled at a specified rate and examined at intervals of 3 °C for flow characteristics. The lowest temperature at which movement of the specimen is observed is recorded as the pour point [21].

2.7.3 Passivator

Passivators, also known as metal deactivators, react with reactive metal surfaces and dissolved metals such as copper and reduce their rate of reaction with compounds in the insulating liquid (Fig. 2.46).

The harmful effects of corrosive sulfur compounds in transformer oil are well known by now. These failures occur because corrosive sulfur in the insulating liquid reacts with copper to form copper sulfide, a conductive compound. The copper sulfide can form at the copper surface or with copper ions in the insulating liquid and paper. The use of metal passivators to block reactions between sulfur and copper is a well-established mitigation technique. Passivators are composed of two basic types, sulfur based, and nitrogen based. Passivator that has been used in electrical insulating liquids are nitrogen based and have been predominantly benzotriazole or its derivatives (Figs. 2.47, 2.48 and 2.49). One passivator that has recently been suggested for use for suppression of corrosive sulfur reactions is that product, named Irgamet 39 (Fig. 2.49).

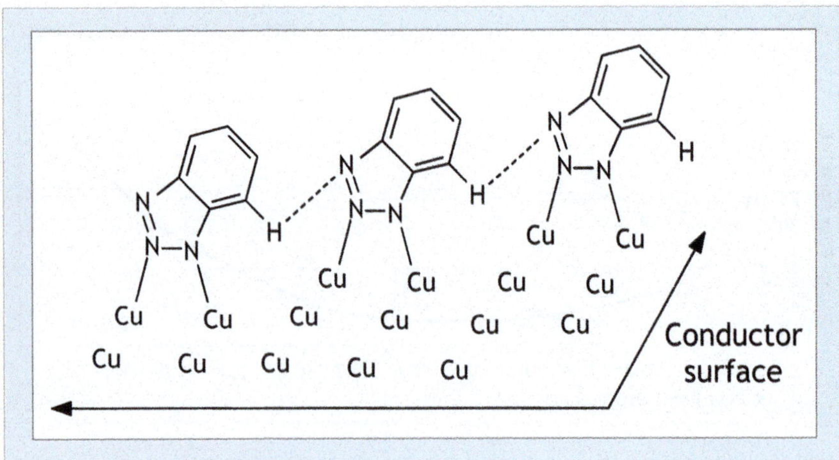

Fig. 2.46 Passivator protects copper surface

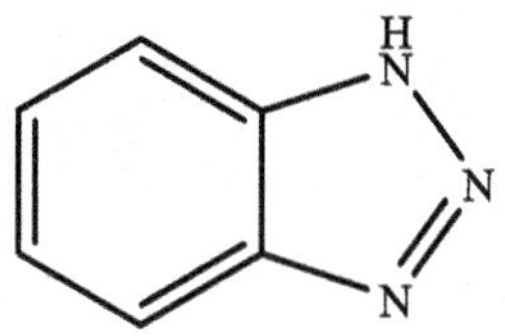

Fig. 2.47 Benzotriazol

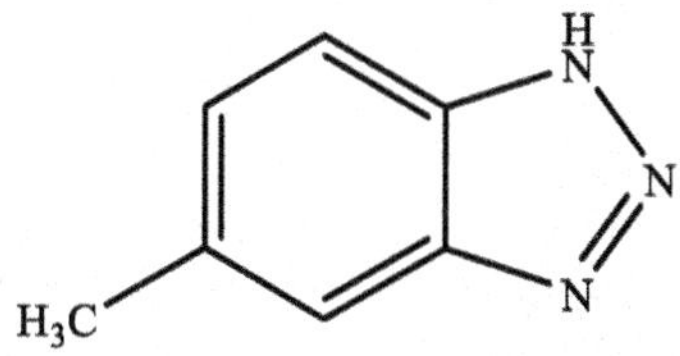

Fig. 2.48 5-methyl-1H-benzotriazol

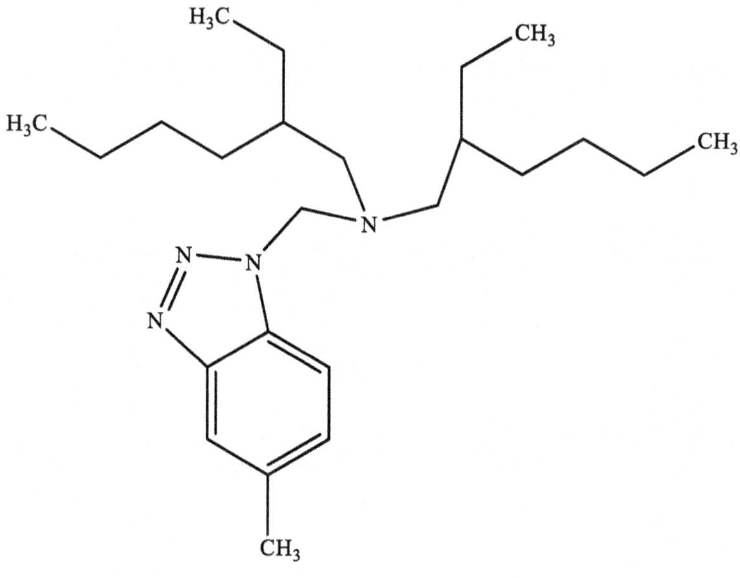

Fig. 2.49 Benzotriazol—Irgament 39 (N,N-Bis(2-ethylhexyl)-5-methyl-1H-benzotriazole-1-methylamine)

The action of the passivator is to have the nitrogen group of the benzotriazole molecule bind with copper and other reactive metal surfaces. This process is a chemical bonding that, given the right circumstances can be reversed. The passivator molecule is attracted to the metal surface and is held to the reactive site so that same site cannot be occupied (attacked) by a corrosive molecule [22]. Passivator molecules can also bind free copper ions or particles present in the bulk of insulating liquid.

For new dielectric insulating liquids the best approach is to use a liquid that does not have any amounts of corrosive sulfur compounds. Today most dielectric insulating fluids do not contain a passivator. In laboratory tests Pagger and his team were able to detect a passivator in a synthetic ester (Midel 7131) that has not yet been announced [23].

2.7.3.1 Passivator—Test Method

IEC 60666 [24]: This test method covers the determination of passivators of the family of derivatives of benzotriazole like Benzotriazol, 5-methyl-1H-benzotriazol, and N,N-Bis(2-ethylhexyl)-5-methyl-1H-benzotriazole-1-methylamine (Irgament 39) in liquids by high performance liquid chromatography (HPLC). A weighed portion of liquid is diluted with pentane and passed under vacuum through a silica gel cartridge, previously rinsed with methanol and pentane. The residue of non-polar liquid constituents retained by the solid phase is then eluted with a further volume of pentane and discarded. The analytes are eluted with a known volume of methanol. The solution is injected into a HPLC system equipped with a reverse-phase column, and passivator detected with a UV detector at a wavelength of 260–270 nm.

References

1. Shah, Z.H., and Q.A. Tahir. 2011. *Dielectric Properties of Vegetable Oils*. JSR Publications.
2. Martin, R. 2014. Fluid Maintenance and Dissolved Gas Analysis for Ester-based Transformer Liquids. In *5th Workshop, Oil/Paper and Gas-Insulated Systems in Components of Electrical Power Supply*. Graz, June 2014.
3. Pagger, E. 2013. *Alternative Insulating Liquids Compared to the Classic Mineral Oil*. Doctoral Thesis, Graz University of Technology, May 2013.
4. Colombo, C., et al. 2017. *Macauba: A Promising Tropical Palm for the Production of Vegetable Oil, OCL*. EDP Sciences. https://doi.org/10.1051/ocl/2017038.
5. Gunstone, F. 2004. *The Chemistry of Oils and Fats, Sources, Composition, Properties and Uses*. Blackwell Publishing Ltd., ISBN 1-4051-1626-9.
6. Gomna, A., et al. 2019. Review of Vegetable Oils Behavior at High Temperature for Solar Plants: Stability, Properties and Current Applications. *Elsevier, Solar Energy Materials and Solar Cells* 200.
7. Hammond, E.W. 2003. *Vegetable Oils, Types and Properties, Encyclopedia of Food Sciences and Nutrition (Second Edition)*. Elsevier.
8. Totzauer, P., and P. Trnka. 2019. Different Ways to Improve Natural Ester Oils. In *13th International Scientific Conference on Sustainable, Modern and Safe Transport (TRANSCOM 2019)*. Elsevier.
9. Bartz, W.J. 2006. *Basics of Lubricants*. Fuchs Academy, 03/2006.
10. Evangelista, Jr., et al. 2017. Development of a New Bio-Based Insulating Fluid from Jatropha crcas Oil for Power Transformers. *Advances in Chemical Engineering and Science* 7: 235–255.
11. Kumara, S., et al. 2017. Comparison of Coconut/Sesame/Castor Oils and Their Blends for Transformer Insulation. *Conference Paper*. https://doi.org/10.1109/ICHNFS.2017.8300410.
12. Suwarno, A., and I.S. Darma. 2008. Dielectric Properties of Mixtures Between Mineral Oil and Natural Ester from Palm Oil. *Wseas Transaction on Power Sytems* 3 (2), February 2008.

13. Sitorus, H.B., et al. 2014. Physiochemical and Electrical Properties of Jatropha Curcas Methyl Ester Oil as a Substitute for Mineral Oil. In *18th International Conference on Dielectric Liquids (ICDL)*. Bled, Slovenia.
14. Kurzweil, P., et al. 2021. Environmental Impact and Aging Properties of Natural and Synthetic Transformer Oils under Electrical Stress Conditions. *Advanced Sustainable Systems, Wiley-VCH GmbH*.
15. Opinion of the Scientific Panel on Food Additives, Flavorings, Processing Aids and Materials in Contact with Food on a request from the Commission related to *tertiary*-Butylhydroquinone (TBHQ). 2004. *The EFSA Journal* (2004) 84.
16. Thanigaiselvan, R., et al. 2015. Investigation on Eco friendly Insulating Fluids from Rapeseed and Pongamia Pinnata Oils for Power Transformer Applications. *Journal of Electrical Engineering and Technology*. https://doi.org/10.5370/JEET.2015.10.6.2348.
17. ASTM D2668-07. 2013. Standard Test Method for 2,6-di-tert-Butyl- p-Cresol and 2,6-di-tert-Butyl Phenol in Electrical Insulating Oil by Infrared Absorption.
18. Pisareva, S., et al. 2001. Determination of Antioxidant Ionol (2,6-Di-*ter*-Butyl-4-Methylphenol) in Transformer Oils by a Kinetic Method and IR Spectrometry. *Journal of Analytic Chemistry* 56 (10).
19. ASTM D4768-11. 2019. Standard Test Method for Analysis of 2,6-Ditertiary-Butyl Para-Cresol and 2,6-Ditertiary-Butyl Phenol in Insulating Liquids by Gas Chromatography.
20. Modi, P., et al. 2020. An Overview of Different Pour Point Deppresant Synthesized and Their Behavior On Different Crude OilS. *IJCRT* 8 (8), August 2020.
21. ASTM D97-17b. 2017. Standard Test Method for Pour Point of Petroleum Products.
22. Lewand, L. 2006. *Passivators—What They are and How They Work*. Double Engineering Company, NETA WORLD, Spring.
23. Pagger, E., and Scala, M. 2014. Diffusion and Transport Processes—"Post-Mortem and Laboratory Tests". In *5th Workshop, Oil/Paper and Gas-Insulated Systems in Components of Electrical Power Supply*. Graz, June 2014.
24. IEC 60666. 2010. Detection and Determination of Specified Additives in Mineral Insulating Oils.

Chapter 3
Production Process of Dielectric Liquids

This chapter deals with the most important processes to produce the different insulating liquids. The largest part is dedicated to vegetable oils and other oils, which can be used as insulating liquid. The production of the various dielectric insulating liquids is very different and has in some cases a lot of steps to obtain high purity insulating liquids. All the considerations are made with reference to mineral oil which serves as the basis insulating oil most commonly used.

3.1 Mineral Oil

The raw material for mineral oil is crude oil. Crude oil is currently the most important raw material and contains many chemical compounds used for many different purposes. Table 3.1 shows the range of elements most frequently found in petroleum. Petroleum is a mixture of many hydrocarbons; a distinction being made between straight chain (Fig. 3.1), branched chain (Fig. 3.2) alkanes (paraffins) and cyclic, saturated (Fig. 3.3) cycloalkanes (naphthene) and cyclic unsaturated (Fig. 3.4) aromatics.

After several process steps, a base oil from crude oil is obtained for further processing (Fig. 3.5).

Naphthenic refined oil and its binary mixtures with alkyl benzene (ethyl benzene and cumene) at different concentrations have been employed as insulating oil in a wide variety of electrical equipment. Petroleum is chemical very stable in the absence of oxygen, proves its underground storage over millions of years. However, it is a long way from the raw material crude oil to the finished insulating liquid. Extensive process steps such as distillation, extraction, absorption, and hydrogenation are necessary to achieve the finished product from the raw material. During the procedural process,

E. P. Pagger et al., *Biological Insulating Liquids*,
https://doi.org/10.1007/978-3-031-22460-7_3

Table 3.1 Crude oil—chemical composition [%]

Carbon	83.0–87.0
Hydrogen	10.0–14.0
Sulfur	0.05–6.0
Nitrogen	0.1–2.0
Oxygen	0.05–1.5
Metals	0.00–0.14

Fig. 3.1 Straight-chain aliphatic molecule

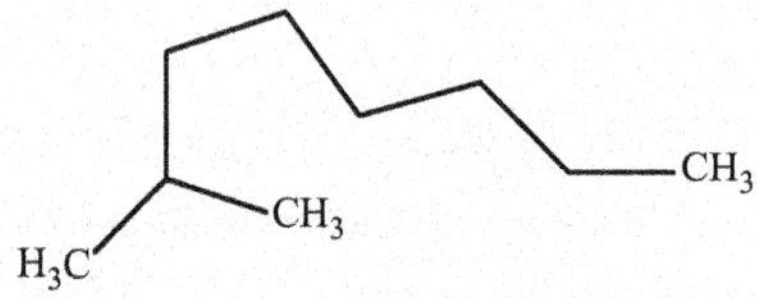

Fig. 3.2 Branched-chain aliphatic molecule

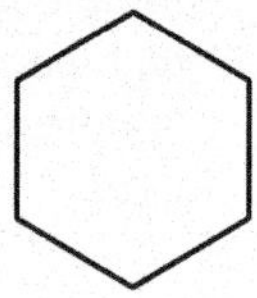

Fig. 3.3 Cycloalkane (left hand side)

there are possibilities to intervene in the process in a regulatory manner and thus to influence the quality of the product. A basic distinction is made between

- inhibited
- and non-inhibited mineral oil.

3.2 Ester Liquids

The chemical structure of ester molecules and their origin is completely different compared to mineral oil. The functional ester-group comes from the chemical linkage which is formed from the reaction of an alcohol and a carboxylic acid in the presence of sulfuric acid by elimination of water (Eq. 3.1).

Fig. 3.4 Aromatic molecule—ring structure (right hand side)

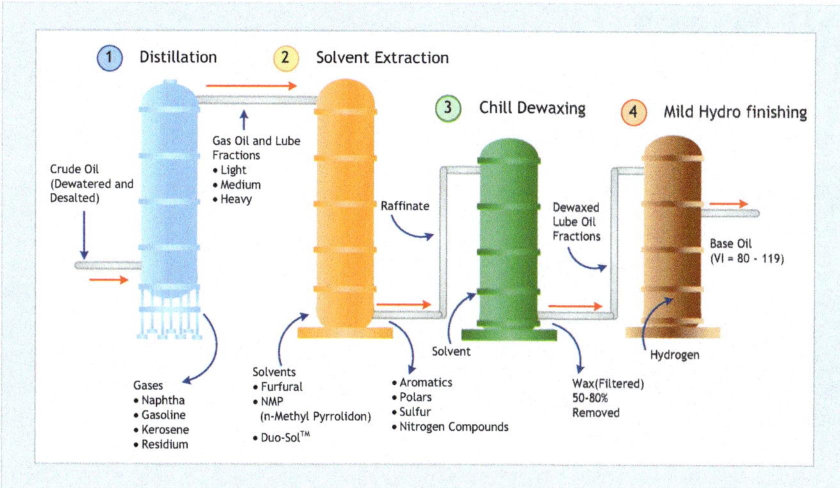

Fig. 3.5 Base oil production

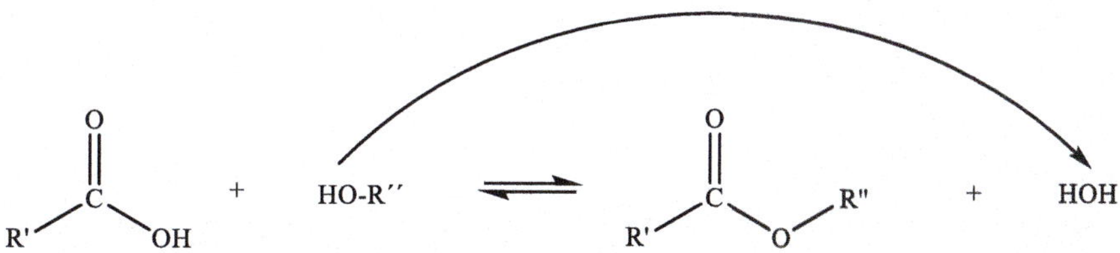

(3.1)

$$K = \frac{[c_C] * [c_D]}{[c_A] * [c_B]}$$

In this reaction, there is an equilibrium between starting materials and products, which can be shifted in the desired direction by the excess of one of the components. The law of mass action, where K is a constant, shows that for a large recovery of ester (C) as much water as possible must be eliminated. Substances such as sulfuric acid are used as water absorbents. The ester linkage occurs in both synthetic- and natural ester. There can be up to four ester-groups in one molecule. Every day we are confronted with various esters as fruit-aromatic substances (Figs. 3.6, 3.7, 3.8, 3.9, 3.10 and 3.11).

Vegetable, renewable oils form the basis for the biological insulating fluids, while carboxylic acids and pentaerythritol stemming from petroleum crude oil form the

Fig. 3.6 Ethyl formate (smells like rum)

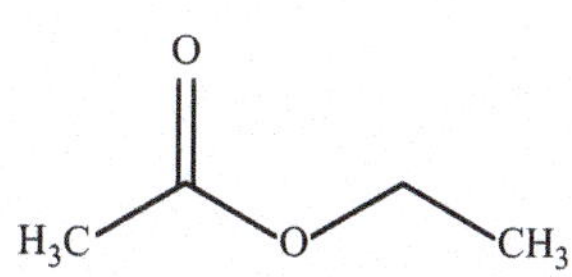

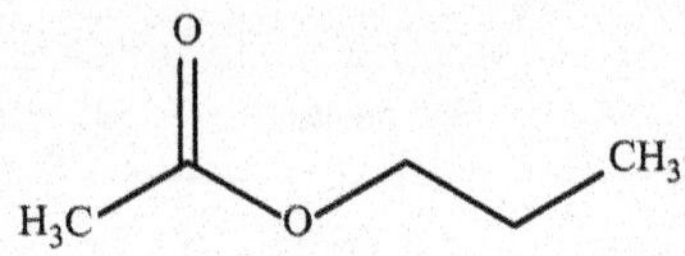

Fig. 3.7 Propyl ethanoate (smells like pear)

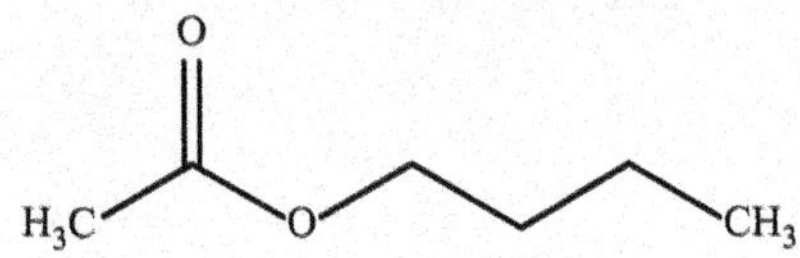

Fig. 3.8 Butyl acetate (smells like pear, banana)

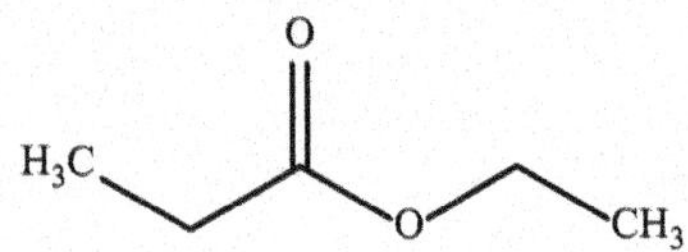

Fig. 3.9 Ethyl propionate (smells like rum)

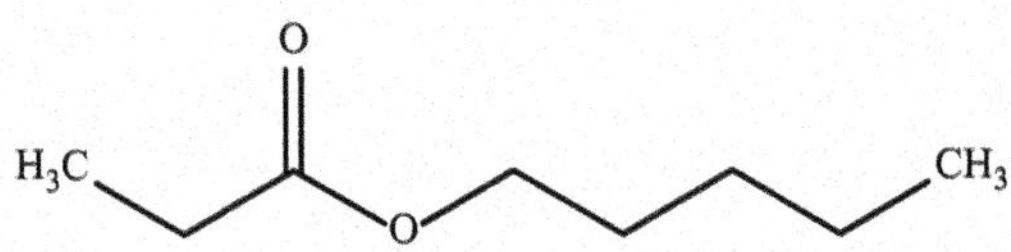

Fig. 3.10 Pentyl propionate (smells like apple)

basis for the synthetic esters. Compared to mineral oil, biological insulating liquids have a lower oxidation stability and not suitable for free breathing transformers.

3.2.1 Synthetic Ester

Synthetic esters are derived from petrochemicals and made in a lab. The raw materials used to make esters can be linear, branched, saturated, unsaturated, monofunctional, difunctional, or polyfunctional. For insulating liquids, they are usually the product of a polyol (a molecule with more than one alcohol functional group—mostly pentaerythritol) with carboxylic acids. The acids can be derived from synthetic- or natural origin. The main chemical-physical differences are achieved by varying the molecular chains R^x (Eq. 3.2).

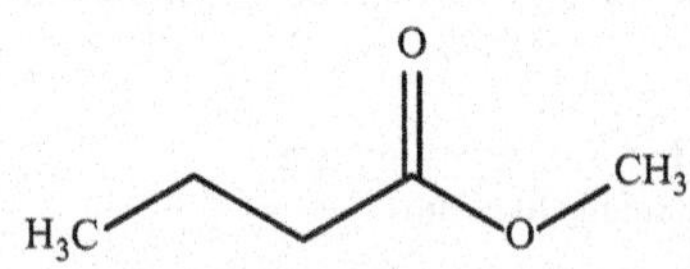

Fig. 3.11 Methyl butonate (smells like ananas)

Fig. 3.12 Autoclave vessel

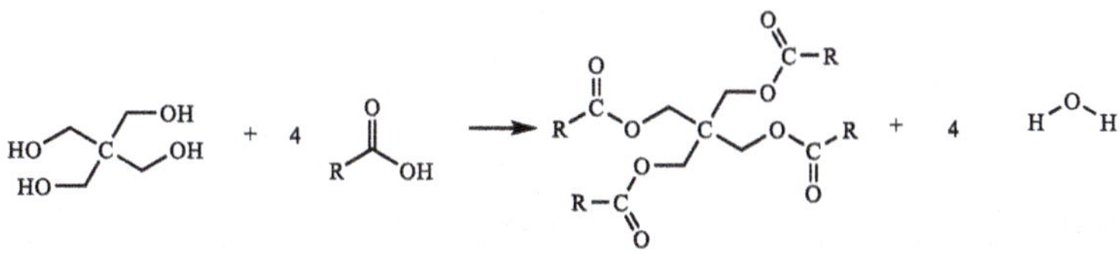

(3.2)

While these are fixed for the natural esters depending on the type of plant, they can be freely selected for the synthetic ester. Hence, for synthetic ester the acids used are usually saturated (no C=C double bonds) in the chain, giving them a very stable chemical structure with a reduction possibility of attacks by oxygen. Synthetic esters are produced at elevated temperatures in pressurized autoclave vessels (Fig. 3.12).

3.2.2 Biological Insulating Liquids (Natural Esters)

Biological insulating liquids are produced from vegetable oils, manufactured from renewable plant crops. Plants produce these liquids as part of their natural growth cycle. Such oils have been part of human culture for millennia. The first oil producer was primitive man who employed a stone mortar to squeeze fallen olives and noticed that oil was released (ca. 5000 BC). Edible vegetable oils are also used in food, both in cooking and as supplements. Many oils, edible and otherwise, are burned as fuel, such as in oil lamps and as a substitute for petroleum-based fuels. There are several types of plant oils, distinguished by the method used to extract the oil from the plant. To produce vegetable oils, the oil first needs to be removed from the oil-bearing plant components, typically seeds. Crushing, pressing, and oil extraction are upstream processes for getting natural crude oils (Fig. 3.13 [1]). This can be done via mechanical extraction using an oil mill or chemical extraction using a solvent. The extracted oil can then be purified and, if required, refined, or chemically altered. The vegetable crude oil will usually contain at least 95% of triacylglycerols.

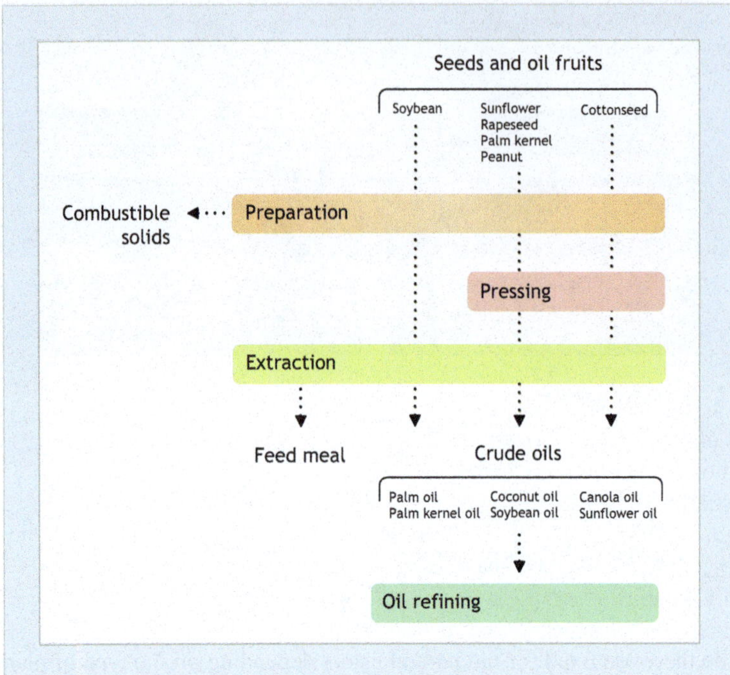

Fig. 3.13 From seeds and fruits to the natural crude oil [1]

For mechanical extraction, the relevant part of the plant may be placed under pressure to extract the oil, giving a pressed oil. The improvement of oil processing has been influenced by technological evolution over the years. The pressing system has been continuously improved by the introduction of electrically driven hydraulic pumps, cage, and column presses and, more recently, open mono-block super-presses that permit reaching pressures of 350–500 atmospheres [2]. Nowadays, mostly there are three extraction systems in use—pressure, percolation, and centrifugation. Depending on the source material (seeds, cereal, fruit), mechanical crushing steps are carried out beforehand. Centrifugation is a process that can separate oil from the other materials (water, solids, etc.) based on Stoke's law that determines the speed at which non-miscible liquids are separated under centrifugal force. Separation of the oil from the rest of the liquor can also be achieved by setting tanks based on gravity. Trying should be done by vacuum dryer. All produced biological insulating liquids should be protected from oxidation (sealed container, nitrogen blanketing). Another method for separating the vegetable oil from the plant is solvent extraction. This process produces higher yields, is quicker and less expensive. The most common solvent is petroleum-derived hexane. When the seeds are not very rich in oil, pre-pressing is not undertaken and the seeds, suitable adjusted to size and moisture content, are extracted directly with solvent. The solvent is recovered for re-use, but there is a loss of 8–10 L/tone of meal and 2 L/tone of oil. The oil yields

Table 3.2 Yields of oil and of meal obtained by extraction of the major oilseeds [3]

Oilseed	Oil (%)	Meal (%)
Soybean	18.3	79.5
Cottonseed	15.1	57.4
Groundnut	40.3	57.2
Sunflower	40.9	46.9
Rapeseed	38.6	60.3
Palm kernel	44.6	54.0
Copra	62.4	35.4
Linseed	33.3	64.2

of some common vegetable sources are given in Table 3.2. This technique is used for most of the "newer" industrial oils such as soybean and corn oils. Supercritical carbon dioxide can be used as a non-toxic alternative to other solvents. Industrial plants usually combine mechanical and solvent extraction processes to extract oil from seeds.

Generally, crude vegetable oils, extracted from oil seeds, have a dark color, and contain solid constituents such as proteins and fibers. These vegetable oils are treated to obtain a refined, bleached, and deodorized oil which is the starting material used to develop a candidate fluid (Fig. 3.14).

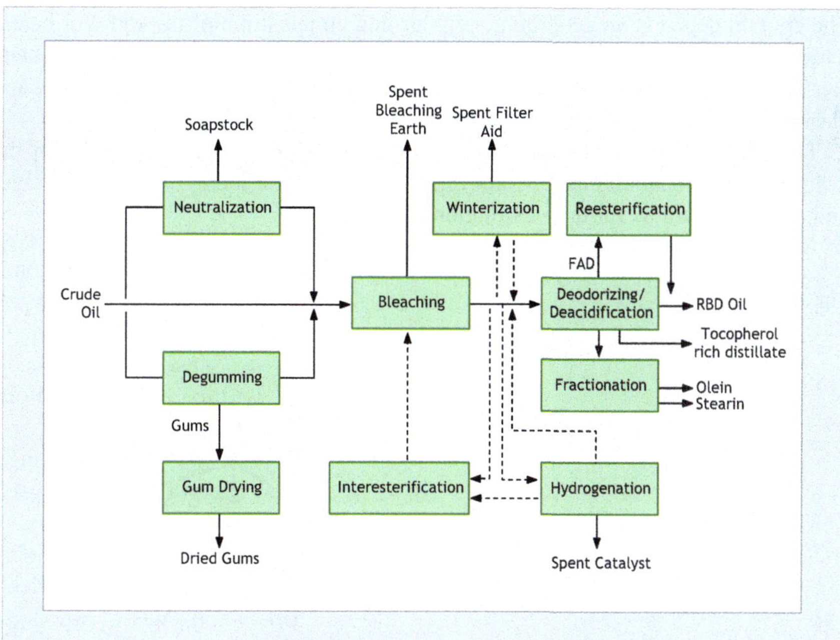

Fig. 3.14 The way from crude oil to the refined, bleached, deodorized (RBD) oil

The procedure to refine the vegetable oil consists of different steps like alkaline refinement, bleaching and deodorization. To deodorize and remove pigments, soap, and other impurities, the clarification is commonly applied in edible oil refining. For the oil clarification a combination of activated carbon and alkaline earth oxide is useful. The clarified oil should be filtered under vacuum with quantitative filter paper to remove the solids.

In general, vegetable oils are constituted by triacylglycerol molecules, whose hydrolysis produces free fatty acids, which are responsible for the increase of the acidity of these oils. For this reason, these acids must be neutralized and removed. Alkaline refinement is the first step to eliminate free fatty acids in vegetable oil. The neutralization can be done by sodium hydroxide (Eq. 3.3).

$$RCOOH + NaOH \rightarrow RCOONa + H_2O \tag{3.3}$$

Bleaching is the second step to eliminate coloring materials. In this step, clay filter press which further purify the oil are usually used. Finally, deodorization which is a high-temperature, high-vacuum steam-distillation process to remove volatile and odoriferous materials is performed. The drying process should remove water without increasing the acidity. Additionally, a winterization process may also remove easily freezing saturated fats and improve the pour point of the base vegetable oil. After refining the content of triacylglycerol will rise to 97–99% depending on the level of (unsaponifiable) material insoluble in aqueous alkali after hydrolysis [3].

Jatropha curcas crude oil contains toxins known as phorbol esters (Fig. 3.15). The contain depends on geographic regions and composition of the soil. For health reasons, the toxin should be removed from the oil. Most part of phorbol esters can be extracted with methanol and also the processing steps of neutralization and clarification led to significant reduction of phorbol esters [4].

The biological insulating liquids have an advantage of local manufacturing by the availablity of a local oil plant. Moreover, they require less energy for their treatment and simple apparatus for their extraction.

Fig. 3.15 Phorbol

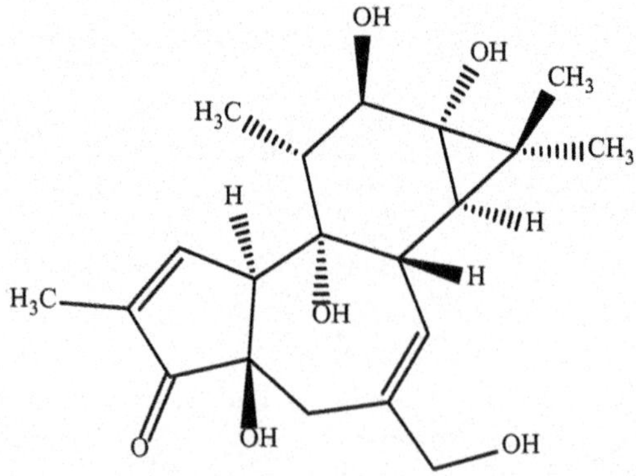

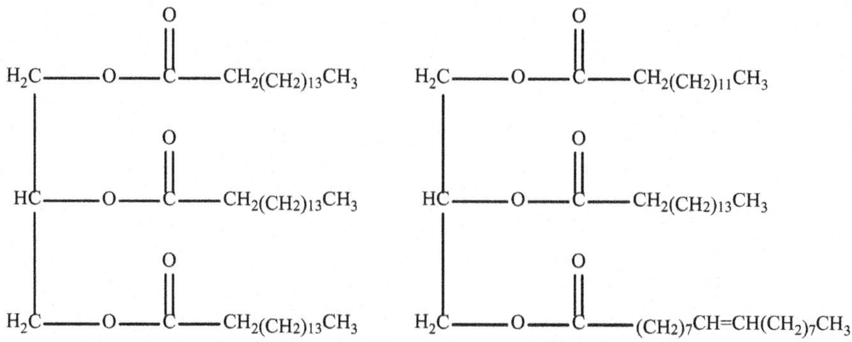

Fig. 3.16 Structures of natural ester

The structure of the biological insulating liquids is based on a glycerol backbone which is bonded to three naturally occurring fatty acid groups. These fatty acids may be the same or different (Fig. 3.16).

There is one key difference to synthetic esters (Sect. 3.2.1): many of the chains present in biological insulating liquids are unsaturated (contain C=C double bonds) and this means that they are less oxygen stable. They are therefore considered mostly suitable for hermetically or with rubber bag sealed applications and not recommended for free breathing equipment. Hydrogenation can reduce the double bonds and make the vegetable oil more stable to oxidative attacks, but this is combined with increasing viscosity.

3.2.2.1 Effect of Vegetable Oil Treatment

Table 3.3 shows the effect of the different treatment processes on the electrical properties of coconut, castor, and sesame oil under FDS (Frequency Domain Dielectric Spectroscopy) [5]. Usually, raw oil contaminates with various impurities including moisture during oil processing. In general, raw oil treatment resulted in a significant reduction (more than 40%) of the conductivity in all three vegetable oil types. It is very clear that the conductivities reduced because of removal of contamination and moisture. Similar reductions were observed in loss tangent curves, but loss tangent values at 50 Hz did not show any significant variations. As far as the relative permittivity is concerned, castor oil has relatively higher permittivity than the other two oil types. In the raw castor oil getting from Ricinus plants (Fig. 3.17), the hydroxyl group (polar) in the molecular structure (Fig. 2.31) increases the permittivity value. However, after the treatment the permittivity is reduced drastically and determines in similar range as other vegetable oils. It is reasonable to state that the hydroxyl (OH^-) groups in castor oil might have been removed during neutralization in addition to breakdown of usual COOH bonds in the triglyceride chains. After transesterification castor oil is mostly used to produce biodiesel.

Table 3.3 Change of electrical properties under FDS measurements

Oil type	Permittivity at 1 kHz	Conductivity at 1 MHz (pS/m)	Tan δ at 50 Hz (%)
Coconut (new)	2.91	546	6.9
Coconut (treated)	2.69	270	3.7
Castor (new)	5.21	120	0.8
Castor (treated)	2.37	72.5	0.9
Sesame (new)	3.04	1041	18
Sesame (treated)	3.61	436	51

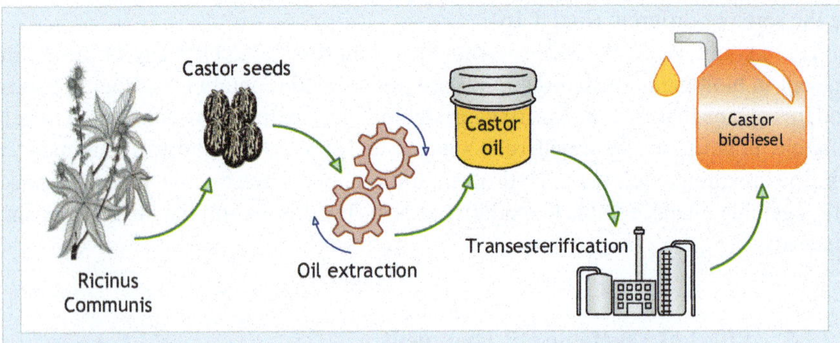

Fig. 3.17 Production of biodiesel from castor oil

The oxidation stability of vegetable oils depends not only on the feedstock, it depends also on the refining process and processing steps of the oil. The refining process consists of several steps that reduce the concentration of free fatty acids, waxes, metals, coloring pigments and odors but also contribute to the reduction of the concentration of natural antioxidants as tocopherols [6].

3.2.3 *Methyl Ester Liquids Stemming from Vegetable Oils*

A method to reduce the higher viscosity of vegetable oils is the process of transesterification. This process is used on a large scale in biodiesel production. This process aims to convert free fatty acids and triglycerides into methyl esters and glycerol (see

Table 3.4 Methyl ester contents in JMEO

Methyl ester	Percentage (%)
Methyl tetra decanoate/myristic acid methyl ester (Fig. 3.18)	0.2
Methyl palmitoleate/palmitoleic acid methyl ester (Fig. 3.19)	1.6
Methyl palmitate/palmitic acid methyl ester (Fig. 3.20)	26.2
Methyl oleate/oleic acid methyl ester (Fig. 3.21)	65.4
Methyl stearate/stearic acid methyl ester (Fig. 3.22)	6.0
Others	0.4

Sect. 2.3.2). Transesterification is the process of cutting C–O bond between triglycerides and methanol on the carbonyl carbon. The production of insulating liquid using transesterification has not yet progressed beyond the laboratory scale.

Sitorus et al. performed the transesterification of crude jatropha curcas oil (CJCO) using methanol (CH_3OH) with the ratio between oil and methanol in 1:6 mol. They used potassium hydroxide (KOH) as catalyst and a reaction temperature of 63 °C [7]. After chemical treatment and setting, the mixture was separated to become two layers where the fatty acid methyl ester (FAME) was on the top and the glycerol on the bottom, respectively. The jatropha curcas methyl ester (JMEO) obtained from transesterification process of jatropha curcas oil needs further treatment to be useable as dielectric insulating liquid. The JMEO must be washed, dried under vacuum, and finally pass through an absorption process to reach the water content which is appropriate for the IEEE standard C.57.147 [8]. After this long procedure the resulting insulating liquid has a composition as shown in Table 3.4 [7].

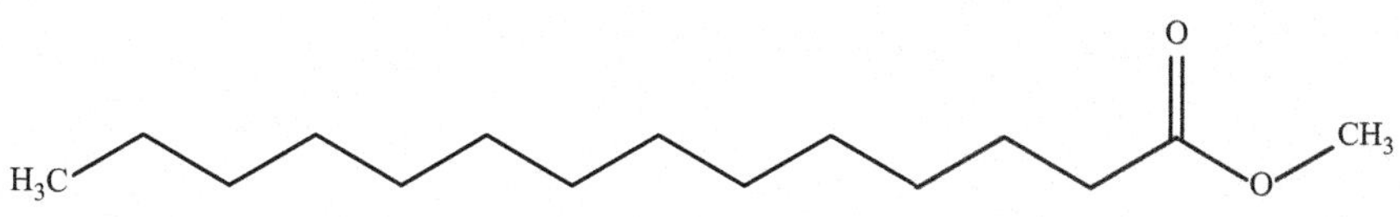

Fig. 3.18 Myristic acid methyl ester

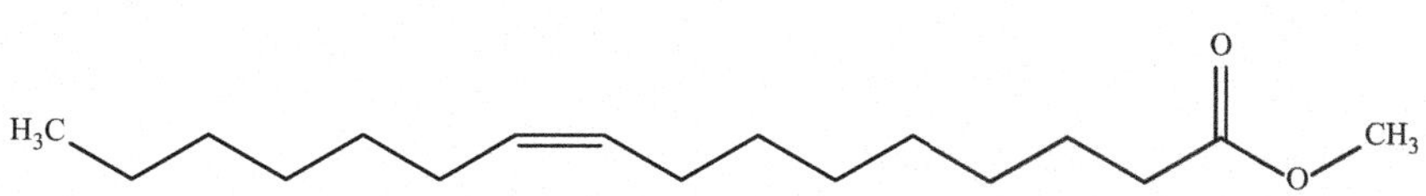

Fig. 3.19 Palmitoleic acid methyl ester

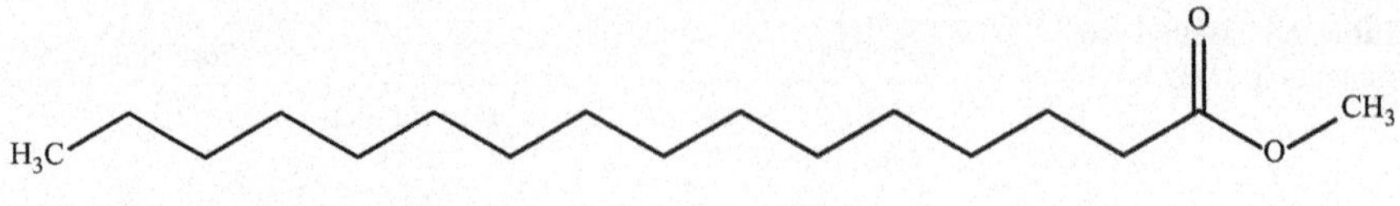

Fig. 3.20 Palmitic acid methyl ester

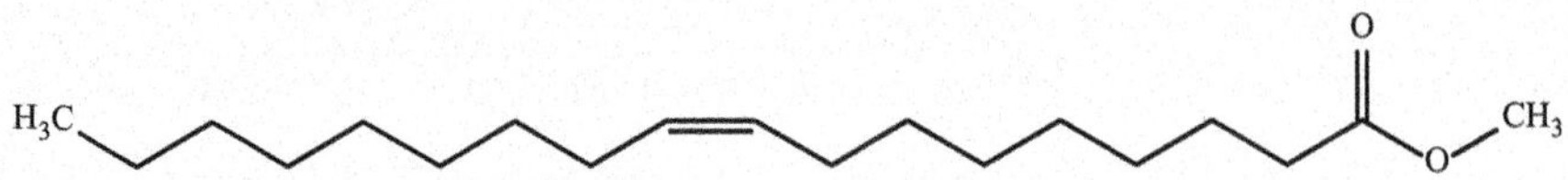

Fig. 3.21 Oleic acid methyl ester

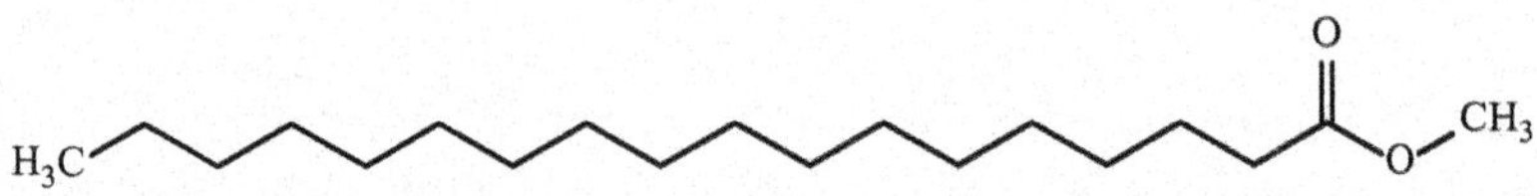

Fig. 3.22 Stearic acid methyl ester

The question is, if the chemical process used to convert the original vegetable oil (CJCO) into JMEO makes ecological sense. In Table 3.5 the physicochemical properties of CJCO and JMEO are presented. To reduce the kinematic viscosity, a chemical process is necessary. However, this chemical treatment also reduces the flash and fire point, which is a serious disadvantage. Acidity could be reduced by neutralization, and water content by drying or absorption. After drying, the breakdown voltage increases more than 100%. That means, drying improves the breakdown voltage and not the transesterification. Evangelista et al. show in [9] a jatropha curcas insulation liquid with acceptable properties (see Chap. 4) without transesterification.

Table 3.5 Physicochemical properties of CJCO and JMEO [7]

Property	Standard	CJCO	JMEO
Relative density at 15 °C (g/cm^3)	ASTM D1298 [10]	0.90–0.95	0.896
Kinematic viscosity at 40 °C (mm^2/s)	ASTM D445 [11]	30–55	10.45
Water content before adsorption (mg/kg)	ASTM D1533 [12]	1000–2000	1160
Water content after adsorption (mg/kg)	ASTM D1533 [12]	–	64.91
Total acid number (mg KOH/g)	ASTM D664 [13]	0.90–1.50	0.071
Flash point (°C)	ASTM D92 [14]	≥ 240	191
Breakdown voltage before treatment (kV)	IEC 60156 [15]	35–85	42.6
Breakdown voltage after treatment (kV)	IEC 60156 [15]	–	87.0
Iodine number (%wt)	AOCS Cd. 1-25 [16]	–	41.4

3.3 Bio-based Hydrocarbon Liquids

The Nytro® Bio 300X from Nynas is a fully bio-based hydrocarbon dielectric liquid, produced using severe hydro processing combined with isomerization. The processing technology enables the utilization of a variety of feedstocks. Bio-based residues and by-products make up most of the mainly plant-based feedstock. Crops and vegetable oils are also suitable for using as starting materials and hydrocarbon provider.

3.4 Silicone Oil

Quartz (SiO_2) is reduced to Si with carbon in an electric furnace and reacted with methyl chloride and a catalyst at 300 °C to form chloromethylsilanes. The catalyst used in this Müller-Rochow process consists of copper, copper oxide and a number of various metal oxides as promotors. While a broad variety of chloromethylsilanes are formed the dichlorodimethylsilane (Me_2SiCl_2) will be the predominant part (Eq. 3.4).

$$\mathrm{Si} + \mathrm{MeCl} \xrightarrow[\text{yield}]{\substack{\text{Catalyst}\\ 300\,^{\circ}\mathrm{C}}} \underset{>80\%}{\mathrm{Me_2SiCl_2}} + \underset{>5\%}{\mathrm{MeSiCl_3}} + \underset{\sim 4\%}{\mathrm{Me_3SiCl}} + \underset{2-3\%}{\mathrm{MeSi(H)Cl_2}} \cdots \qquad (3.4)$$

$$\mathrm{Me{=}H_3C}$$

The hydrolysis from dichlorodimethylsilane forms dihydroxy dimethyl silane and hydrochloric acid (Eq. 3.5).

$$\mathrm{Me_2SiCl_2} + 2\mathrm{H_2O} \rightarrow \mathrm{Me_2Si(OH)_2} + 2\mathrm{HCL} \qquad (3.5)$$

The base material for the silicone oil is obtained with the subsequent polycondensation. Silicone oil is a polydimethylsiloxane, a polymer in which dimethyl silane groups are linked to one another via oxygen bridges to form a chain—simplified representation (Eq. 3.6). In the case of silicone oil, X is in the range of 100.

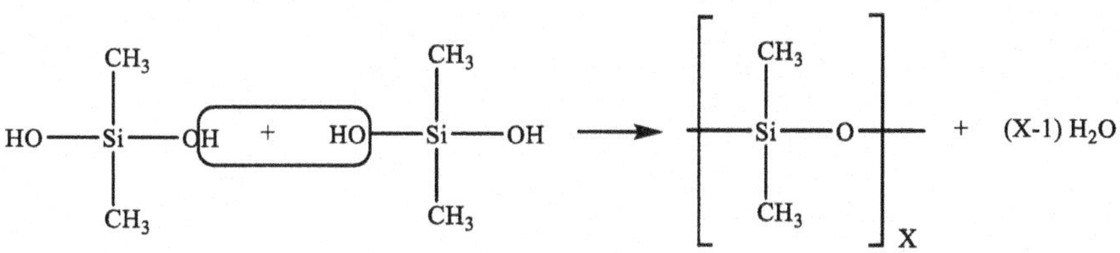

(3.6)

References

1. Air Liquid Engineering & Construction. 2017. Extraction and Refining for High-Quality Natural Oils.
2. Garcia-González, D.L., and A. Ramón. 2021. Olive Oil, AOCS Lipid Library®.
3. Gunstone, F. 2004. *The Chemistry of Oils and Fats, Sources, Composition, Properties and Uses*. Blackwell Publishing Ltd. ISBN 1-4051-1626-9
4. Evangelista Jr., M.G., et al. 2017. Development of a New Bio-based Insulating Fluid from Jatropha Curcas Oil for Power Transformers. *Advances in Chemical Engineering and Science* 7: 235–255.
5. Kumara, S., et al. 2017. Comparison of Coconut/Sesame/Castor Oils and Their Blends for Transformer Insulation. *Conference Paper*. https://doi.org/10.1109/ICHNFS.2017.8300410.
6. Diestre, E. et al. 2011. Oxidation Stability of Non-Inhibited Vegetable Transformer Liquids, CIRED. In *21st International Conference on Electricity Distribution*, Frankfurt.
7. Sitorus, H.B., et al. 2014. Physiochemical and Electrical Properties of Jatropha Curcas Methyl Ester Oil as a Substitute for Mineral Oil. In *18th International Conference on Dielectric Liquids (ICDL)*, Bled, Slovenia.
8. IEEE C57.147. 2018. IEEE Guide for Acceptance and Maintenance of Natural Ester Liquids in Transformers.
9. Evangelista, Jr. M.G., et al. 2017. Development of a New Bio-based Insulating Fluid from Jatropha Caracas Oil for Power Transformers. *Advances in Chemical Engineering and Science* 7: 235–255.
10. ASTM D1298-12b. 2017. Standard Test Method for Density, Relative Density (Specific Gravity), or API Gravity of Crude Petroleum and Liquid Petroleum Products by Hydrometer Method.
11. ASTM D445-21. 2021. Standard Test Method for Kinematic Viscosity of Transparent and Opaque Liquids (and Calculation of Dynamic Viscosity).
12. ASTM D1533-20. 2020. Standard Test Method for Water in Insulating Liquids by Coulometric Karl Fischer Titration.
13. ASTM D664-18e2. 2018. Standard Test Method for Acid Number of Petroleum Products by Potentiometric Titration.
14. ASTM D92-18. 2018. Standard Test Method for Flash and Fire Points by Cleveland Open Cup Tester.
15. IEC 60156. 2018. Insulating Liquids—Determination of the Breakdown Voltage at Power Frequency—Test Method.
16. AOCS Official Method Cd 1-25. 2019. Iodine Value of Fats and Oils Wijs Method.

Chapter 4
Properties of New Insulating Liquids and Main Differences

This chapter describes and compares the essential properties of the various insulating liquids. This applies to insulating liquids that are already available to use, but also to those biological liquids that have the potential to serve as dielectric insulating liquids in the future. The main chemical, physical, electrical, and ecological differences of the different insulating liquids are described. When different vegetable oils are compared, their chemical and electrical properties are significantly very depending on the carbon chain length, as well as their saturation/unsaturation nature. Besides the chemical properties, also the physical properties like density, temperature, viscosity, surface tension and water capacity play an important role for their use as insulating material. Especially the temperature behaviour, as insulating liquids are also used for cooling, and the water absorption capacity are very important. Other factors are the electrical properties, which must withstand the requirements of the electrical stresses and electrical operations. Recently, the interests in environmental and economic properties are rising.

4.1 Chemical Properties

The chemical structure of ester molecules (natural ester and synthetic ester, respectively) is very different from mineral oil molecules. Esters have the general chemical formula (Fig. 4.1), with R^1 as a carboxylic acid and R^2 as an alcoholic rest. Their wide distribution in nature, and their numerous practical applications make them perhaps to the most important carboxylic acid derivatives.

Table 4.1 shows typical chemical properties (neutralization number, sulfur content) of selected dielectric insulating liquids issued by the manufactures.

E. P. Pagger et al., *Biological Insulating Liquids*,
https://doi.org/10.1007/978-3-031-22460-7_4

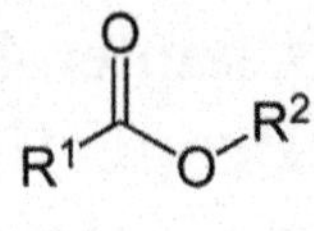

Fig. 4.1 Ester—general formula

4.1.1 Neutralization Number

The hydrolysis of ester bonds of biological insulating liquids release fatty acids. The free fatty acids introduced by this process are long chain organic, hardly dissociated acids. Most of these have 18 carbons in length, while a few have 16 carbon sequences (see Sect. 2.3.2), other carbon sequences are very rare. In principle, the strength of organic acids is inversely related to the chain length (Table 4.2) and the whereabouts of the carboxylic acids in the water/oil system depends on the number of methyl groups in the molecule. The probability of the acid to stay in the water-phase decreases linearly with the logarithm of the number of methyl groups in the acid [27]. Short chain organic acids such as formic acid and acetic acid are middle strong and, in sufficient amount, can be detrimental to the condition of other materials in contact with the insulating liquid. These acids are above all generated during the degradation of cellulose via hydrolysis. Acids can dissociate in water to form hydronium ions. A measure of the easiness of an acid to dissociate is the acid dissociation constant, K_a or pK_a. pK_a is the negative logarithm of the acid dissociation constant K_a. Long chain acid such as stearic acid are weak (Table 4.2) and have not been associated with any detrimental effects. Hence, the acids can be sub-divided into the two groups of low molecular (LMA) and high molecular acids (HMA). From literature, three LMA are found in transformers liquid: formic, acetic and levulinic acid [28]. LMA molecules are defined by their ease of solubility in water and have a typical pK_a value in the region of 3.7–4.7.

LMA have been found to be more active in the degradation of paper insulation. LMA are mainly generated from the chemical hydrolytic processes in the paper and due to their polar and hydrophilic nature, they tend to accumulate in the paper insulation when using mineral oil as insulating liquid. Additionally, because these molecules are small, they will also partition into the transformer liquid. Other factors that affect partitioning are: partial concentrations, polarity and temperature. LMA that are assumed to originate from the paper insulating aging have a molar mass in the range from 46 to 116 g/mol. HMA have much larger molar masses that are typically greater than 240 g/mol. HMA have been found to be generated mainly from chemical oxidation or hydrolytic processes relating to the transformer liquid, and due to their non-polarity and large molecular size they tend to stay in the transformer liquid [29].

In [29] Yan et al. tested the extraction of low molecular weight acids from transformer liquids in water. They varied the stirring duration of the samples from 1 to 12 h. The results show that for mineral oil the extraction process is completed within the first hour, whereas for the synthetic ester Midel 7131 this takes approximately up to three hours. The short period of time for extraction in the mineral oil is most likely due to its non-polar nature. The polar LMA molecules are more likely to diffuse into

Table 4.1 Typical initial chemical properties

Liquid	Literature	Neutralization number		Sulfur content	
		mg KOH/g	Test method	%	Test method
Nynas Nytro® 4000X	[1]	< 0.01	IEC 62021 [2, 3]	< 0.01	ISO 14596 [4]
Nynas Nytro® BIO 300X	[5]	< 0.01	IEC 62021 [2, 3]	< 0.01	ISO 14596 [4]
Envirotemp™ 360 Synthetic Ester	[6]	0.01	IEC 62021 [2, 3]	–	
Midel® 7131	[7]	< 0.03	IEC 62021 [2, 3]	–	
BecFluid® 9902	[8]	< 0.03	IEC 62021 [3]	–	
Nycodiel® 1233	[9]	0.01	IEC 62021 [2]	–	
Nycodiel® 1244	[9]	0.01	IEC 62021 [2]	–	
Nycodiel® 1258	[9]	0.01	IEC 62021 [2]	–	
MATROL-BI® FDE01A	[10]	0.03	IEC 62021 [2]	–	–
Envirotemp® FR3™ Fluid[a]	[12]	0.022	ASTM D974 [13]	Non-corrosive	ASTM 1275 [14], IEC 62697 [15]
Midel® eN 1204	[16]	≤ 0.04	ASTM D974 [13], IEC 62021 [17]	Non-corrosive	ASTM 1275 [14], IEC 62697 [15]
Midel® eN 1215	[18]	≤ 0.04	ASTM D974 [13], IEC 62021 [17]	Non-corrosive	ASTM 1275 [14], IEC 62697 [15]
BIOTEMP®	[19]	0.03	ASTM D974 [13], IEC 62021 [17]	Non-corrosive	ASTM 1275 [14], IEC 62535 [20]
NeuGen Plus	[21]	≤ 0.06	ASTM D974 [13]	–	

(continued)

Table 4.1 (continued)

Liquid	Literature	Neutralization number		Sulfur content	
		mg KOH/g	Test method	%	Test method
Paryol Electra 7426®	[22]	0.1–0.6	IEC 62021 [17]	Non-corrosive	ASTM 1275 [14], IEC 62535 [20]
MATROL-BI® FDV01A	[23]	0.03	IEC 62021 [17]	Not present	62535 [20]
Jatropha Curcas Oil	[24]	0.04	ASTM D974 [13]	Non-corrosive	ASTM 1275 [14]
Powersil® Fluid TR 50	[25]	< 0.01	IEC 60836 [26]	–	

[a] pH value:5.8; test method EPA 9045C [11]

Table 4.2 pH value of different liquids

Number of carbon atoms in the molecule	Acid	pK_a	pH-value (concentration 1 mol/l)
–	Hydrochloric acid	≈ −6	≈ 0
1	Formic acid	3.75	≈ 2.2
2	Acetic acid	4.75	≈ 2.8
5	Levulinic acid	4.65	≈ 2.3
6	Caproic acid	4.85	≈ 4.0
8	Caprylic acid	4.85	≈ 4.0
10	Capric acid	4.90	≈ 4
12	Lauric acid	5.3	≈ 5.2
14	Myristic acid	4.95	≈ 6.1[a]
18	Stearic acid	6.9 (10.15)	≤ 7[a]
18	Oleic acid	6.4 (9.85)	≤ 7[a]
18	Linoleic acid	(9.24)	≤ 7[a]
18	Linolenic acid	6.2 (8.28)	≤ 7[a]

[a] Limited information, as up to C_{14}, these acids are practically insoluble in water

water rather than staying in the mineral oil. For the synthetic ester MIDEL 7131 the attraction between LMA and ester can be stronger due to their polar nature, which is probably why it takes longer for LMA to be extracted into water from the synthetic ester. In addition, a residual amount of LMA was found in MIDEL 7131 even after 12 h of stirring. This means that there will be a small amount of LMA that cannot be separated by using water, which is trapped by the polar synthetic ester molecules via hydrogen bonding. That means, the ultimate extraction of acid in an ester will always be less than that in the mineral oil at the same initial LMA concentration due to its polar chemical nature.

Not all acids are created equal. The strength (pH Value) of an acid depends on its concentration and degree of dissociation and can be calculated by Eqs. 4.1 and 4.2.

In general: Acids, which can arise during the chemical breakdown of biological insulating fluids, are part of daily life in pharmaceuticals, cosmetics, and food.

$$c\left(H_3O^+\right) = \sqrt{K_a * c_{Liquid}} \tag{4.1}$$

$$pH = -\log c(H_3O^+) \tag{4.2}$$

K_a Constant of equilibrium
C Concentration (mol/l).

The exact pK_a values of biologically important long-chain free fatty acids (FFA) are mostly unknown. The few reported values scatter widely: e.g., for stearic or oleic acids from 6 to 10.5 [30, 31]. In Table 4.2 the data in the bracket stems from [31]. However, there is consensus that pK_a of the carboxyl group, which is ≈4.75 in water, is shifted to higher values in a hydrophobic environment.

Most short-chain carboxylic acids have a pK_a value of ca. 4.8. For example, when acetic acid (CH_3COOH) and propionic acid (CH_3CH_2COOH) are dissolved in water, their pK_a values are found to be 4.74 and 4.87, respectively [31]. The pK_a also be raised by increasing the carbon chain length of the carboxylic acid. However, because electronic effects are not felt beyond two to three carbons, when the chain length is increased beyond about four carbons, the pK_a tends to level off. Increasing the chain length from pentanoic to hexanoic acid, for example, increases the pK_a from 4.82 to 4.83. Therefore, we know that intramolecular interactions (i.e., the effects on the carboxylate anion by the rest of the carbon chain) become negligible beyond four carbons in the alkyl chain [31]. The pK_a values of long-chain fatty acids depend on the chain length of the fatty acid molecule and can increase to values much higher than those for shorter chain analogs. The chain length of the fatty acid affects its pK_a value. It is known that, as the chain length increases, van der Waals interactions between the chains of adjacent molecules increase, bringing these molecules closer to each other. When this happens, the carboxylic acid groups of the fatty acids are packed closer, shielding the hydrogen atom between the two oxygen atoms. The closer the molecules, the more strongly shielded the hydrogen atom and consequently the higher the pK_a value [31]. On the other hand, pK_a value decreases with increasing number of double bonds. For example, linolenic acid contains many kinks in its chain caused by three cis double bonds. These kinks prevent the molecules from packing closely. Alterations of the local relative dielectric permittivity ε of the medium are not responsible in the change of pK_a values, because neither chain length nor the level of unsaturation influences the position of the FFA's carboxylic group relative to the aqueous bulk phase [30].

The fact that short- and middle-chain carboxylic acids (C2–C10) dissociates into H^+ ions explain their corrosiveness with copper and steel, since H^+ ions govern corrosion mechanism. Short- and middle-chain fatty acids could be therefore used as markers to give an early indication of insulating liquid quality.

4.1.1.1 Neutralization Number—Test Methods

There are several standards [2, 3, 13, 17, 32, 33] for determination of the acidy in insulating liquids. For colored sample the potentiometric method (IEC 62021-1 [2], ASTM D664-18e2 [32]) is preferable.

4.1.2 Aging Stability

In addition to the good breakdown voltage and good heat transfer characteristics, the selected liquid must have a good aging stability. This is combined with a high stability against oxidation. It is well established that a high degree of unsaturation causes poor thermal and oxidative stability of biological insulating liquids. Another concern is the tendency of vegetable oils to hydrolysis. Therefore, the main reason for degradation of biological insulating liquids are moisture, oxygen, and temperature. The possible positions of vulnerable chemical attack are shown in Fig. 4.2.

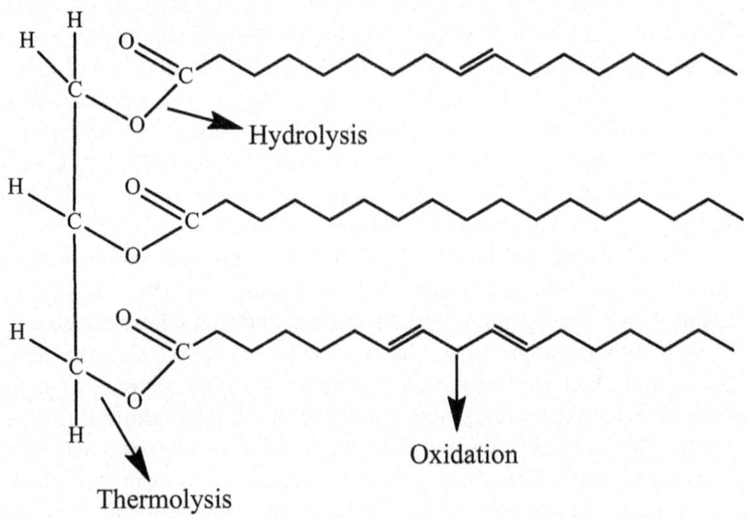

Fig. 4.2 Vulnerable positions for chemical attack of biological insulating liquids

4.1.2.1 Oxidation Stability

Oxygen constitutes as one of the most influential factors in the aging of every insulating liquid by oxidation. The oxygen molecule exists in two forms—in its normal ground state it is a triplet form ($^3O_2$) and in the excited state it exists in a singlet form ($^1O_2$). Both react with olefinic systems and while they share some similarities, there are some important differences between the reactions of these two forms of oxygen. Triplet oxygen is a diradical ·O–O· and reacts mainly at allylic centers to give allylic hydroperoxides. In contrast, singlet oxygen is an electrophilic substance reacting with electron-rich olefinic systems, but also producing allylic hydroperoxides. It is more reactive than triplet oxygen by 93.6 kJ/mole and has a lifetime of only 50–700 μs [34].

Oxidation stability of insulating liquids is a crucial factor since this is highly necessary that insulating liquid should not be oxidizing with the passage of time. The oxidation stability of insulating liquids is a key apprehension of end consumers. All insulating liquids oxidize, and so to all intents and purposes different oxidation byproducts may affect the performance of a transformer. The oxidation stability of the insulating liquid is a feature that specifies its resistance to oxidation during operation. A large value of oxidation stability is desired to certify extended service lifetime of insulating liquid. Mineral oil also suffer from the oxidation process throughout service resulting in the development of sludge, but adding antioxidants may overdue the process. Furthermore, as biological insulating liquids are more biodegradable than mineral oil ones, they tend to have lower oxidation stability. Biological insulating liquids are not as resistant to oxidation as mineral oil liquids. For biological insulating liquids, ASTM D6871 [35] does not yet recommend oxidation stability testing methods or limits. Varied laboratory oxidation stability tests have been developed to be used as control tests for evaluating the relative oxidation stability of dielectric liquids in a laboratory environment. Oxygen can be naturally present or my ingress into the insulating liquid from the environment. Therefore, the oxidation stability of insulating liquids is more important for free breathing transformer. Increasing temperature influence the oxidation stability negative and some metals such as copper act as a catalyst.

The oxidation stability study has been done by different authors who firstly have aged the insulating liquid and later have measured acidity and tan δ, as they constitute the most suitable properties to evaluate the oxidation [36]. The results of Darwin et al. have shown that the relative stabilities of insulating liquids to oxidation are silicone oil > synthetic ester > mineral oil > natural ester (biological insulating liquid) [37]. The values listed in Table 4.3 largely confirm this statement, even a comparison is very difficult due to the very different test times. The test times for natural esters are only a tenth of those of oxidation-stable insulating liquids. No results are available for silicone oil. However, it can be assumed that they have a very high oxidation stability.

Biological insulating liquids indicate lower oxidation stability than their counterparts. The existence of carbon–carbon dual bond converts dielectric liquids susceptible to oxidation. In biological insulating liquids, additional carbon–carbon dual

Table 4.3 Oxidation stability (148 h, 120 °C) according to IEC 61125C [36], ASTM D2440 [38]

Liquid	Total acidity (mg KOH/g)	Viscosity at 40 °C (mm^2/s)	Dissipation factor at 90 °C (tan d)	Sludge (%per mass)
Nynas Nytro® 4000X [1]	< 0.01*	–	< 0.01*	< 0.01*
Nynas Nytro® BIO 300X [5]	0.01*	–	< 0.01*	0.01*
Envirotemp™ 360 Synthetic Ester [6]	0.21**	–	–	0.004**
Midel® 7131 [7]	0.02	–	–	< 0.01
BecFluid® 9902 [8]	< 0.3	–	–	0.005
Nycodiel® 1233 [9]	0.09	–	–	0.004
Nycodiel® 1244 [9]	0.08	–	–	0.007
Nycodiel® 1258 [9]	0.04	–	–	0.002
Envirotemp® FR3™ Fluid [40]	0.1***	17.1% increase***	0.1***	–
Midel® eN 1204 [16]	< 0.1***	8% increase***	< 0.1***	–
Midel® eN 1215 [18]	< 0.07***	15.7% increase***	0.05***	–
BIOTEMP® [19]	0.36	–	–	0.12
NeuGen plus [21]	0.08***	12.5% increase***	0.0782***	–
Paryol Electra 7426® [22]	0.1–0.6***	1–2% increase***	≤ 0.5***	–

*After 500 h
**After 800 h
***After 48 h

bonds are existent compared to mineral oil liquids, and they are largely prone to oxidation. The oxidation procedure is irreparable and because of reaction action, oxygen is expended. Nevertheless, the process of oxidation in biological insulating liquids varies from mineral oil. In case of natural ester liquids, constant contact to oxygen does not develop any solid precipitates. A fine layer of gel is developed in biological insulating liquids with prolonged contact to oxygen.

Oxidation stability of biological insulating liquids relies on the dissemination of fatty acids, filtering action, existing organic/natural antioxidants. It is likewise known that the oxidation stability of biological insulating liquids reduces with rise in unsaturated fatty acid level. Oxidation stability of oleic acid (one double bound) is stated tenfold larger than linoleic acid (two double bounds) whereas linoleic acid is twice more stable than linolenic acid (three double bounds) [39].

The formation of oxidation products obeys to a chain mechanism consisting of initiation, propagation, branching and termination steps [41]. The key event initiation is the formation of a lipid radical, R· (Eq. 4.3).

$$RH \rightarrow R\cdot + H\cdot \tag{4.3}$$

This can occur by thermal or photochemical cleavage of an R-H bond or by hydrogen atom abstraction from R-H by an initiator free radical. Metals can act as catalysts. Free radicals rapidly react with oxygen to form peroxy radical ROO· (Eq. 4.4)

$$R\cdot + O_2 \rightarrow ROO\cdot \tag{4.4}$$

The peroxy radical can then attack another lipid molecule to remove a hydrogen atom to form a hydroperoxide ROOH and another free radical R· (Eq. 4.8), propagation the oxidation process.

$$ROO\cdot + RH \rightarrow ROOH + R\cdot \tag{4.5}$$

These produced hydroperoxides are unstable and may degrade to radicals RO· and OH· (Eq. 4.6) that accelerates propagation of the reactions. The RO· radical will react with more oxygen to form new hydroperoxides. However, not all free radicals propagate the oxidation process, some may interact with each other or degrade to give non-radical products.

$$ROOH \rightarrow RO\cdot + \cdot OH \tag{4.6}$$

The oxidized compounds formed by hydroperoxide decomposition lower surface tension in the insulating liquids.

Mark Lashbrook writes in [42] that in synthetic ester Midel 7131 there are no double bonds in the R chains between carbons. This increases its oxidation resistance and thermal stability. The oxidation stability of insulating liquids for transformers is a critical issue because the oxidation of dielectric liquids generates the formation of by-products such as acids, sludge, etc., when using mineral oil. This may cause problems in heat transfer in a transformer and reducing the dielectric properties of the insulation.

In particular, the base oils of the biological insulating liquids usually contain naturally unsaturated compounds that facilitate the possibility of oxidative attack, or form longer molecules through polymerization, which has a strong influence on the viscosity. The characteristics of biological insulating liquids may be stabilized by selecting a suitable configuration of saturated and unsaturated triglycerides during preparation. Nevertheless, as mentioned above, it's combined with disadvantages like higher pour point and viscosity. To achieve durability and the highest possible stability, antioxidants additives can be added as inhibitors to these insulating liquids that decelerate oxidation, thus enhancing service life expectancy. Large transformer designs frequently use nitrogen in headspace, or those with insulating liquid conservators need a membrane wall between insulating liquid and outer venting. Nevertheless, it is recommended to use biodegradable insulating liquids only in hermetically sealed devices. Therefore, airtight closing against ambient atmosphere is the finest solution to take advantage of the features of biological insulating liquids.

4.1.3 Gassing Tendency

The gassing tendency of insulating liquids is measured commonly according to ASTM Test Method D2300 [43]. In a test cell the liquid–gas interface is saturated with a gas (usually hydrogen), then subjected to a radial electrical stress of 3–4 kV/mm (the nominal test calls for an applied stress voltage of 10 kV over a 3-mm liquid gap). The gas phase above the liquid is ionized and the oil–gas interface subjected to ion bombardment. The amount of gas evolved or absorbed in the liquid is measured in a burette and expressed in the units of volume per unit of time, at a given temperature.

Poor drying or oil impregnation of the paper insulation or both may also lead to bubbles of air or water, especially in overlapping areas of thick layer insulation. Bubbles of air or nitrogen may be formed because of oversaturation in insulating liquid due in temperature or pressure changes, and bubbles of water vapor or carbon monoxide because of hot spot in the paper insulation. Rates of gas formation generally increase with increasing temperature or applied voltage. Nascent hydrogen is also formed, either by the decomposition of dielectric liquid vapors, or through reactions at the dielectric liquid–gas interface, and may recombine to form hydrogen gas. This nascent hydrogen can also react with the liquid molecules (hydrogenation of fatty acids in biological insulating liquids), and bubble formation is reduced or eliminated; the liquid is said then to be gas-absorbing.

The appearance of partial discharges results in the production of different gases due to the decomposition of insulating liquid molecules. The main formed gas is hydrogen (H_2). But there is also formed methane (CH_4), ethane (C_2H_6), ethylene (C_2H_4), and acetylene (C_2H_2). Such gases can endanger the safety of the device. Thus, the use of low gas-evolving or high gas-absorbing insulating liquid is preferred, especially if the system is sealed or hermetic closed (not in contact with the atmosphere). With a high gas-absorbing oil, partial discharge activity can continue over a long period of time without harm.

There are differences between mineral oil liquids and biological insulating liquids gassing tendency. Biological insulating liquids have inherently lower gassing tendency than mineral oil liquids, well below the lower range of mineral oil liquids and generally fairly negative (gas absorbing). The typical range for biological insulating liquids is – 50 μl/min to – 90 μl/min (IEEE C57.147 [44]). An insulating liquid with a negative gassing tendency reduces hydrogen gas bubbles which result from electrical and thermal fault conditions in the transformer. A negative gassing tendency also reduces the risk of transformer failure and explosion due to hydrogen gas generation, providing an extra margin of safety.

4.1.3.1 Stray Gassing

Increased awareness of stray gas production has led to industry studies and the development of standardized tests such as ASTM D7150 [45] to characterize various insulating liquids.

Stray gassing means the formation of gasses when dielectric insulating liquids are heated at relatively low temperatures (90–200 °C). In natural ester, at comparatively low temperature (80–250 °C), a significant formation of stray gases such as hydrogen and ethane are witnessed for a specific period (weeks to months) subsequently the transformer is activated [38]. This may be due to chemical reactions within the transformer, possibly influenced by catalytic materials. Martin et al. suggested in [46] that stray gassing in natural ester Envirotemp FR3 liquid may occur at temperatures between 40 and 50 °C. Such unexpected gas formation at low temperature can lead to confusion in the dissolved gas analyzes interpretation. Then, the generation of these gases could be linked to an electrical fault in the equipment whereas it is linked to the degradation of the insulating liquid. The biological insulating fluids show a higher tendency towards stray gassing. Especially, in terms of ethane, compared to other insulating liquids. Martin shows in [47] that the type of base vegetable oil can have an influence on the amount of ethane produced. In this experiment the content of dissolved gases after a period of one day at an elevated temperature in the presence of oxygen is compared (Table 4.4). It is interesting to note that the level of ethane is significantly higher in the biological liquids having a higher polyunsaturated content.

The interference from this data is that low-oleic biological oils are likely to produce higher levels of stray gassing than high biological oleic oils. This does not mean that the biological insulating liquids which have higher unsaturated content are less preferable. It just means that it is useful to know the oil base type when evaluating DGA data, especially when high levels of ethane are indicated.

In [48] Deng et al. present experimental research on gassing characteristics of mineral oils (Shell Diala BG and Nynas 10XN) tested under normal operating temperature (70–140 °C). Comparing the decomposition of transformer mineral oil Shell Diala BG and transformer oil Nynas 10XN they found that the hydrocarbon gases

Table 4.4 Dissolved gas analyzes (DGA) results for various vegetable oils

	High oleic Sunflower	Peanut	Soyabean	Flaxseed
Polyunsaturated content	0.2%	0.2%	7%	53%
H_2 (ppm)	357	282	316	708
CH_4 (ppm)	21	10	10	17
C_2H_6 (ppm)	4	26	563	2371
C_2H_4 (ppm)	8	16	7	16
C_2H_2 (ppm)	0	0	0	0
CO (ppm)	203	389	408	977
CO_2 (ppm)	876	2232	1330	3212

produced by the two transformer oils were mainly CH_4 and C_2H_6 at the heating temperature range. The gas production and gassing rate of the transformer oils increased with increase of heating temperature. Gas production and gassing rate of Shell Diala BG was significantly higher than those of Nynas 10XN at the same heating temperature and time. Shell Diala BG always produced a certain amount of CH_4 and C_2H_6, while the gas production of Nynas 10XN was almost zero when the heating temperature was reduced to 100 °C and below, which indicated that the thermal stability of Shell Diala BG was obviously worse than that of Nynas 10XN. In general, the maximum gassing rate of the mineral oils was in the order: CH_4–C_2H_6–C_2H_4–C_2H_2 which is except for C_2H_6 similar as for vegetable oils (Table 4.4). The chemical composition (the content of paraffins and aromatics in Shell Diala BG is larger than that in Nynas 10XN) caused the difference in the thermal stability between the two mineral oils, which in turn lead to the difference in the gas production characteristics.

4.1.4 Gas Solubility

For the non-mineral liquids, densities differ a bit (Table 4.6) and viscosities are higher (Table 4.7) compared to mineral oil. This fact influences the gas solubility. Müller et al. tested the gas solubility of nitrogen, hydrogen, and air in different insulating liquids [49]. On many transformers, nitrogen is the gas with the highest concentration dissolved in the insulating liquid. However, it depends on the sealing principle of the transformer conservator tank, e.g., free breathing or nitrogen blanket. Table 4.5 shows the maximum solubility of nitrogen and hydrogen for the analyzed insulating liquids.

Nitrogen solubility increases with increasing temperature. The absolute values are different for the liquids, but for nitrogen the temperature dependent change is

Table 4.5 Maximum solubility of nitrogen and hydrogen at atmospheric pressure [49]

Insulating liquid	Max. solubility of nitrogen (%)		Max. solubility of hydrogen (%)
	25 °C	90 °C	25 °C
Mineral Oil Nytro Lyra X	8.5	11.0	5.15
Mineral Oil Diala DX	9.0	11.1	4.49
Natural Ester Envirotemp FR3	7.0	9.3	3.68
Synthetic Ester Midel 7131	7.4	9.2	4.35
Silicone Oil Powersil Fluid TR 50	13.9	14.7	–

similar. Silicone oil behaves different with almost double of solubility and a much smaller increase with temperature. A lower viscosity leads to higher gas solubility and vice versa, except for silicone oil. Hydrogen has a smaller maximum solubility compared to nitrogen.

4.2 Physical Properties

Clear differences between the insulating liquid groups can be seen here, which must also be incorporated into the construction, design, and processing of the equipment.

4.2.1 Density

The relative density (specific gravity) is not significant in determining the quality of an insulating liquid. All dielectric insulating liquids show a decrease of density (r) with increase of temperature. This is because of the volume expansion with increasing temperature. The volume (V) expansion reduces the density since the mass (m) of the dielectric insulating liquids is constant. This can be expressed as (Eq. 4.7).

$$\rho = \frac{m}{V} \tag{4.7}$$

Table 4.6 shows the values of density issued by the manufacturers. Nynas Nitro Bio 300X has by far the lowest density, while the densities of the synthetic esters are just below 1 g/cm^3 at a temperature of 20 °C. The biological insulating fluids are almost all in the range from 0.91 to 0.925 g/cm^3 at this temperature.

Because of the increase of volume when temperature rises the density decrease (Eq. 4.7). Figure 4.3 shows the coefficients of thermal expansion [54] of dielectric insulating liquids from which data where available. Except for NeuGlen, the gradient of the curves is very similar for the different insulating liquids. Compared to the other insulating liquids, Nytro Bio 300X has a significantly higher starting value. Evangelista specifies the thermal expansion coefficient for Jatropha curcas in [24] as 7×10^{-4} per °C without giving a reference temperature.

The change in density, as a function of temperature is inversely proportional to the volume expansion and thus is decreasing linearly (Eq. 4.8). The values for volume expansion per °C are in the range of $-6.64 * 10^{-4}$ and $-7.10 * 10^{-4}$.

$$\rho(T) = y_0 + a * T \tag{4.8}$$

In Fig. 4.4 the curves of the density as a function of the temperature, as far as they have been announced by the manufacturers, are depicted.

Table 4.6 Density of different dielectric insulating liquids

Liquid	Density			
	(g/cm^3) at 15 °C	(g/cm^3) at 20 °C	(g/cm^3) at 25 °C	Test method
Nynas Nytro® 4000X [1]	–	0.868	–	ISO 12185 [50]
Nynas Nytro® BIO 300X [5]	–	0.785	–	ISO 12185 [50]
Envirotemp™ 360 Synthetic Ester [6]	–	0.960	–	ISO 12185 [50] ISO 3675 [51]
Midel® 7131 [7]	–	0.970	–	ISO 12185 [50] ISO 3675 [51]
BecFluid® 9902 [8]	–	0.970	–	ISO 3675 [51]
Nycodiel® 1233 [9]	–	0.950	–	ISO 12185 [50]
Nycodiel® 1244 [9]	–	0.980	–	ISO 12185 [50]
Nycodiel® 1258 [9]	–	0.970	–	ISO 12185 [50]
MATROL-BI® FDE01A [10]	0.944	–	–	ISO 3675 [51]
Envirotemp® FR3™ Fluid [12]	–	–	0.920	ASTM D1298 [52]
Midel® eN 1204 [16]	–	0.920	–	ISO 3675 [51]
Midel® eN 1215 [18]	–	0.920	–	ISO 3675 [51]
BIOTEMP® [19]	–	–	0.910	ISO 12185 [50] ASTM D1298 [52]
NeuGen Plus [21]	0.910	–	–	ASTM D1298 [52]
Paryol Electra 7426® [22]	–	0.910–0.925	–	ISO 3675 [51]
MATROL-BI® FDV01A [23]	0.917	–	–	ISO 3675 [51]
Jatropha Curcas Oil [24]	–	0.912	–	ASTM D1298 [52]
Castor Oil [53]	–	–	0.960	ASTM D1298 [52]
Powersil® Fluid TR 50 [25]	–	0.96	–	ISO 3675 [51]

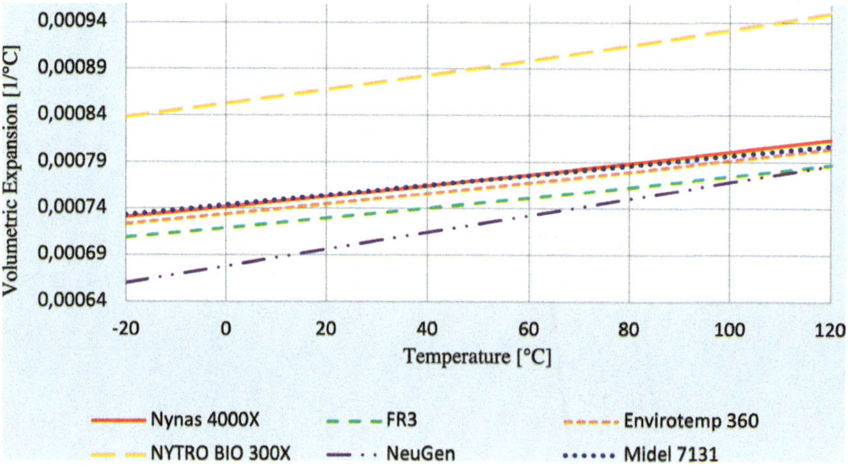

Fig. 4.3 Coefficient of thermal expansion

4.2.1.1 Density—Test Methods

Density is defined as the mass of liquid per unit volume at a defined temperature. There are different methods for determining the density of a liquid

- ISO 12185: Gives a method for the determination, using an oscillation U-tube densitometer, of the density of crude petroleum and related products within the range 600–1100 kg/m^3 which can be handled as single-phase liquids at the test temperature and pressure [50].
- ISO 3675: Method for laboratory determination of the density by hydrometer at 15 °C [51].
- ASTM D1298: This test method covers the laboratory determination using a glass hydrometer in conjunction with a series of calculations [52].

4.2.1.2 Coefficient of Thermal Expansion—Test Methods

The coefficient of thermal expansion of liquid is the change in volume per unit volume per degree of temperature. It can be calculated by determining the specific gravities of liquid at any two temperatures. The average coefficient of expansion over the given temperature range is Eq. 4.9.

$$\lambda = \frac{\rho_0 - \rho_1}{\rho_0 * (T_1 - T_0)} \tag{4.9}$$

λ Coefficient of thermal expansion (K^{-1})
ρ_0 Specific gravity at higher temperature (g cm^{-3})
ρ_1 Specific gravity at higher temperature (g cm^{-3})

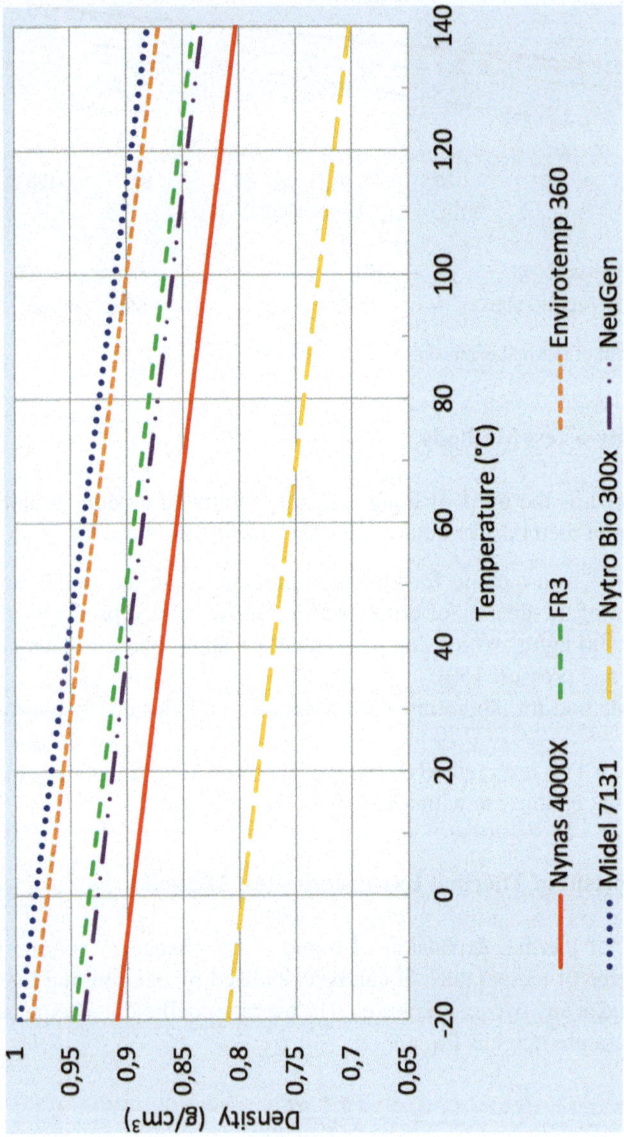

Fig. 4.4 Density versus temperature

T_0 Lower temperature (K)
T_1 Higher temperature (K).

4.2.2 *Viscosity*

The absolute viscosity of liquids is its resistance to flow and shear due to internal friction and is measured with SI units of Pa*s (pascal-second) or cP (centipoise). In contrast, the kinematic viscosity of liquid is its resistance to flow and shear due to gravity and it is measured with SI units of m^2/s or cSt (centistokes). Fasina and Colley present the viscosity of different vegetable oils (Table 4.7) in dependence of temperature in [55].

The kinematic viscosity of oil can be obtained by dividing the liquid absolute viscosity with its corresponding density [56]. The viscosity of a liquid plays a key part in determining the cooling capability of the liquid because this property represents the resistance to the flow. For this reason, high viscosity may slower the flow of liquid in the winding cooling ducts and increase the operation temperature of the transformer. This is not necessarily a critical issue but should be considered for the design of a power transformer. Kinematic viscosity is the most important factor for heat transfer. In insulating liquids for high voltage applications, this feature is narrowly associated with heat transport efficiency. The heat transmission capability of a liquid is hugely affected by its viscosity. Liquids with high viscosity has considerably decreased cooling performance. Low viscous liquids enhance cooling efficiency by easy circulation. Besides, high viscosity also needs to be considered during liquid impregnation of solid cellulose in the transformer manufacture process. The more

Table 4.7 Viscosity (cP) of vegetable oil at different temperatures

	Sample temperature (°C)									
Oil source	35	50	65	80	95	110	120	140	160	180
Almond	43.98	26.89	17.62	12.42	9.15	7.51	6.54	5.01	4.02	3.62
Canola	42.49	25.79	17.21	12.14	9.01	7.77	6.62	5.01	4.29	4.65
Corn	37.92	23.26	15.61	10.98	8.56	6.83	6.21	4.95	3.96	3.33
Grape Seed	41.46	25.27	16.87	11.98	9.00	10.37	9.18	7.50	6.10	4.78
Hazelnut	45.55	27.40	17.83	12.49	9.23	7.56	6.69	5.25	4.12	3.48
Olive	46.29	27.18	18.07	12.57	9.45	7.43	6.49	5.29	4.13	3.44
Peanut	45.59	27.45	17.93	12.66	9.40	7.47	6.47	5.14	3.75	3.26
Safflower	35.27	22.32	14.87	11.17	8.44	6.73	6.22	4.77	4.11	3.44
Sesame	41.14	24.83	16.80	11.91	8.91	7.19	6.25	4.95	4.16	3.43
Soybean	38.62	23.58	15.73	11.53	8.68	7.17	6.12	4.58	3.86	3.31
Sunflower	41.55	25.02	16.90	11.99	8.79	7.38	6.57	4.99	4.01	3.52
Walnut	33.72	21.20	14.59	10.51	8.21	6.71	5.76	4.80	3.99	3.46

viscous a liquid, the slower flow speed inside the capillaries will be. Consequently, the liquid with low viscosity is preferred during impregnation process. Increasing the temperature of the impregnation can help to reduce the impregnation time. It has been well established that temperature has a strong influence on the viscosity of liquids, with viscosity generally decreasing with temperature.

The viscosity of mineral oil and bio-based liquid are at normal operating temperature of a power transformer (25–80 °C) significantly lower compared to the ester liquids (Table 4.8). With parameter quotient Q (Eq. 4.10) the variation of viscosity versus temperature is determined, where ν_1 and ν_2 are the values of viscosity at temperatures T_1 and T_2, respectively ($T_1 < T_2$). In general, the synthetic and natural esters show a stronger temperature dependence compared to insulating liquids with lower viscosities. Above all, environmentally friendly ester-based liquids show superior viscosity values in comparison to mineral oil, being this variance more significant at lower temperature. The viscosity of the natural esters and the viscosity of the synthetic esters are almost in the same range, independent of the producers (Fig. 4.5).

At the normal operating temperature of power transformers, the viscosity of biological insulating liquids is higher than that of mineral oil, but lower than that of silicone oil. Viscosity of biological insulating liquids decreases with the degree of unsaturated bonds in the molecule increase and depends on either the chain length of the fatty acid or the alcohol moiety. Thus, castor oil having hydroxyl groups (polar) shows the highest viscosity. The characteristic of the curve of the silicone oil is very different. The curve runs much flatter. Relatively speaking, their viscosity is little changed by changes in temperature. This viscosity stability results from the ease of rotation about the Si–O–Si bond, which prevents close packing of the molecules and reaction between the weak intermolecular forces or dipoles.

$$Q = \frac{\nu_1(T_1)}{\nu_2(T_2)} \tag{4.10}$$

In addition to the already known dependence on temperature, Pagger examined in [61] whether moisture, which can change during operation of the transformer, influences the viscosity and how the different liquids behave. McShane et al. reports [62] that the viscosity of the soya-based natural ester Envirotemp FR3 in a free-breathing transformer increased by 8.6% after seven years of operation.

According to Möller and Nasser [63], the viscosity increases when water-in-oil emulsions are formed according to the empirical equation (Eq. 4.11).

$$\eta_E = \eta_L + \eta_L * 0.025 * y \tag{4.11}$$

η_E mPa s (dynamic viscosity of solution/water-in-oil emulsion)
η_L mPa s (dynamic viscosity of the insulating liquid)
y % water content

Analyzes done by [61] at three different temperatures and different water contents could not confirm this statement (Fig. 4.6).

An increase in viscosity as described in [63] could only be found for mineral oil at temperatures 20 and 50 °C, and for silicone oil at temperatures 50 and 80 °C. The other dielectric insulating liquids showed a decrease in viscosity in all areas. Because of the low viscosity of water (Table 4.9) [64] compared to the insulating liquids, it would be expected anyway that the viscosity decreases as the water content increases.

Equation 4.11 is not applicable for alternative insulating liquids. The results of mineral oil show still the best conformity. Generally, the results show that the expected change in viscosity due to moisture absorption during operation of the devices is so small that it does not have to be considered, when designing the devices.

Table 4.8 Viscosity of different insulating liquids

Insulating liquid	Viscosity (mm^2/s)					Q
	(0 °C)	(20 °C)	(40 °C)	(100 °C)	Test method	T_1 = 40 °C; T_2 = 100 °C
Nynas Nytro® 4000X [1]	45.8	16.1	7.9	2.1	ISO 3104 [57]	3.8
Nynas Nytro® BIO 300X [5]	12	6.1	3.8	1.4	ISO 3104 [57]	2.7
Envirotemp™ 360 Synthetic Ester [6]	–	–	34.0	7.0	ISO 3104 [57]	4.9
Midel® 7131 [7]	233	74.7	29.5	5.3	ISO 3104 [57] ASTM D445 [58]	5.6
BecFluid® 9902 [8]	–	–	25.0	–	ISO 3104 [57]	–
Nycodiel® 1233 [9]	–	–	16.1	3.8	ISO 3104 [57]	4.2
Nycodiel® 1244 [9]	–	–	21.6	4.6	ISO 3104 [57]	4.7
Nycodiel® 1258 [9]	–	–	27.2	5.2	ISO 3104 [57]	5.2

(continued)

Table 4.8 (continued)

Insulating liquid	Viscosity (mm^2/s)					Q
	(0 °C)	(20 °C)	(40 °C)	(100 °C)	Test method	T_1 = 40 °C; T_2 = 100 °C
MATROL-BI® FDE01A [10]	–	–	21.4	4.7	ISO 3104 [57]	4.6
Envirotemp® FR3™ Fluid [40]	190	–	32–34	7.7–8.3	ISO 3104 [57] ASTM D445 [58]	4.2–4.1
Midel® eN 1204 [16]	232	–	37	8.3	ISO 3104 [57] ASTM D445 [58]	4.5
Midel® eN 1215 [18]	206	–	32	7.6	ISO 3104 [57]	4.2
BIOTEMP® [19]	276	–	42	9.0	ISO 3104 [57] ASTM D445 [58]	4.7
NeuGen Plus [21]	225	80	39	8.5	ASTM D445 [58]	4.6
Paryol Electra 7426® [22]	–	–	37–43	8.0–10.0	ISO 3104 [57] ASTM D445 [58]	4.6–4.3
MATROL-BI® FDV01A [23]	–	–	42	9.0	ISO 3104 [57]	4.7
Jatropha Curcas oil [24]	208	–	39.7	8.3	ASTM D445 [58]	4.8
Pongamia Pinnata oil [59]	–	–	38	–	ASTM D445 [58]	–

(continued)

Table 4.8 (continued)

Insulating liquid	Viscosity (mm^2/s)					Q
	(0 °C)	(20 °C)	(40 °C)	(100 °C)	Test method	T_1 = 40 °C; T_2 = 100 °C
Powersil® Fluid TR 50 [25]	–	–	40	–	ISO 3104 [57]	–
Coconut oil [60]	–	–	29	–	ASTM D445 [58]	–
Sesame oil [60]	–	–	36	–	ASTM D445 [58]	–
Castor oil [60]	–	–	265	–	ASTM D445 [58]	–

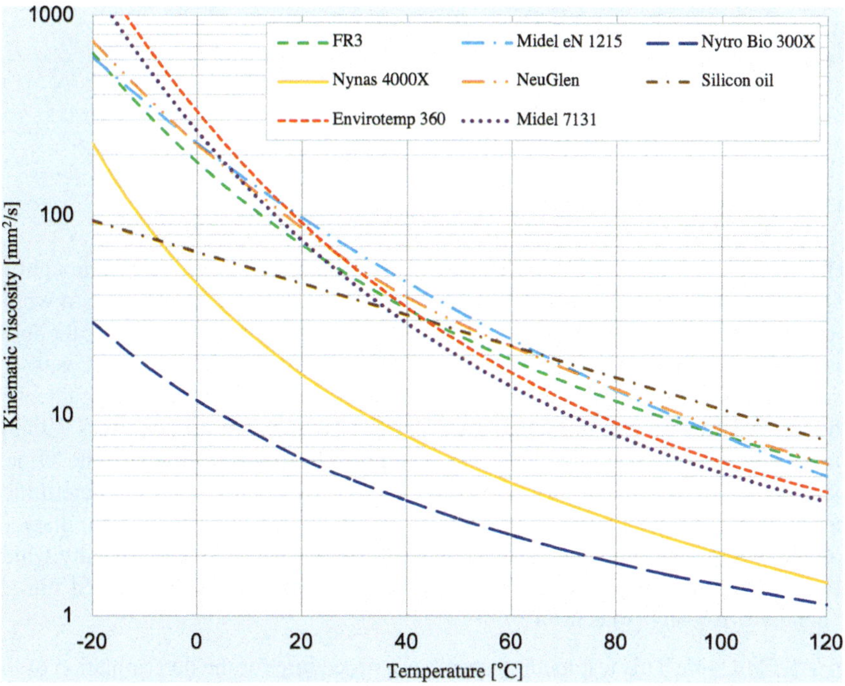

Fig. 4.5 Viscosity versus temperature

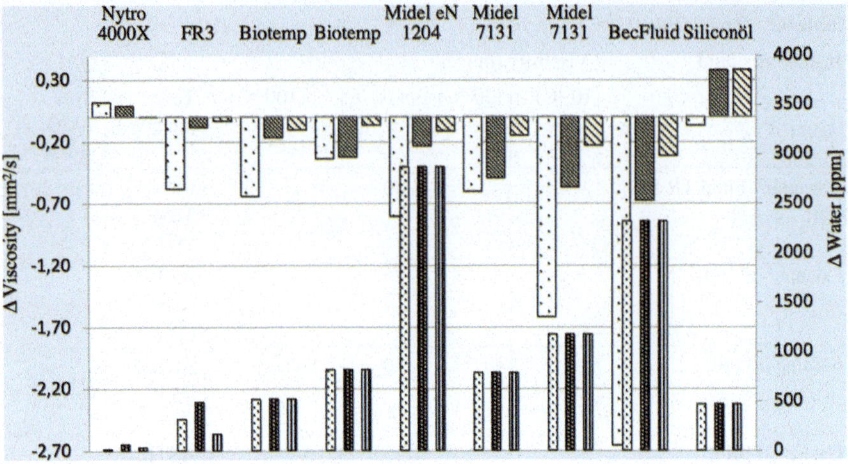

Fig. 4.6 Change in viscosity as a function of moisture

Table 4.9 Viscosity of water versus temperature

Temperature (°C)	20	50	80
Viscosity (mm^2/s)	1.004	0.553	0.365
Viscosity (mPa s)	1.001	0.548	0.355

4.2.2.1 Viscosity—Test Methods

The viscosity of a liquid is a measure of its resistance to flow. It can be thought of a type of internal friction which prevents the oil from moving as freely as it would normally. There are several ways of measuring and reporting viscosities, the most used being dynamic viscosity and cinematic viscosity. The dynamic viscosity (η) is defined as the force per unit area required to maintain a unit velocity gradient in the liquid. If the ratio of the force to the velocity gradient remains constant with the increasing velocity (rate of shear), the liquid is said to be Newtonian. The SI unit of dynamic viscosity is N s m^{-2} or kg s^{-1} m^{-1} or Pa s. Since viscosity determined normally by the time of flow of a liquid under its own weight in a capillary of known dimensions, its value is also reported in terms of cinematic viscosity which is defined as the as the dynamic viscosity divided by the density. The SI unit of cinematic viscosity (ν) is m^2 s^{-1}.

- ASTM D445: This test method specifies a procedure for the determination of the kinematic viscosity, ν, of liquid petroleum products, both transparent and opaque, by measuring the time for a volume of liquid to flow under gravity through a calibrated glass capillary viscometer. The dynamic viscosity, η, can be obtained by multiplying the kinematic viscosity, ν, by the density, ρ, of the liquid [58].
- ISO 3104: This standard corresponds to test method ASTM D445.

4.2.3 *Interfacial–Surface Tension*

Interfacial tension is the tension at the phase boundary (interface) of two immiscible liquids and surface tension is the tension at the phase boundary between the liquid and the gas phase.

The surface tension of liquids is composed by (σ_P) polar and disperse parts (σ_D) (Eq. 4.12).

$$\sigma = \sigma_P + \sigma_D \tag{4.12}$$

The polar part (σ_P) is caused by

- Dipole–dipole interaction
- Hydrogen bonds
- Lewis acid–base interaction

and the disperse part of the surface tension through Van der Waals interaction.

While the surface tension of water has a very high polar component, it is practically zero in the case of aliphatic hydrocarbons which are liquid at room temperature. In addition to density and viscosity, surface tension is the third essential property of liquids. A molecule that is located at the phase boundary is in a different energetic state than a molecule inside the phase. Inside the liquid, the resulting force on the molecule is zero, while on the surface the resulting force is directed inwards. To increase the boundary surface by dA, work by dW is necessary. From this the following dimension can be derived (Eq. 4.13).

$$\sigma = \frac{dW}{dA} = \frac{J}{m^2} = \frac{Nm}{m^2} = \frac{N}{m} \tag{4.13}$$

Since the interfacial tension represents the tension between the two condensed phases, it is tempting to determine it by calculating the difference between the surface tension. However, since there are always interactions between the liquids (even if immiscible), the mere subtraction of the surface tensions to determine the interfacial tension is not permissible. Because of the interactions between the liquids, the interfacial tension between two liquids is always smaller than the difference between the surface tensions of the two liquids. Oil aging products such as water, organic acids, and polar oil oxidation products, which are mostly hydrophilic in character, lower the interfacial tension of the oil in the water/oil system. The interfacial tension is determined with a tensiometer (ring method) [65, 66]. The surface tension is determined using the ring method, and in addition, via the height of rise in the capillary. In this way, the contact angle that occurs in the event of incomplete wetting can be determined. With the capillary method, the surface tension is determined from the capillary height (h), the density (ρ), the acceleration due to gravity (g) and the tube radius (r). If the surface tension, determined by means of a tensiometer, and the capillary height are known, the contact angle (θ) can be calculated as follows Eqs. 4.14, 4.15, and 4.16.

$$\sigma_K = \frac{h * r * g * \rho}{2 * \cos\theta} \tag{4.14}$$

$$\sigma_K = \sigma_T \tag{4.15}$$

$$\theta = \cos^{-1}\left(\frac{h * r * g * \rho}{2 * \sigma_T}\right) \tag{4.16}$$

4.2.3.1 Test Results

Table 4.10 shows the interfacial tension (water/insulating liquid) of new, not yet used insulating liquids at 20 °C. The interfacial tensions of the natural esters are the lowest. The surface tension was determined by means of a tensiometer at different temperatures and the values were extrapolated to a temperature of 80 °C (Fig. 4.7). Silicone oil has the lowest surface tension (Table 4.10 and Fig. 4.7).

There is always an interaction at the interface between two immiscible liquids. The more molecules enter into the interaction, the more intense it will be. A qualitative assessment consists in subtracting the interfacial tension water/insulating liquid from the difference (surface tension of the water minus surface tension of the insulating liquid) expressed in Table 4.11.

If there were no interaction between the insulating liquid and the water, the value in the right column of Table 4.11 should be zero. The more hydrophilic substances there are, the higher the value.

Table 4.10 Interfacial tension and surface tension at 20 °C

Insulating liquid	Interfacial tension (σ_I) (mN/m) at 20 °C	Surface tension (σ_S) (mN/m) at 20 °C
Water		71.6
Nynas Nytro 4000X	39.0	30.2
Envirotemp® FR3™ Fluid	24.0	34.7
Biotemp®	23.9	31.8
Midel® eN	25.0	34.3
Midel® 7131	30.9	28.4
BecFluid® 9902	32.2	31.4
Powersil® Fluid TR 50	35.0	22.5

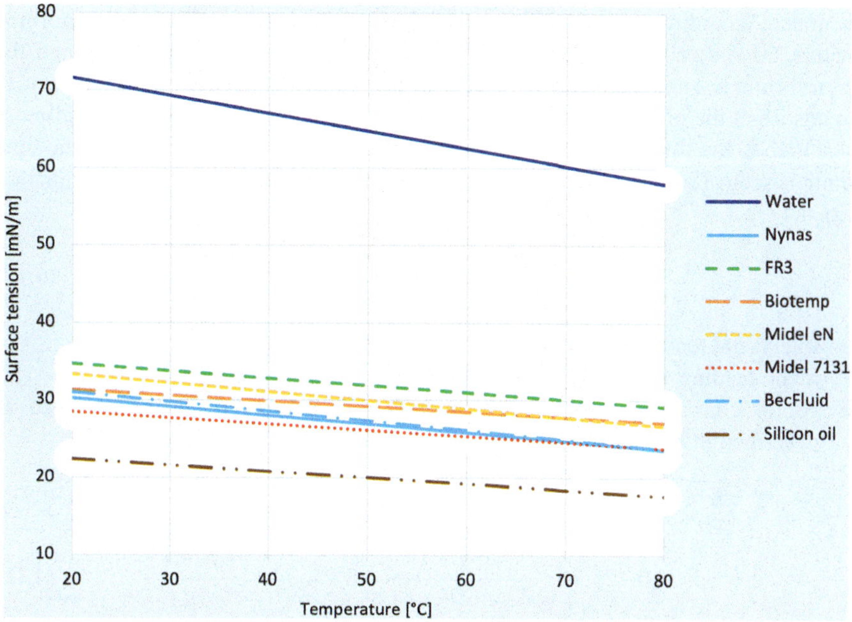

Fig. 4.7 Surface tension versus temperatures determined with a tensiometer

Table 4.11 Differences in the surface tension compared to water and to the interfacial tension at 20 °C

Insulating liquid	$\sigma_{S,Water} - \sigma_{S,IL}$ (mN/m) at 20 °C	$(\sigma_{S,Water} - \sigma_{S,IL}) - \sigma_{I,IL}$ (mN/m) at 20 °C
Nynas Nytro 4000X	41.4	2.4
Envirotemp® FR3™ Fluid	36.9	12.9
Biotemp®	39.8	15.9
Midel® eN	37.4	12.4
Midel® 7131	43.2	12.3
BecFluid® 9902	40.3	8.0
Powersil® Fluid TR 50	49.2	14.2

$\sigma_{S,Water}$ Surface tension of water
$\sigma_{S,IL}$ Surface tension of insulating liquid
$\sigma_{I,I}$ Interfacial tension of insulating liquid

4.2.3.2 Interfacial Tension—Test Methods

The principle is to determine the force which is necessary to pull out a ring of the interface between insulating liquid and water in the direction of the measured insulating liquid (Standard IEC 62.961 [65] and ASTM D971-20 [66]). The most significant difference of the two standards is the time point at which the measurement is done. According to the ASTM standard, the measurement should be performed within 60 s after the oil–water interface has been formed. At this time point, the equilibrium has not been reached, as it generally takes much longer for the impurities to adsorb at the oil–water interface. This makes the method less accurate and more sensitive to the timing of the measurement. With the standard IEC 62.961, the time point is set to 180 ± 30 s. The interfacial tension is calculated by following equation Eq. 4.17.

$$\sigma = P * F \tag{4.17}$$

σ Interfacial tension (mN/m)
P Scale reading when film ruptures (mN/m)
F Factor converting scale reading in mN/m to interfacial tension obtained as described in Eq. 4.18

$$F = 0.725 + \sqrt{\frac{1.452 * P}{C^2 * (D - d)} + 0.04534 - \frac{1.679 * r}{R}} \tag{4.18}$$

where:

C Circumference of ring (mm)
D Density of water (g/ml)
d Density of insulating liquid (g/ml)
R Radius of ring (mm)
r Radius of wire of ring (mm).

4.2.4 Water Absorptive Capacity

The ester linkages, present in both synthetic and natural ester, make these liquids polar and are responsible for the very different interaction with water comparing to mineral oil. This part of the molecular structure is somewhat polar as water. This leads to the possibility for water molecules to be attracted to esters. This interaction is known as hydrogen bonding. This gives esters a much higher moisture saturation limit and allows far higher dissolved water content. Consequently, esters have a particular affinity for water molecules, in a way that mineral and silicone oil liquids cannot. The different water absorption and adsorption capacity of the insulating liquids influences the physical, chemical, and electrical properties of the liquids as described in [61]. The equilibrium curves between cellulose and the insulating liquid [67] depend on the water content of the liquid, which influences paper ageing [68]. The water saturation of the different liquids depends on the temperature and can be calculated by Eq. 4.19 or Eq. 4.20.

$$Saturation\,(T) = 10^{A_{10} - \frac{B_{10}}{273.1+T}} \tag{4.19}$$

$$Saturation\,(T) = e^{A_e - \frac{B_e}{273.1+T}} \tag{4.20}$$

T	Temperature in °C
$A_{10},\ B_{10}$	Moisture saturation coefficient expressed in its decimal logarithm form
$A_e,\ B_e$	Moisture saturation coefficient expressed in natural logarithmic form.

Equations 4.21 and 4.22 can be used to convert the moisture saturation coefficient values (Table 4.12).

$$A_e = A_{10} * \ln(10);\ B_e = B_{10} * \ln(10) \tag{4.21}$$

$$A_{10} = A_e * \log(e);\ B_{10} = B_e * \log(e) \tag{4.22}$$

Table 4.12 shows the coefficients for calculating the curves in Fig. 4.8. Mineral oil has the lowest water absorption capacity followed by the bio-based liquid. Silicone oil has a slightly higher water absorption capacity compared to mineral oil. Between the biological insulating liquids (natural esters) there is very little difference in water absorption capacity. There is a huge difference between biological insulating liquids and mineral oil in water solubility. Biological insulating liquids can absorb many times more moisture than mineral oil or bio-based liquids at different temperatures before saturation happens. This is, because moisture molecules are easy to be bonded to hydrophilic groups of biological insulating liquid molecules. This phenomenon is not found in mineral or bio-based insulating oils from Nynas. The IEEE C57.147 maintenance guide for natural esters [44] suggests that the relative water content of the insulating liquid should be maintained at the same level as for mineral oil. In this case the ppm value limit can be calculated as far as the saturation limit is known, based on the fluid temperature.

Table 4.12 shoes the water content in ppm at 20% relative humidity. The IEC 61.203 maintenance standard for synthetic ester [69] stipulates that in use, water contents should be below 400 ppm.

Table 4.12 Coefficients of moisture tolerance

Insulating liquid	A	B	Typical water content—new liquid (ppm) [70, 71]	Value at 20% saturation (ppm)
Nynas Nytro 4000X [1, 61]	7.0895	1587	< 20	12
Nynas Nytro Bio 300X [5, 72]	7.2387	1592	< 20	16

(continued)

Table 4.12 (continued)

Insulating liquid	A	B	Typical water content—new liquid (ppm) [70, 71]	Value at 20% saturation (ppm)
Envirotemp™ 360 Synthetic Ester [6]	5.6588	734	< 50	314
Midel® 7131 [7, 73]	5.2788	542	50	578
BecFluid® 9902 [8]	–	–	50	–
Nycodiel® 1233 [9]	–	–	55	–
Nycodiel® 1244 [9]	–	–	32	–
Nycodiel® 1258 [9]	–	–	30	–
MATROL-BI® FDE01A [10]	–	–	50	–
Envirotemp® FR3™ Fluid [40, 61]	5.3318	684	4–50	218
Midel® eN 1204 [16, 61]	5.3422	687	50	218
Midel® eN 1215 [18]	–	–	50	–
Biotemp® [74]	5.5633	751	< 50	221
NeuGlen Plus [21]	5.3367	679	< 100	229
Paryol Electra 7426® [22]	–	–	70–150	–
MATROL-BI® FDV01A [23]	–	–	50	–
Jatropha curcas oil [24]	–	–	177	–
Powersil® Fluid TR 50 [25, 61]	5.9142	1063	< 50	45

– No data available

The solubility of water in all these dielectric liquids increase with temperature (Fig. 4.8). It could be observed that the most polar liquids (synthetic esters) absorb more water across the temperature range. Biological insulating liquids have 3 ester linkages per molecule, whilst synthetic esters may have mostly 4 linkages per molecule. The more ester linkages the more hydrogen bridges, the more water can be dissolved. Clearly, more polar ester linkages can absorb a lot more water across the temperature range. These differences become evident considering the amount of water that will dissolve in these insulating liquids (Fig. 4.8).

This better water solubility decreases the humidity influence on insulation strength and dries the paper. This could increase the life of a transformer, as its life is in first approximation controlled by the state of the paper. At elevated temperatures, ester liquids can undergo hydrolysis, consuming available water from the paper, thereby improving paper ageing characteristics. Moreover, the esterification of the reactive OH sites on cellulose with bulky ester groups stabilizes the molecule and protect them for further attacks and extends thermal life of the solid insulation. Long alkyl chains in fatty acid produced by hydrolysis with hydrophobic characteristic are arranged

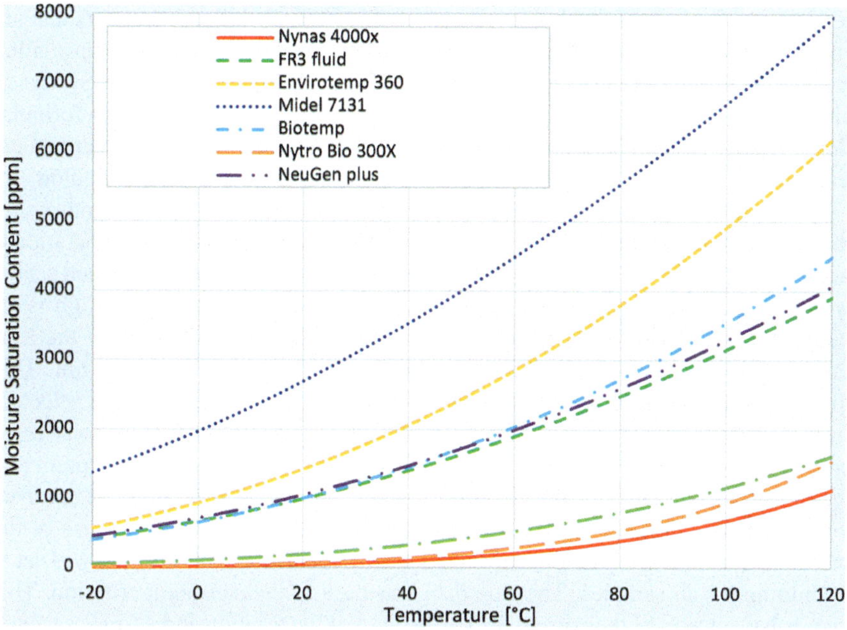

Fig. 4.8 Moisture saturation content of different insulating liquids

in nearly parallel with cellulose, which leads to a water barrier on the surface of cellulose, avoiding extensive water penetration into the pressboard. On the other hand, due to this higher water solubility, ester liquids recapture moisture rapidly and special care must be taken during handling compared to mineral oil. In contrast, because of the low polarity of mineral oil, the mutual attraction between water and cellulose is relatively much stronger, and thus more water tends to be absorbed by the cellulose, leading to further deterioration of the pressboard.

4.2.4.1 Water Content—Test Method

Principle: If clear and bright, and free from both water droplets and particulate matter on swirling, a weighed portion is injected into the titration vessel of a coulometric Karl Fischer apparatus, in which iodine for the Karl Fischer reaction is generated coulometrically at the anode. When all the water has been titrated, excess iodine is detected by an electrometric end-point detector and the titration is terminated. Based on the stoichiometry of the reaction, one mole of iodine reacts with one mole of water, thus the quantity of water is proportional to the total integrated current according to Faraday's Law [71, 75]. This test is significant in that it can indicate the presence of water even if the presence of water may not be evident from electrical test.

The determination of water in oils and fats requires a working medium that dissolves these substances to release the water to be titrated. For this purpose, only the ready-to-use reagent Hydranal® was used in this work [76]. Some reagents that are suitable for mineral insulating liquids may not be suitable for biological insulating liquids. As stated in ASTM D1533 [71], Annex A1 "Alternative Solvent Systems," alternative reagents may be needed for certain biological insulating liquids formulations. The water content of the solid samples is determined by thermal expelling and transferring the water vapor to the titration vessel of the Karl Fischer Coulometer [77]. It is important to ensure that all the moisture is expelled without decomposing the insulating material. The advantage of this method is that no sample preparation is necessary. The samples can be weighed directly into the headspace vials and sealed with PTFE-coated septa. The sample heated in the oven module at 140 °C gives off its moisture as water vapor, which is transferred to the measuring cell of the Karl Fischer Coulometer with the help of a gas stream via a heated transfer line. Due to the sensitivity of the samples to oxygen, an inert carrier gas must be selected. In this case, argon was used as the carrier gas because of its higher molar mass compared to nitrogen. Since the needle only reaches shortly below the septum and the heavier argon sinks to the bottom, the moisture can immediately be removed via the needle casing. The short puncture length had to be chosen because of the pressboard samples, which filled the vials quite a lot. The argon flow rate was set at 50 ml/min for all samples. The stop drift was used as the end-point criterion. This drift value consists of the current drift at the start of the titration and a freely selected relative drift. This drift was set to 30 μg/min for the pressboard samples, 15 μg/min for the Nomex samples and 5 μg/min for the paper samples. Under these conditions, titration times of up to eight hours were still achieved with the pressboard samples.

4.2.5 *Miscibility of Alternative Insulating Liquids*

Although, in most cases, different types of insulating liquids are miscible (except for silicone liquid), such mixtures should typically be avoided in transformers and liquid processing equipment when possible and practical. This is due to potential negative impact on key environmental, performance, and fire safety characteristics. Obviously, some low percentage contamination cannot be avoided when retrofilling, particularly transformers and other equipment with impregnated cellulose material.

The miscibility of alternative insulating liquids has been used by some authors to improve the properties of mineral oil by mixing. Although silicone oil is miscible with mineral oil, it can cause excessive foaming in mineral oil even in small quantities. Biological insulating liquids are typically miscible and compatible with mineral oil insulating liquids. Because of reducing the flash and fire point, a maximum of 7% mineral oil in the biological insulating liquid is acceptable.

When refilling mineral oil used equipment with alternative insulating liquids, it should be noted that the mixture must not be calculated linearly regarding the flash point and fire point. As expected, these parameters are first dominated by the product with the lower boiling point. In [61], Pagger produced a mixture of silicone oil and mineral oil, the mineral oil content being 1.2%. In this concentration range, the two liquids are absolutely miscible. After stirring with the magnetic stirrer, only one phase can be determined. A clear influence on the flash point is found. If silicone oil has a flash point > 250 °C and mineral oil has a flash point of 138–150 °C, the mixture had a flash point of 192 °C. The low percentage of mineral oil (1.2%) reduced the flash point of the silicone oil by at least 23%. Additionally, silicone oil is not miscible with ester liquids. However, natural and synthetic esters are miscible with mineral oil in all proportions.

For the theoretical value of the mixture, the value can be calculated according to the mixture law (proportionality law), given by the following relationship (Eq. 4.23)

$$\Psi_C = X * \frac{\Psi_A}{100} + Y * \frac{\Psi_B}{100} \tag{4.23}$$

where, at a given temperature and in the same conditions, X is the % by volume of an oil A with characteristic ψ_A, Y is the % volume of an oil with characteristic ψ_B and C is the mixture of A and B with characteristic ψ_C.

Gockenbach et al. studied the properties of a mixture of synthetic ester liquid (Midel 7131) and a mineral oil (Shell Diala D) [78–80]. Their work demonstrated that some electrical and physical properties of the mixed liquids (Table 4.13) were not inferior to those of mineral oil, particularly for mixtures with less than 20% ester content. For the mixture with 50% of ester liquid, the density and the viscosity exceed the limiting values suggested by the standards for mineral oil [81]. It can be said that for the property values that do change, the change may or may not be proportional to the ratio of the content of the liquids.

The results in Table 4.14 shows that the liquid mixtures do not follow the mixture law expressed by (Eq. 4.23), due to chemical interactions between the liquids, exactly.

Table 4.13 Some properties of mineral oil mixed with synthetic ester (experimental results)

Property		Mineral oil 100%	90% Mineral oil 10% Synthetic ester	80% Mineral oil 20% Synthetic ester	50% Mineral oil 50% Synthetic ester	Synthetic ester 100%
Density exp. (kg/dm^3)	At 23 °C	0.856	0.881	0.890	0.918	0.960
	At 90 °C	0.810	0.841	0.851	0.876	0.915
Water solubility (ppm)	At 20 °C	45	100	310	830	2700
	At 100 °C	650	940	1600	2900	7200
Viscosity (mm^2/s)	At 20 °C	16	17.4	19.44	28.65	63
	At 90 °C	2.3	3.45	3.76	4.80	7.70
Breakdown voltage (kV)		> 55	> 55	> 55	> 55	> 55
Dielectric constant at 25 °C		2.2	> 2.2	> 2.3	> 2.6	3.3

Table 4.14 Some properties of mineral oil mixed with synthetic ester (calculated results)

Property		Mineral oil 100%	90% Mineral oil 10% Synthetic ester	80% Mineral oil 20% Synthetic ester	50% Mineral oil 50% Synthetic ester	Synthetic ester 100%
Density (kg/dm^3)	At 23 °C	0.856	0.867	0.877	0.908	0.960
	At 90 °C	0.810	0.821	0.831	0.863	0.915
Water solubility (ppm)	At 20 °C	45	311	576	1373	2700
	At 100 °C	650	1305	1960	3925	7200
Viscosity (mm^2/s)	At 20 °C	16	20.7	25.4	39.5	63
	At 90 °C	2.3	2.84	3.38	5.15	7.70

Pierre et al. tested mixtures consisting of mineral oil and two other kinds of insulating liquids (namely silicon oil and synthetic ester) [82]. These three liquids, as well as their mixtures, are analyzed and compared based on the main properties (heat transfer, breakdown voltage, aging stability, electrostatic charging tendency) required for an insulating liquid for power transformers. Because of a technical and economical point of view, the emphasis has been put on the mixture based on 80% of mineral oil with 20% of another liquid. It should be mentioned that Pierre et al. used non-inhibited naphthenic mineral oil for their tests.

Suwarno and Darma used in their experiment [83] mixtures of conventional mineral oil with methyl ester. Mineral oil used in this experiment is Shell Diala B. This is a highly refined base oil which is hardly biodegradable. Ester used in this experiment is methyl ester made from palm oil using methanol for transesterification according to Chap. 2, Sect. 2.3.2. The concentration of ester in the mixture ranges from 0, 25, 50, 75 to 100%.

The experimental results of breakdown voltage show that in term of breakdown voltage, the increase of methyl ester content improved the quality of the mixture. It is also in all samples clearly seen that breakdown voltage increase with temperature. Suwarno and Darma found a more or less linearly correlation of the mixtures (BV_E) between breakdown voltage and content of palm methyl ester (E) in % (Eq. 4.24). BV_0 is the breakdown voltage without any palm methyl ester.

$$BV_E = BV_0 + 0.085 * E \tag{4.24}$$

Dielectric losses factor (tan δ) slightly increases with the methyl ester content. Losses factor is influenced by the degree of polarization of the liquid. Chemical structure indicates that mineral oil has a symmetric structure while methyl ester contains an unbalanced structure. This unbalanced structure of the methyl ester enhances its losses factor. The dielectric losses factor of pure methyl ester of palm oil is much higher than mineral oil. The resistance of dipoles during polarization process increases the losses in the ester oil sample.

The work of Suwarno and Darma also shows that dielectric constant slightly increases with the content of methyl ester and temperature. As expected, the density and the acidity of the oil mixture increases with the ester content.

Although in many cases different types of less flammable insulating liquids are miscible, such mixtures should generally be avoided in transformers and insulating liquid filled equipment as practical, unless such mixtures are done purposely for certain applications or to achieve certain properties.

4.3 Thermal Properties

Fuel, air, and temperature are three elements for combustion. The combustion cannot happen so long as one of them is failed. Under the normal circumstances, insulating oil of the transformer and other insulating systems are isolated from air and the temperature is not high for normal operation (< 105 °C), so the combustion will not happen. The combustion in transformers mainly refers to the combustion of insulating liquid ejected from the transformer in the air, so from the perspective of combustion and fire resistance, the combustion of liquid-immersed transformers refers mainly to the combustion of transformer liquid.

The flammability of high voltage equipment is the most important security worry of recent times. There are numerous instances of high voltage apparatus outbursts leading to flames that are hard to quench and extend to nearby regions as liquid outflows. The flash point defines the tendency of liquid insulation to establish a combustible compound with an atmosphere under ordered laboratory environments. This experiment provides a characteristic for assessing a material's overall flammability susceptibility. Flash point is applied in shipment and security guidelines to describe flammable and combustible materials. Flash point may designate the potential existence of highly impulsive and flammable materials.

Fire safety is a key concern for today's users of dielectric insulating liquids. This is especially so when considering their use in areas where a fire would be highly damaging and difficult to control, for example in public buildings, apartment blocks, industrial application such as steel works, in subway tunnels, or in financial centers.

Petroleum-based mineral oils have been widely used as insulating liquid since the end of the nineteenth century. Because of their good performance and reasonable costs, mineral oil was the liquid of the choice. However, low flash- and fire-point of these liquids fall far short of the safety requirement, hence combustion and explosion accidents constantly occur. Flash and fire points are the most critical factor to consider when specifying transformer fire safety, as it is the property that most directly prevents a fire in the event of a transformer failure. Flash- and fire-point are acknowledged in IEC 61.039 [84] and IEEE C57.1200 [85]. Power transformers filled with mineral oil are one of the most dangerous electrical devices due to the large amounts of insulating oil and the direct contact with high-voltage elements. Electrical faults that lead to arcing can cause the mineral oil to lose its dielectric properties. The mineral oil evaporates and creates a pressure increase in the transformer. If the pressure inside the

Table 4.15 Fire classification of liquids

Class	Fire point	Class	Net calorific value (MJ/kg)
O	≤ 300 °C	1	≥ 42
K	> 300 °C	2	< 42 and ≥ 32
L	No measurable fire point	3	< 32

transformer exceeds the mechanical strength of the container, an explosion occurs. A fire is very likely the result. Although the electrical transformers are equipped with protective mechanisms that detect excessive energy flows, they are often too slow to switch off the energy supply in good time before the damaging event occurs, although the response times are in the millisecond range [86]. One of the most terrible accidents from this title happened in 2010 in Dhaka. In this transformer fire, the event and its consequences killed over one hundred people [87] and over two hundred were admitted to hospitals with critical injuries—mainly burns [88]. The extent of the accident was so dramatic because the transformer was set up in a tightly built-up area surrounded by a plastic factory. Even if this event, on this scale, must be viewed as extremely dramatic and more of a singular event, fatal accidents with transformers filled with mineral oil happen again and again [89].

Mineral oil insulated equipment is the main source of substation fire. Nearly all natural- and synthetic ester and silicone oil are thermal stable and have a fire point > 300 °C and they can be K2 to K3 classified (IEC 61.039 [84]). This standard designates a class K for insulating liquids with a fire point above 300 °C and defines a system for classifying insulating liquids according to the fire point and net calorific value. The characteristics on which the system is based are given together with limiting values in Table 4.15.

The higher bond energy of the ester linkage makes it stable to heat. Its fire resistance qualities and high fire point are traits that benefit customers by protecting them against the harmful effects of transformer failure-caused fire. This feature greatly reduces the likelihood of a disastrous fire occurring in association with a failed transformer in highly populated areas. This reduces the need for ancillary fire-fighting equipment and the transformers are suitable for indoor, outdoor, and underground installations. Additionally, IEC 61.039 [84] and IEC 61.936-1 [90] allows smaller distances to buildings and adjacent transformers. To reduce fire risk, it is advisable to replace more flammable insulating fluids (typically mineral-based) with ester fluids. Table 4.16 shows a comparison of the flash-, fire point and heating value of the different insulating liquids. Biological insulating liquids have significantly higher flash and fire points than that of conventional mineral oil liquids. The flash point of an insulating liquid is specified for safety reasons. It is the lowest temperature of a flammable liquid at which the vapor pressure is sufficient to form a flammable mixture with air near the surface of the liquid. IEC specifies the Pensky-Martens closed cup method [91, 92]. In the USA the Cleveland open cup test (ISO 259) [93] and (ASTM D92) [94] is used, which gives a 5–10 °C higher flash point value. The flash point depends on the light part of the insulating liquid and is extremely sensitive to contaminants from lighter oils, such as gas oil or gasoline. Even though both methods yield relatively poor reproducibility, the closed cup method is preferred because it provides better repeatability.

The fire point is the lowest temperature at which a liquid is heated in an open container and attains sufficient combustible vapors to ignite and sustain a fire for 5 s. Shengwei et al. showed in [95] that compared with mineral insulating oil, the

Table 4.16 Thermal properties of insulating liquids

Insulating liquid	Flash point (°C)	Flash point (°C)	Fire point (°C)	IEC classification	Heating value (MJ/kg)
Test method	ISO 2719 [91] ASTM D93 [92]	ISO 2592 [93] ASTM D92 [94]	ISO 2592 [93] ASTM D92 [94]	IEC 61039 [84]	ASTM 240 [97]
Nynas Nytro® 4000X [1]	146	–	165–170 [61]	O1	–
Nynas Nytro® BIO 300X [5]	145	–	–	O	–
Envirotemp™ 360 Synthetic ester [6]	277	–	316	K	–
Midel® 7131 [98]	260	275	316	K3	30.8
BecFluid® 9902 [8, 99]	274	–	303	K2	36.8
Nycodiel® 1233 [9]	248	–	284	O2	32.0
Nycodiel® 1244 [9]	255	–	304	K3	30.6
Nycodiel® 1258 [9]	255	–	310	K3	31.8
MATROL-BI® FDE01A [10]	260	–	310	K	–
Envirotemp® FR3™ Fluid [40]	260	320–330	350–360	K	–
Midel® eN 1204 [16, 98]	> 260	> 315	> 350	K2	37.5
Midel® eN 1215 [18, 98]	> 260	> 315	> 350	K2	37.2
BIOTEMP® [19]	–	340	360	K	–
NeuGen plus [21]	–	318	360	K	–
Paryol Electra 7426® [100]	–	≥ 300	≥ 350	K2	< 42
MATROL-BI® FDV01A [23]	260	–	330	K	–
Jatropha curcas oil [24]	–	310	> 340	K	–
Castor oil [53]	229	–	–	–	–
Powersil® Fluid TR 50 [25]	> 240	–	> 340	K	–

natural ester insulating oil is not easily ignited by external fire sources and it has a certain self-extinguishing function. This is associated with a high ignition point and boiling point and low saturated vapor pressure of the natural ester insulating oil. Their test results of the fire tray also presented that the mineral insulating oil can be easily ignited under the oxyacetylene flame, while the plant insulating oil cannot be easily ignited. This indicates that the natural ester insulating oil is inflammable and not easy to explode. In the event of a strong external fire, a natural ester insulating oil transformer cannot catch fire for 30 min, which provides a relatively long time for a firefighting operation, and therefore has a certain fire resistance property.

Low values of either flash or fire points of biological insulating liquids are an indication of contamination with lower flash and fire point materials, such as conventional mineral oil. Hydrolysis and oxidation breakdown triglycerides, producing free fatty acids lowering the flash- and fire-point. Hoffmann studied in [96] thermal stability of rapeseed oil at 210 °C by measuring the flash point during isothermal tests of 720 and 2160 h with various thermal energy storage materials. The flash point evolution was correlated to acid value (a) by Eq. 4.25. Where a is the acid value in KOH mg/g and T the temperature in degree Celsius.

$$T_{Flash\ point} = 327.589 - 11.504a + 0.481a^2 \tag{4.25}$$

To improve the fire resistance of the whole biological insulating liquid filled transformer system, it is recommended to improve the fire resistance of the transformer using the fire-resistant high-pressure casing and sealing materials, improving the structure and threshold value of pressure relief valve, improving the mechanical strength of liquid tank, and taking other measures.

Both biological insulating liquids (natural ester) and synthetic ester can offer a high degree of fire safety, due to their high fire points. Smoke density, the actual damage from the transformer fire but also damage from pool fires, needs to take into consideration.

Apart from the calorimeter test, Hellebuyck carried out a pool fire test [101]. In this case, the insulating liquid was heated by a cooking stove and a propane-butane mixture burner until a sustained fire was reached. After reaching a sustained fire in the pan, the content of the pan were poured into a large tray filled with the same nature of insulating liquid, recreating a sudden burst in the transformer container. The liquid level in the tray was 2 mm. The results of the investigation were:

Mineral Oil The fire spread across the surface of the tray resulting in a big pool fire.

Natural Ester When pouring the natural ester heated up till to 360 °C in the tray filled with natural ester at 20 °C, the fire extinguished.

Synthetic Ester As expected, the flame height was lower than in the natural ester test as the heat of combustion is lower for the synthetic ester. When pouring the ester in the tray filled with synthetic ester at 20 °C, the fire extinguished.

Silicon Oil When pouring the silicon liquid in the tray filled with silicon oil at 30 °C, the fire did not fully extinguish but kept on burning around the crust areas and produced white smoke.

Summarizing all insulating liquids burned when enough heat was applied to them. Whenever the upper layer of the insulating liquid reached its fire point, sustained burning was reached. For mineral oil liquids, this happens very quickly. A couple of seconds with the burner was enough. The other insulating liquids required preheating on the stove, combined with reasonably long heating with the burner. Pouring the burning insulating liquid in a pool of cold liquid induced heat losses to the extent that except for mineral oil liquid, their temperature went below their respective fire points, resulting in the extinguishment of the fire.

Laboratory tests show a lower smoke density when burning ester liquids. The oxygen present in the ester liquids (see Sect. 2.3), 4–10% of the required one for combustion, is available in the core zone of the flame, which is not the case with mineral oils.

Up to date, no cases of transformer fires have been reported involving biological insulating liquids [102]. Silicone oil forms a crust on the liquid's surface which impedes combustion (see Sect. 6.1.2.3).

In IEC 60695 [103] it is concluded that by measuring the fire point and heat release rate of insulating liquids, and the corrosion damage, smoke opacity and toxic hazard effects of fire effluent from burning insulation liquids, the hazards associated with insulating liquids used in electrotechnical devices can be assessed, based on the principles that

- the higher fire point, the more difficult is ignition.
- if ignition occurs, the lower the heat release rate and production of fire effluent, the lower is the expected hazard and difficulty of firefighting.

4.3.1 Flash Point—Test Method

The flash point of insulation liquid is the temperature whereat liquid surface is blistered and creates an adequate vapor which would develop a combustible mixture with atmosphere that can be ignited. Flash point values are a function of the apparatus design, the condition of the apparatus used, and the operational procedure carried out. Flash point can therefore only be defined in terms of a standard test method, and no general valid correlation can be guaranteed between results obtained by different test methods, or with test apparatus different from that specified.

- ASTM D92 [94] or ISO 2592 [93]: This test method describes the determination of the flash and fire point of petroleum products by a manual or an automated Cleveland open cup apparatus. This test method is applicable to all petroleum products with flash points above 79 °C and below 400 °C except fuel oils which are most commonly tested by the closed cup procedure.

- ASTM D93 [92] or ISO 2719 [91]: These test methods cover the determination of the flash point of combustible liquids, liquids with suspended solids, liquids that tend to form a surface film under the test conditions, biodiesel and other liquids in the temperature range from 40 to 360 °C (ISO 2719: 370 °C) by a manual or an automated Pensky-Martens closed-cup apparatus.

4.3.2 Fire Point—Test Method

ASTM D92 [94] or ISO 2592 [93] are also used to determine the fire point, which is a temperature above the flash point, at which the test specimen will support combustion for a minimum of 5 s.

4.3.3 Heating Value—Test Method

The heating value of an insulating liquid is the energy created by the combustion of the flammable material.

- ASTM D240: A weighted amount of liquid is burned in an oxygen bomb calorimeter under controlled conditions. The heat released by the combustion of a unit mass of liquid under defined oxygen pressure provides the heat of the sample [97].

4.3.4 Specific Heat Capacity and Thermal Conductivity

The specific heat capacity is the energy required to raise the temperature of the unit mass of a given substance by 1 K. Therman conductivity measures the ability of a material to conduct heat or the rate of heat transfer through a unit thickness of the material per unit area and per unit temperature difference. In addition to the viscosity, the specific heat capacity (Figs. 4.9 and 4.10) and the thermal conductivity (Figs. 4.11 and 4.12) are those parameters that influence the cooling performance of the insulating liquid. For Fig. 4.10 the polynomial second order and the coefficients of [41] are used. Both parameters are temperature dependent. Evangelista specifies the specific heat for Jatropha curcas [24] with 1.92 J/g without giving a reference temperature. It can be observed that the heat capacity of all mentioned insulating liquids increases with temperature and the thermal conductivity decreases. Specific heat capacity increases with the chain length and degree of saturation of vegetable oils [104].

Gomna et al. give in [41] several correlations of specific heat capacity versus temperature of vegetable oils with their coefficients. Figure 4.10 presents the fitted curves of the polynomial correlation with T in °C calculated by equation Eq. 4.26.

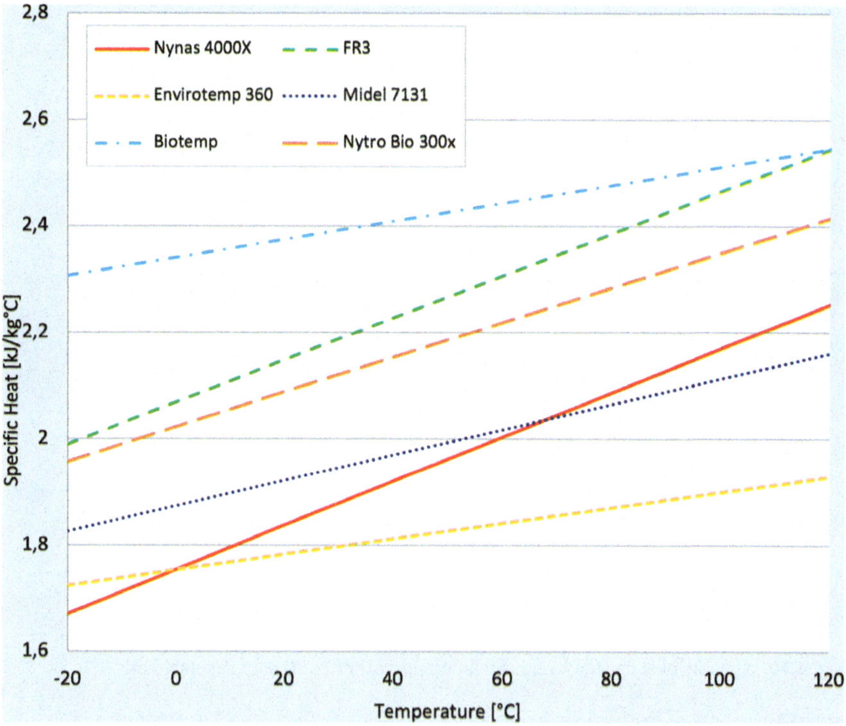

Fig. 4.9 Specific heat capacity versus temperature of different insulating liquids

There is no great difference within the vegetable oils.

$$Cp = a * T^4 + b * T^3 + c * T^2 + d * T + e \quad (4.26)$$

In [41], Gomna et al. show that vegetable oils show different behavior of thermal conductivity depending on fatty acid composition. Saturated oils have stronger decrease of thermal conductivity than unsaturated oils.

Studies reported changes in thermal characteristics of vegetable oils after being used for frying [41]. The decrease was attributed to the disappearance of triglycerides and degradation products formed.

Heat transfer is an important function that insulating liquid must ensure in transformers. This function is realized by both thermal conductivity and convection. The convection represents all the properties which lead to the heat transfer by fluid displacement (viscosity, specific heat, thermal expansion coefficient) whereas the conduction depends on the conductivity of the fluid. The most influential parameter for the heat transfer is the kinematics viscosity (Fig. 4.5). The viscosity shows that the heat transfer in a transformer by convection will be less efficient with biological insulating liquids and least efficient with silicone oil. In contrary Figs. 4.9, 4.10, 4.11

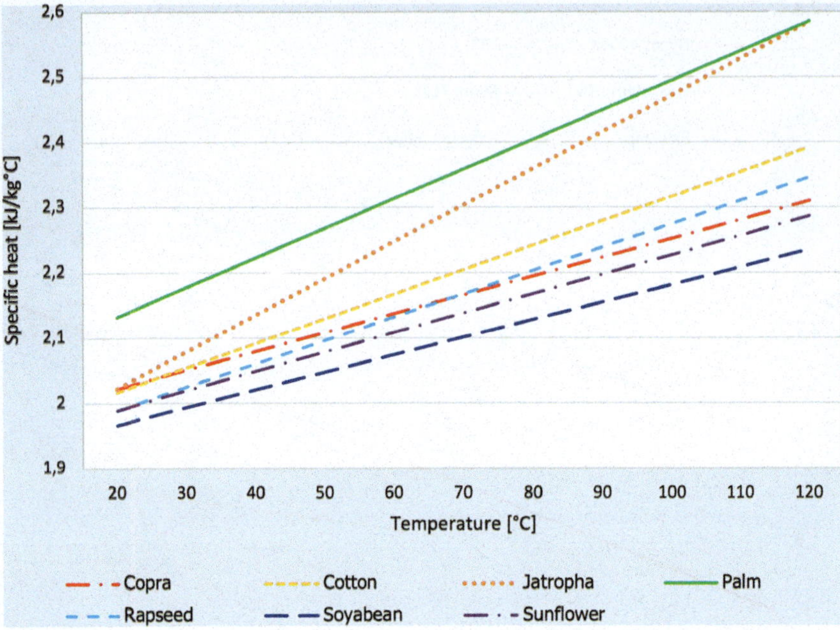

Fig. 4.10 Specific heat capacity versus temperature of different pure vegetable oils

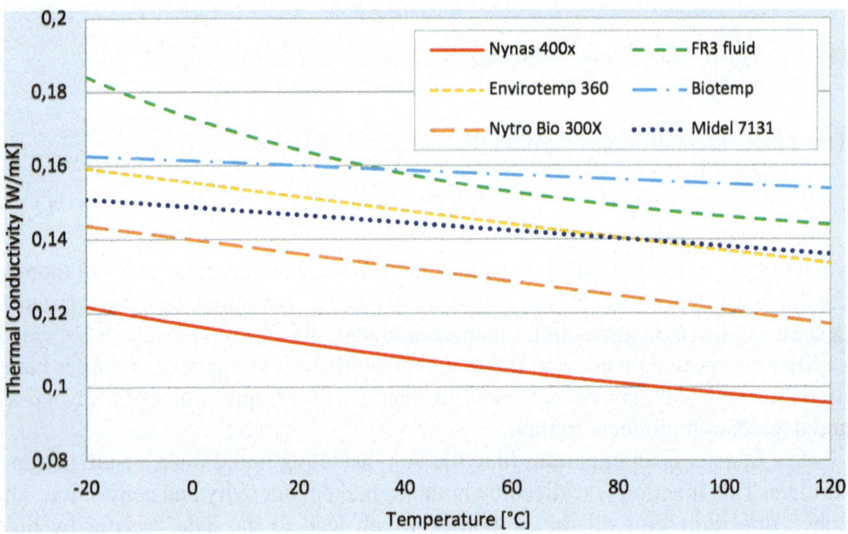

Fig. 4.11 Thermal conductivity versus temperature of insulating liquids

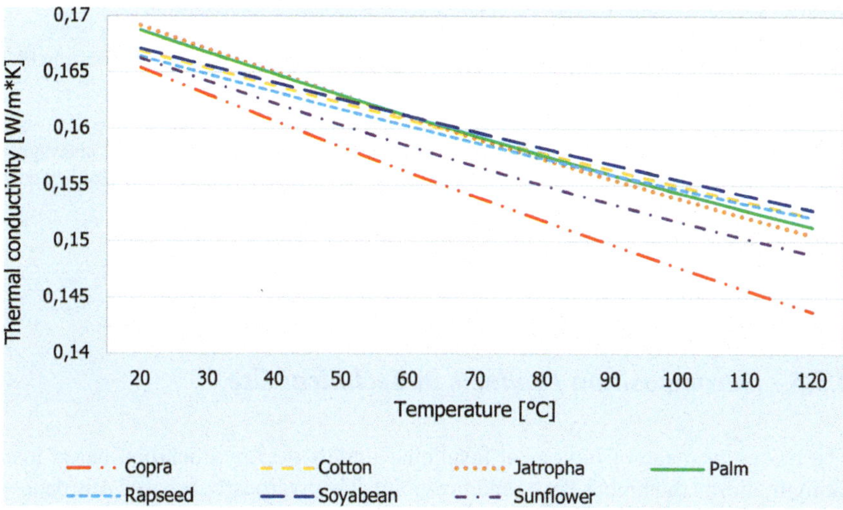

Fig. 4.12 Thermal conductivity versus temperature of different pure vegetable oils

and 4.12 show that biological liquids have in case of specific heat and thermal conductivity better heat transport behavior. This is not necessarily a critical issue, but care must be taken especially when designing the cooling system of power transformers.

4.3.4.1 Specific Heat Capacity—Test Method

The specific heat capacity of liquid is the amount of energy required to produce a given temperature change within a quantity of liquid ($J*K^{-1}*g^{-1}$).

- ASTM E1269: This test method covers the determination of specific heat capacity by differential scanning calorimetry [105].

4.3.4.2 Thermal Conductivity—Test Method

The thermal conductivity of a liquid is a measure of its ability to transfer energy as heat in the absence of mass transport phenomena.

- ASTM D2717: The thermal conductivity can be determined by measuring the temperature gradient produced across the liquid by a known amount of energy introduced into a glass cell by heating a platinum resistance element. It is expressed in such units as $W*m^{-1}*K^{-1}$ or $J*s^{-1}*m^{-1}*K^{-1}$ [106].

Table 4.17 Decomposition products/temperature

Insulating liquid	Nynas Nytro Taurus	Nynas Nytro Bio 300x	Midel 7131	Midel eN 1204	Envirotemp FR3
Decomposition products	Broken chains, alkanes	Alkanes, acids, ketenes	Alkanes, methyl ester	Ester cleavage, C=C aldehydes	Ester cleavage, C=C aldehydes
Decomposition temperature (°C)	174	169	309	386	380

4.3.5 Decomposition Products in Technical Use

The most advantage of biological insulating liquids over hydrocarbon-based insulating liquids is the higher flash- and fire-point. The hydrocarbon-based oils decompose above 150 °C, while synthetic, natural esters and silicon oil are stable up to 300 °C. Kurzweil et al. studied the thermal decomposition behavior of different insulating liquids by thermogravimetric analyses under a nitrogen atmosphere [107]. For determining the qualitative composition of the cracked gases, they used a nitrogen atmosphere in a closed crucible, and for the decomposition temperature an open crucible, respectively (Table 4.17). A clear relation with the flash and fire point can be seen.

4.3.6 Pour Point

Insulating liquids generally circulate within the high voltage equipment for the intention of customary natural cooling. Therefore, the appropriate flow of this dielectric liquid insulation is significant in providing the required cooling of the equipment. The pour point of the liquid insulation refers to that lowermost temperature at which insulating liquid merely initiate to flow/pour easily, when examined under prescribed conditions. The pour point of liquid insulation is specified as temperature whereat liquid rejects to pour/flow under specified experiment circumstances. Information about the pour point of the insulating liquid is vital to evaluate its appropriateness for application in specific weather. Insulating liquids having higher pour points may not be suitable for using in power transformers, especially in low-temperature climates. Practical application of coconut oil in transformers in tropical climates has worked well despite its higher pour point. Pour point is a valuable measure to know how insulating liquid will execute at low temperatures specifically, while this is crucial to startup a transformer in extremely cold conditions. When the temperature of insulating liquid falls under the pour point, it inhibits conventional flowing and hinders the cooling of the transformer. Furthermore, the movement of tap changer may be impacted [38]. Although biological insulating liquids have a relatively high pour

Table 4.18 Pour points of new different insulating liquids [°C]

Nynas Nytro® 4000X [1]	Nynas Nytro® BIO 300X [5]	Envirotemp™ 360 Synthetic ester [6]	Midel® 7131 [98]
– 57	– 60	– 48	– 50
BecFluid® 9902 [8]	Nycodiel® 1233 [9]	Nycodiel® 1244 [9]	Nycodiel® 1258 [9]
– 50	– 66	– 45	– 60
MATROL-BI® FDE01A [10]	Envirotemp® FR3™ Fluid [40]	Midel® eN 1204 [98]	Midel® eN 1215 [98]
– 48	– 21	– 31	– 18
Biotemp® [19]	NeuGen plus [21]	Paryol electra 7426® [22, 111]	MATROL-BI® FDV01A [23]
– 15 to – 20	– 15	– 15 to – 27	– 21
Coconut oil [60]	Sesame oil [60]	Castor oil [60]	Powersil® Fluid TR 50 [112]
23	– 9	– 27	< – 50

point, they have a very good service record in transformers installed in cold and hot climates. Rapp et al. demonstrated in [108] that freezing at the natural ester liquids had no effect on their physical, chemical, or electrical properties. Thanks to the mixture of esters in the biological insulating liquids, they do not suddenly freeze or thaw, and remain in a liquid state even at very low outside temperatures. Their tests also showed that a transformer can be energized at solid phase temperature without adverse effects. Regarding synthetic esters, they have a pour point close to that of conventional mineral oil (Table 4.18). The test method for determining the pour point is ISO 3016 [109] and ASTM D97 [110], respectively.

Flow improving additives, alkyl-methacrylate polymers, and alkylated-polystyrenes, act as surface active agents, preventing crystals from growing into long interlocking chains, dispersing them rather into clusters around the additive molecule acting as pour point depressants.

The pour points of the vegetable oils are mainly determined by their structures determined by the saturated and unsaturated nature. Saturated fatty acids, having linear zig-zag hydrocarbon chains, have very close intermolecular interactions, causing a higher melting point. Together unsaturation and branch off in fatty acid possess an encouraging effect on dropping pour point. Unsaturated fatty acids are additionally efficient in decreasing pour point than branched fatty acids of identical carbon length. The existence of aromatic assemblies in natural ester is also favorable to maintaining lower pour point [38].

For coconut oil with the highest pour point and the sesame oil, it should be mentioned that these are commercially machine processed produced oils without a refining, bleaching and deodorization process.

4.3.6.1 Point—Test Methods

The pour point temperature of liquids is the lowest temperature at which the liquid is observed to flow when cooled and examined under the prescribed conditions of

- ASTM D97: After preliminary heating, the sample is cooled at a specified rate and examined at intervals of 3 °C for flow characteristics. The lowest temperature at which movement of the specimen is observed is recorded as the pour point [110].
- ISO 3016: Corresponds to ASTM D97.

4.4 Electrical Properties

Insulating characteristics of insulating liquids differ depending on several dynamic interactions, for example moisture content, temperature, applied voltage, frequency level, polarity, electrode configuration, particles, etc. Of interest here is the one which considers a material's resistance to the flow of electricity as a guide. For example, materials having a specific resistance of less than one microohm per centimeter are called conductors, those having less than one giga-ohm per centimeter are called semiconductor, and those having higher resistivities are referred to as insulators or dielectric materials. It is the last group, that comprises above all the dielectric insulating liquids that is of interest here in this book.

To choose appropriate insulating liquids, it is essential to identify the dielectric characteristics such as breakdown voltage, dielectric constant, dielectric dissipation factor, etc. Table 4.19 shows the dielectric breakdown voltage value and dissipation factor from new dielectric insulating liquids announced by the manufactures. The company IMCD (NeuGen) only uses the ASTM standard for the dissipation factor. A comparative representation of the parameters resistivity and dielectric constant is not possible because of insufficient amount of data. There is no big difference in the electrical properties of the described insulating liquids. Above all, the biological insulating liquids have excellent electrical behavior.

A property closely related to the polar nature of the ester linkage is the DC volume resistivity of the biological insulating liquids. These liquids have a DC volume resistivity, which is typically an order of magnitude lower than that of mineral oil. Because of using different standards and sources, it is not so clearly seen in Table 4.19. This does not affect the liquid's ability to operate as an effective AC dielectric, but what users will notice is that measurements of insulation resistance on a transformer filled with synthetic ester or biological insulating liquid will be lower than would be expected for mineral oil. The magnitude of difference will depend on factors such as the volume of paper insulation and the dielectric structure.

Table 4.19 Dielectric breakdown voltage and dissipation factor of different insulating liquids

Insulating liquid	Dielectric Breakdown (kV)	Dissipation Factor (90 °C, 50 Hz)	Volume resistivity (90 °C) (GΩ m)
Test method	IEC 60156 [113]	IEC 60247 [114] ASTM D924[a] [115]	IEC 60247 [114] ASTM D1169[b] [116]
Nynas Nytro® 4000X [1]	40–60	< 0.001	45.4–495.5 [61]
Nynas Nytro® BIO 300X [5]	40–60	< 0.001	–
Envirotemp™ 360 Synthetic ester [6]	> 60	0.004	24
Midel® 7131 [98]	> 75	< 0.008	3.0–9.5 [61]
BecFluid® 9902 [8, 99]	99	0.005	6 (2.7–4.4) [61]
Nycodiel® 1233 [9]	65	0.01	10
Nycodiel® 1244 [9]	> 70	0.01	7.5
Nycodiel® 1258 [9]	> 70	0.01	10
MATROL-BI® FDE01A [10]	77	0.008	6.2
Envirotemp® FR3™ Fluid [40]	> 50	0.02	4.0–4.6 [61]
Midel® eN 1204 [16, 98]	> 75	< 0.03	9.5 [61]
Midel® eN 1215 [18, 98]	> 75	< 0.03	–
BIOTEMP® [74]	> 75	< 0.02	25 °C 150[b] (2.0–3.4) [61]
NeuGen Plus [21]	≥ 70	25 °C 0.05[a] 100 C < 4%[a]	100 °C 0.77[b]
Paryol Electra 7426® [22]	≥ 60	0.02–0.05	–
MATROL-BI® FDV01A [23]	80	0.015	–
Jatropha Curcas oil [24]	55[++]	–	–
Pongamia Pinnata oil [59]	51–77	–	–
Powersil® Fluid TR 50 [25]	> 40	< 0.001	> 100 (148) [61]

[++] ASTM D877/D877M-19, Standard Test Method for Dielectric Breakdown Voltage of Insulating Liquids Using Disk Electrodes, 2019
[a] ASTM D924
[b] ASTM D1169

4.4.1 Dielectric Constant and Refractivity

The relative permittivity (dielectric constant) is one of the most important parameters of insulating materials. The electrical field strength can be derived from Coulomb's law (Eqs. 4.27–4.28) [117]

Table 4.20 Test results—permittivity, refraction index at 90 and 20 °C [61]

Insulating fluid	ε_r (50 Hz, 90 °C)	ε_r (50 Hz, 20 °C)	Δ%[a]	n (20 °C)	n^2	$\varepsilon_r - n^2$	Water content (ppm)
Nynas Nytro 4000X	2.08	2.18	4.6	1.4741	2.173	0.01	8.1
Nynas Nytro 4000X	–	2.17	–	1.4722	2.167	0.00	31.3
Nynas Nytro 4000X	–	2.17	–	1.4727	2.169	0.00	17.2
Nynas Nytro 4000X	–	2.17	–	1.4725	2.168	0.00	16.6
Envirotemp® FR3™ Fluid	2.86	3.15	9.2	1.4739	2.172	0.98	67.1
Biotemp®	2.83	3.11	9.0	1.4702	2.161	0.95	102.0
Midel® eN	2.84	3.12	9.0	1.4729	2.169	0.95	189.4
Midel® 7131	2.97	3.18	6.6	1.4529	2.111	1.07	94.1
BecFluid® 9902	3.02	3.24	6.8	1.4529	2.111	1.13	84.7
Powersil® Fluid TR 50	2.55	2.68	4.9	1.4140	1.999	0.68	84.3
Shell Produkt 4697	2.16	2.26	4.4	1.4815	2.195	0.07	60.9
Shell Diala G	2.15	2.24	4.0	1.4848	2.205	0.04	34.4
Shell Switchgear X	2.08	2.17	4.1	1.4694	2.159	0.01	32.4
Technol 3000 used	2.10	2.19	4.1	1.4753	2.177	0.01	23.7

[a]Deviation between ε_r (20 °C) and ε_r (90 °C) in percentage

$$F = \frac{1}{4 * \pi * \varepsilon} * \frac{Q_1 * Q_2}{r^2} \tag{4.27}$$

$$\varepsilon = \varepsilon_0 * \varepsilon_r \tag{4.28}$$

F Force in N
Q_1, Q_2 Charge in As
ε_0 Absolute dielectric constant of the vacuum $\varepsilon_0 = 8859 * 10^{-12}$ F/m
ε_r Relative dielectric constant[1] (indicates the deviation factor of the dielectricity from the vacuum) for different insulating liquids (Table 4.20)
r Distance between the charges in m.

The electric field strength F_x is inversely proportional to the dielectric constant (Eq. 4.29) [118].

$$\frac{F_1}{F_2} = \frac{\varepsilon_2}{\varepsilon_1} \tag{4.29}$$

The dielectric having the lower dielectric constant is loaded with a higher field strength than the medium having the higher dielectric constant.

[1] The relative dielectric constant is often referred to as "relative permittivity".

The Debye- and Clausius-Mossotti equation (Eq. 4.30) links the macroscopically measurable size of the dielectric constant with the microscopic molar sizes of the electrical polarizability and the permanent dipole moment [119].

$$P_m = \frac{\varepsilon_r - 1}{\varepsilon_r + 2} * \left(\frac{M}{\rho}\right) = \frac{N_A}{3 * \varepsilon_0} * \left(\alpha + \frac{\mu^2}{3 * k_B * T}\right) \tag{4.30}$$

P_m Molar polarizability (m^3*mol^{-1})
M Molar mass ($kg*mol^{-1}$)
r Density ($kg*m^{-3}$)
N_A Avogadro constant ($6.02214076*10^{23}*mol^{-1}$)
a Induced polarization ($T^4*I^2*kg^{-1}$)
m Orientation polarization ($I*m*T^2*s^{-1}$)
k_B Boltzmann constant ($\approx 1.38*10^{-23}*m^2*kg*s^{-2}*T^{-1}$)
T Temperature (K).

In the case of nonpolar liquids, the dielectric constant of the liquid is determined by its electronic and atomic polarizability, molecular weight, and density. Thus, the dielectric constant depends on the electronic structure of the atoms; in addition, if the liquid comprises more than one type of atom, an asymmetrical sharing of electrons may arise as the stronger binding atoms will cause the electron clouds to be displaced eccentrically towards them. Atoms will thereby acquire charges of opposite polarity, whereby the application of an external field will impel a change in the equilibrium positions of the atoms leading to atomic polarization. Atomic polarization need not to be uniquely due to atomic or ionic displacements within the molecule, it may also result from changes in the bond angles between the atoms or ions. For instance, in an n-hexane molecule the C–H group is polarized, with the hydrogen being more negative than the carbon atom; still, the total dipole moment of the molecule is zero. These nonpolar liquids are characterized by a relative dielectric constant ε_r of around 2. If such atoms or molecules are placed in an electric field, a separation of the centers of the positive and negative charge distribution results, and the atoms or molecules become polarized (electronic polarization or induced polarization). The corresponding dipole moment is proportional to the electric field strength. With polar liquids which exhibit a permanent asymmetrical charge distribution between dissimilar atoms, a permanent dipole moment exists even in the absence of an external electrical field. Permanent dipoles are connected with asymmetries of the molecular structure or with the presence of electronegative atoms, like oxygen, in the molecule. Because of their permanent dipole moment, polar molecules are characterized by a nonsymmetrical molecular structure. Evidently, because of their additional permanent dipole contribution, polar liquids are characterized by higher dielectric constants [119].

In castor oil, which consists mainly of ricinoleic acid, the polarity is reinforced by hydroxyl groups in the molecule. Castor oil is important as an electrical insulator for DC and pulse capacitors. The high dielectric constant of 4.5 is favorable for a high energy density of capacitive energy storage. In addition, pulse capacitors with castor oil-paper insulation last about ten times longer than capacitors with mineral oil-paper insulation [118].

The comparison of dielectric constant values of biological insulating liquids (Table 4.20) and cottonseed and corn oil in [120] with that of mineral oil shows that the dielectric constant is higher and in a close-range independent in their origin. This can be attributed to the presence of triglycerides in the biological insulating liquids which have a polar nature. On the other hand, mineral oil mainly consists of refined petroleum, which contains nonpolar alkane molecule and so it has a comparatively lower dielectric constant.

Dielectric constants are not constant, they change with the temperature and the frequency of the electric field. An increase in temperature leads to a decrease in the dielectric constant (Table 4.20). The amount of decrease depends on the molecular structure. The table shows that the biological insulating liquids have the greatest decrease followed by the synthetic ester and silicon oil. The mineral oils show the smallest decrease of the dielectric constant. This decrease in dielectric constant with the increase of temperature is due to the decrease in density, which is directly related to the density of dipoles in the sample. The increase in temperature also causes the increase in kinetic energy of the moving segments, leading to a greater randomness of motion and thus it decrease the dipole orientation, which results in low dielectric constant.

Shah and Tahir show in [120] the frequency dependence of dielectric constant ε for corn oil, cottonseed oil and mineral oil at room temperature. As dielectric dispersion occurred, the dielectric spectra diminished and there is a slight monotonous decrease of dielectric constant values of biological insulating liquids with the increase of frequency. As the frequency increases, it should be noted that the polarization is subject to a mechanical inertia which is greatest for the alignment of large diploes and the smallest for electron polarization. This means that as the frequency increases, the dipoles can no longer follow the field change due to their inertia. This creates a strong frequency dependency (dispersion) of the dielectric constant. Judendorfer pictured the different polarization steps in [121] (Fig. 4.13).

In a transformer, the electrical stress distributes inversely proportional to the permittivity of the material. The weakest material in the solid–liquid insulation is always the liquid. The use of a liquid with a higher dielectric constant brings the dielectric constants of the fluid and solid insulation closer together, thus reducing the stress in the dielectric insulation liquid [122]. Figure 4.14 shows the measured permittivity from different fluids at 90 °C and the water saturation point from these fluids at this temperature, respectively. The polarity helps to keep the water in solution and to increase the permittivity [123].

With fields of higher frequency, the response of the polarization of the dielectric liquid may lag behind the electrical forces. It is subjected to by the electrical field. This lag time usually is called relaxation. It is due to the frictional resistance of the liquid to the change in molecular orientation. The time scale where this relaxation is active in a dielectric liquid depends on the type of polarization and on the structure of the liquid. Relaxation of dipoles in a liquid, which relates to a rotation of the dipole, occurs in a lower frequency range than electronic polarization, which involves the

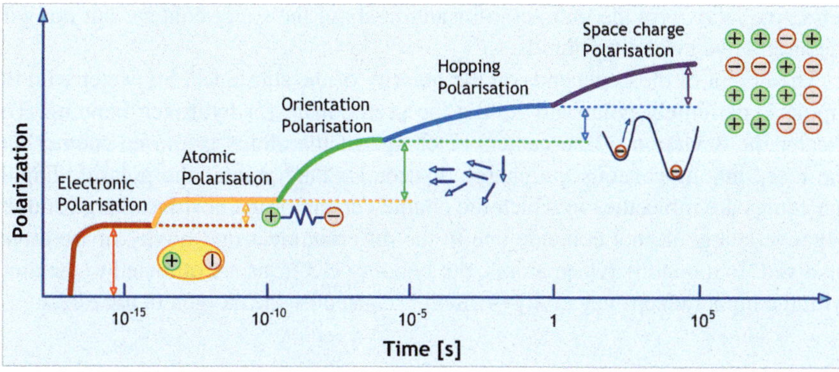

Fig. 4.13 Polarization versus time (s)

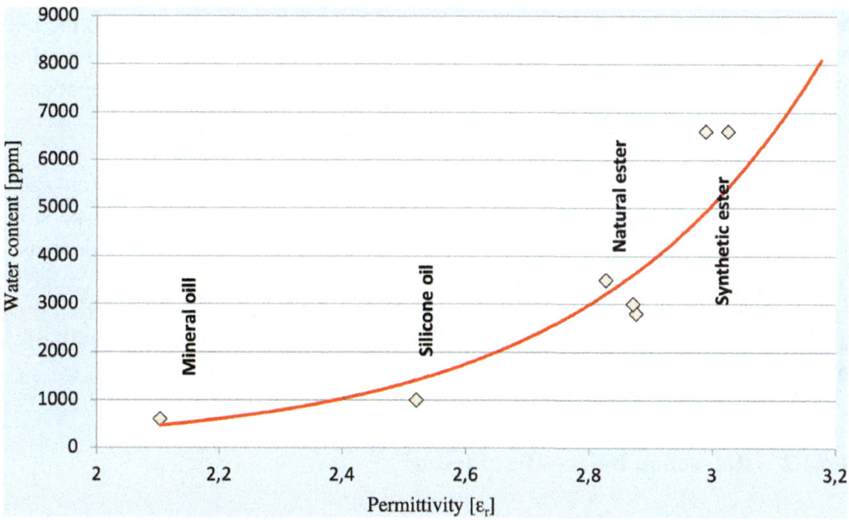

Fig. 4.14 Permittivity of different insulating fluids at 90 °C

displacement of the electronic shell relative to the positive nuclei. Another relaxation is produced by the atomic polarization, the displacement of the atoms in a molecule relative to one another. In [118], a relationship between the dielectric constant ε_r of electrical engineering and the refractive index n from optics is depicted as Eq. 4.31.

$$\varepsilon = n^2 \tag{4.31}$$

Table 4.20 shows the test results taken from [61]. The mineral oil products largely meet the equation, while the alternative insulating liquids show a clearly different behavior. In the case of mineral oils, the determination of the permittivity and the refractive index provide indirect information about the water content, but not with the alternative insulating liquids.

The height of the ε_r depends on the polarity of the substance. So, water with the dipole is extremely polar and fulfills the prerequisite for hydrogen bonding. The greater the deviation of the centers of charge of the atoms from one another and the more inhomogeneous the charge distribution, the greater the polarity. Dipole molecules are molecules in which the charge centers of the positive charge and the negative charge do not coincide due to the different electronegativity of the atoms involved. In a bond between atoms, the bonding electrons are attracted to the atom with the higher electronegativity—which is caused by the oxygen in the esters.

4.4.1.1 Dielectric Constant—Test Method

In the case of constant fields and alternating fields of sufficiently low frequency, the relative permittivity of an isotropic or quasi-isotropic dielectric is equal to the ratio of the capacitance of a capacitor (C_x), in which the space between and around the electrodes is entirely and exclusively filled with the dielectric, to the capacitance (C_0) of the same configuration of electrodes in vacuum Eq. 4.32 [114].

$$\varepsilon_r = \frac{C_X}{C_0} \tag{4.32}$$

The relative permittivity ε_r of dry air, at normal atmospheric pressure, equals 1,00,059, so that in practice, the capacitances C_a of the configuration of electrodes in air can normally be used instead of C_0 to determine the relative permittivity ε_r with sufficient accuracy.

4.4.1.2 Refraction Index—Test Method

The refractive index was determined using the Abbe 2WAJ refractometer with an external water bath for temperature control based on ASTM D1218 [124].

4.4.2 Breakdown Voltage

The dielectric strength of insulating liquids is a measure of its capability to endure electrical stress without collapse. The break down voltage of an insulating structure (liquid and solid insulation) is one of the key factors for the electrical structure of high voltage equipment. It may be defined as the lowest voltage at which there electric conduction occurs that leads to a dielectric rupture. When dielectric material losses its insulation ability, that is, when the barriers to the flow of electricity are suddenly removed by some process, one speaks of electrical breakdown. The dielectric breakdown voltage of an insulating liquid is of importance as a measure of its ability to withstand electric stress. If the insulating material is a solid, such as polyethylene or an epoxy compound, then the electrical breakdown will lead to permanent failure of the system. If, on the other hand, the dielectric is a dielectric insulating liquid, then electrical breakdown will lead only to a short interruption of the functioning of the system since the fluid will recover rapidly and resume its insulating function. This ability to recover from failure renders dielectric insulating liquids very useful in many practical applications. It should be pointed out here that temporary failures naturally cause chemical changes in the liquid, which will lead to a gradual deterioration of the insulating liquid and thus shorten its useful life.

Electrical breakdown always involves the development of a conducting channel in the insulating material. Such channels are of a permanent nature in solids but only of short existence in liquids and gases.

The dielectric strength is a major parameter when choosing oils for use in oil filled equipment such as power transformers. The selected oil must indeed ensure the electrical insulation function as well as a good impregnation of the different solid insulating components of the equipment (paper, polymer, pressboard, wood, etc.) The impregnation eliminates air or other gases, thus avoiding the initiation of partial discharges. One compares the dielectric strength of insulating liquids by their breakdown voltage (BDV) in a given electrode geometry defined by standard specification. BDV measures the efficiency of the insulating liquid as an insulator. BDV is very sensitive to the quality of the insulating liquid, which in turn can be influenced by the presence of different contaminants (conducting and nonconducting particles, moisture and water droplets or other emulsions or gases) [125]. With increasing ambient humidity, the insulation liquids accumulate moisture and the BDV decreases. BDV is a good characteristic for comparing different insulating liquids when the impurity content is well controlled.

Laboratory break down tests, achieved at the ambient temperature according to IEC 60156 [113] on different insulating liquids (Table 4.21; data taken from [61]) and lightening impulse tests of natural ester showed that breakdown voltage for natural ester is similar to mineral oil. Test results from Shah and Tahir show that natural ester oils, especially corn oil, have a somewhat better average BDV, which is 95 kV, than mineral oil [120]. The higher polarity has no negative impact on the electrical stress [126].

4.4.2.1 Effect of Moisture on Breakdown Strength

The electric strength as one of the most important requirements of an insulating liquid is strongly influenced by the amount of contained water. It may be available in two forms: as free moisture or disintegrated water. Polar liquids tend to develop hydrogen bonds with moisture molecules so that water may simply dissolve. Consequently, polar liquids possess exceedingly high humidity endurance. The question is how high the water content of the insulating liquid can rise to achieve a correspondingly

Table 4.21 Breakdown voltage test

Insulating liquid	Breakdown voltage (kV)
Nynas Nytro 4000X	62
FR3™ Fluid	> 90
Biotemp®	90
Midel® eN	81
Midel® 7131	83
BecFluid® 9002	82
Siliconöl Powersil® Fluid TR 50	66

high dielectric strength. Various publications show that the breakdown voltage of mineral oil drops sharply even with a low water content (30–50 ppm) [127, 128]. Pagger showed in [61] that as long as the water content of the ester liquid is less than 200 ppm, there is no direct influence on the breakdown voltage. The same can be said for silicone oil for a threshold of approximately 100 ppm instead of 200 ppm.

If the water content of the insulating liquid exceeds the water saturation and free water is formed, there is a sharp drop in the dielectric strength [127]. While mineral oil liquids compared to ester liquids maintain lower viscosity in cold temperatures, its ability to hold water is very low so the dielectric strength is weakened significantly [129]. In an aging experiment done by Gasser et al. [130] high density pressboard was aged in nine different insulating liquids (vegetable oils, synthetic ester, isoparaffinic and naphthenic mineral oils) at elevated temperatures for twelve months. The water content of the ester liquids decreased after some months, whereas the water content in mineral oil increased in this closed expansion system. That is the reason why the breakdown voltage of the mineral oils aged for two and four months at 150 °C was only about 25% of the initial value. Non-polar insulating liquids are particularly sensitive to the absolute moisture content.

The absolute amount of water that an insulating liquid contains can have a dramatic effect on these electrical properties. Biological insulting liquids (e.g. Midel eN 1204) absorb large amounts of moisture (up to 300 ppm) with no reduction in breakdown voltage [16]. Synthetic ester can even absorb more water (e.g. Midel 7131 up to 650 ppm), before breakdown voltage drops significantly [131].

Alternative insulating liquids show an even more abrupt drop in dielectric strength compared to mineral oil when the critical water content is reached. While the reduction in the strength of mineral oil starts much earlier, the drop is not as rapid. This must be considered when monitoring devices filled with alternative insulating liquids.

Table 4.22 Dielectric breakdown test with different international standards

Standard	Mineral oil (kV)	Synthetic ester (kV)	Natural ester (kV)	Silicone oil (kV)	Low viscosity silicone oil (kV)
IEC 60156 2.5 mm [113]	> 70	> 75	> 75	50	70
ASTM D1816 1 mm [133]	–	–	37	–	–
ASTM D1816 1 mm [133]	60	–	76	–	–
ASTM D877 [134]	55	43	46	43	–

4.4.2.2 Dielectric Breakdown Test with Different International Standards

As standard procedures to perform a breakdown test, different national and international standard are used. Comparison test results are in different new oil products are shown in Table 4.22 [132].

4.4.2.3 Breakdown Voltage—Test Methods

Besides dissolved gas analyses (DGA), the dielectric breakdown test on insulating liquids is the one most used in practice. It offers a quick and easy result on the overall quality of the insulating performance, not only on new but also on used insulating liquids.

A high voltage transformer in the range from 40 kV up to 100 kV is needed to supply the voltage to a test vessel, usually with a content of 50 up to 700 ml. The test cell is equipped with alternatively sphere 12.5 mm diameter or mushroom 36 mm diameter electrode design (Fig. 4.15). The electrode distance varies between 0.5 mm up to 2.5 mm. A constant rate of voltage rises with 0.5 kV/s (ASTM D1816 [133]) or 2 kV/s (IEC 60156 [113]) increases until a sustained breakdown occurs. A series of up to 6 breakdowns is needed to get a statistically valid result. To reduce the scatter of test results, a stirring device shall be used, preferably with a flow in the test cell moving from the base to the top between the electrodes, to minimize the air bubbles remaining between the gap of electrical field.

Care should be taken when filling the test cell with biological insulating liquids to guard against trapping air bubbles that can lead to misleading low breakdown voltages. Due to their higher viscosity, a longer sample rest time (equal to or greater than 15 min at room temperature) is recommended for biological insulating liquids than for mineral oil liquids to allow air bubbles to escape.

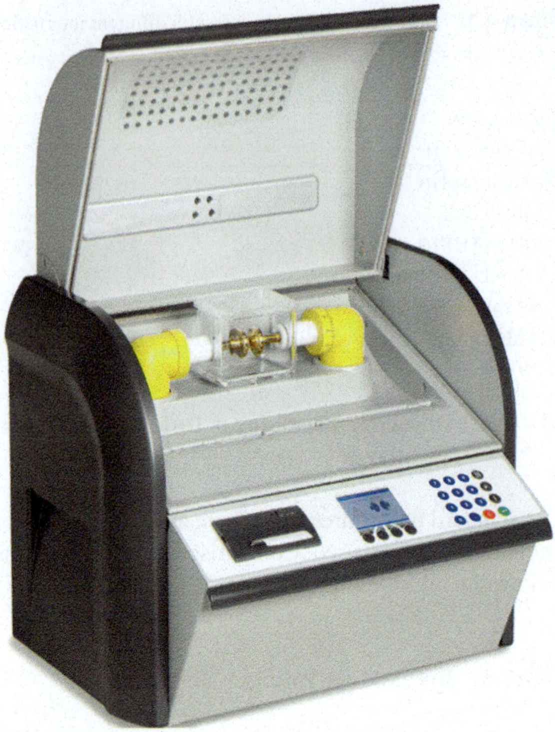

Fig. 4.15 Dielectric Breakdown Tester 100 kV (BAUR ©)

4.4.3 Dielectric Breakdown Voltage Under Impulse Condition

Insulating liquids used in transformers are subjected to transient voltage stresses while being subjected to steady-state voltage stresses associated with continuous operation of the apparatus at commercial power frequencies. The ability of the insulating liquid to withstand transient voltage stresses has become increasingly important to the designer of transformers.

Li et al. present in [135] comprehensive experimental research on both positive and negative lightning impulse breakdown properties in point-plane geometrics with gaps varying from 1 to 50 mm. The breakdown voltages and streamer velocities of different insulating liquids have been measured.

Streamer Velocity The development of streamers in liquid can be divided into two stages: initiation (also called ignition and inception) and propagation. According to the different streamer velocity and shape, streamers in liquids can be classified in first, second, third, and fourth modes depending on their velocity under positive voltage, and in primary and secondary modes under negative voltage as shown in Table 4.23.

The average streamer propagation velocity (v_a) is calculated by the ratio of gap distance (d) to breakdown time (T_b) (Eq. 4.33).

Table 4.23 Streamer development modes of insulating liquids

Polarity	Modes	Streamer velocity
Positive	1st	About 100 m/s
	2nd	1–5 km/s
	3rd	10–30 km/s
	4th	> 100 km/s
Negative	Primary	About 1 km/s
	Secondary	> 5 km/s

Table 4.24 Positive average streamer velocities (km/s) of insulating liquids

Electrode gaps (mm)	Mineral oil	Synthetic ester	Natural ester
1	0.56	0.44	0.33
5	1.63	1.07	1.12
15	1.98	1.52	1.46
25	2.06	1.78	1.60
50	2.29	2.56	1.75

$$v_a = \frac{d}{T_b} \tag{4.33}$$

Table 4.24 presents the results of the average streamer propagation velocity under positive polarity. It can be seen that, at 1–25 mm gaps, the streamer velocity of mineral oil is faster than others of the ester. At 50 mm gap, streamer velocities of mineral oil and synthetic ester are similar and faster than natural ester. This means the streamer velocities are in the 2nd mode under this condition.

Table 4.25 presents the results of the average streamer propagation velocities of insulating liquids under negative polarity. Compared to Table 4.24, the streamer velocity at negative polarity is less than that of positive polarity. The streamer velocities of synthetic ester are mostly lower than those of the other liquids. At 1–5 mm gaps, the streamer velocities of mineral oil are less than those of ester liquids. At 15–25 mm gaps, the streamer velocities of mineral oil are like natural ester. At a 50 mm gap, natural ester has two different streamer velocities, with the slower one is similar to mineral oil and the faster one is faster than any other insulating liquids. This means that there are two modes (primary and secondary) of the streamer under this condition.

Breakdown voltage Under uniform or quasi-uniform electric fields, the breakdown voltages of natural and synthetic esters are close to that of mineral oil liquids. However, it is quite different under heterogeneous fields, such as when point-plane and point-sphere electrodes are applicated. Under the nonuniform field, with positive lightening voltage, at 5–25 mm electrode gaps, the breakdown voltages of natural and synthetic esters are close to those of mineral oils; at 50–150 mm gaps, the breakdown voltages of natural and synthetic esters are lower than that of mineral oil

Table 4.25 Negative average streamer velocities (km/s) of insulating liquids

Electrode gaps (mm)	Mineral oil	Synthetic ester	Natural ester
1	0.14	0.35	0.36
5	0.44	0.66	0.84
15	0.94	0.69	1.04
25	0.98	0.71	1.10
50	1.17	0.87	1.12/3.85

liquids. With negative lightning voltage, at 5–150 mm gaps, the breakdown voltages of natural and synthetic esters are close to that of mineral oils. In addition, the long pulse or switching wave breakdown voltage of natural and synthetic ester is also inferior to that of mineral oil in large gap heterogeneous field, but at 2–50 mm gaps, the initiation voltages of natural and synthetic esters are close to that of mineral oils.

Table 4.26 shows the 50% positive breakdown voltages. The positive lightning breakdown voltage of synthetic ester is lower than others. For the positive breakdown voltage, the dispersion of positive breakdown voltage in 1–15 mm gaps is large, and there are no obvious rules between ester liquids and mineral oils. This implies that the space charge may be involved in the breakdown process. Positive ions may accumulate near the positive needle tip and thus space charges are formed. As the gaps and the voltages in the gaps increase, the space charge tends to be saturated, and its influence on electrical field distribution is reduced.

Table 4.27 shows the negative 50% breakdown voltages. The negative breakdown voltages of mineral oil are always higher than those of ester liquids, and the difference increase with the extension of gaps distance. In addition, the negative breakdown voltages of esters are quite close. Under the same experimental conditions, the differences in breakdown voltages between ester liquids and mineral oil undoubtedly resulted from the difference of their molecular structures. The largest difference between ester liquids and mineral oil are the ester bonds (Fig. 4.1). That means ester bonds play a decisive role in the lower negative breakdown voltage of ester liquids. The synthetic ester has the most ester bonds in the same volume, so its breakdown voltages are the lowest. The ester liquids are polar medium. Thus, a small part of them is separated into ions. Mineral oil is a non-polar medium, making it difficult to separate into ions. Therefore, ester liquids contain more ion pairs than mineral

Table 4.26 The 50% positive breakdown voltages (kV) of insulating liquids

Electrode gaps (mm)	Mineral oil	Synthetic ester	Natural ester
1	28.3	25.6	29.3
5	48.5	31.2	31.8
15	61.7	44.2	42.1
25	67.6	54.4	64.2
50	110.7	83.3	109.3

Table 4.27 The 50% negative breakdown voltages (kV) of insulating liquids

Electrode gaps (mm)	Mineral oil	Synthetic ester	Natural ester
1	29.8	25.9	25.8
5	61.7	37.0	41.1
15	113.2	63.2	69.6
25	149.6	88.9	96.7
50	257.2	149.6	166.0

oil. During the breakdown voltage test, these ions are less stable than molecules, and easier to release electrons, consequently strengthening the process of collision ionization, and reducing breakdown voltage. Therefore, the negative impulse breakdown voltages of ester liquids are lower than those of mineral oil. In the experiment [135], Li showed that double bounds have little effect on the negative breakdown voltage of ester insulating liquid.

The test method (ASTM D3300 [136]) covers the determination of the dielectric breakdown voltage of insulating liquids in a highly divergent field under impulse conditions.

4.4.4 Dielectric Dissipation and Power Factor

Dissipation factor (DF) versus power factor (PF)—what are the differences, and do they matter? In reality, DF and PF are calculated differently (Fig. 4.16, Eqs. 4.34 and 4.35); however, both are calculated from the measurements made in the same test and both describe the inefficiency—or efficiency—of an insulation system. And, for most asset insulation systems that engineers will need to test, the numerical values of DF and PF will be almost the same [137].

$$DF = \tan\delta = \frac{I_{Resisitive}}{I_{Capacitive}} \quad (4.34)$$

$$PF = \cos\varphi = \frac{I_{Resisitive}}{I_{Total}} \quad (4.35)$$

The dielectric dissipation factor (tan δ) is the amount of dielectric loss happening in dielectric fluid when it is exposed to an AC field. The tan δ usually surges with an increasing existence of impurities or aging derivatives, for example, humidity, carbon or additional conducting substances and oxidation derivatives. Power factor (cos φ) designates the dielectric loss in the insulating liquid and therefore the extent of energy dissipated as heat. This power factor test is largely applied as a preemptive preservation experiment for the approval of dielectric liquid. In the end, it is the

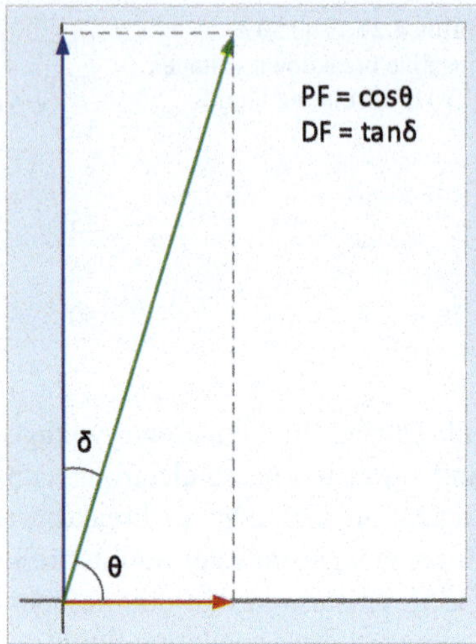

Fig. 4.16 Power and dissipation factor

Table 4.28 Dielectric dissipation factor

Insulating liquid	Dissipation factor
Nynas Nytro 4000X	0.0011–0.0061
Envirotemp® FR3™ Fluid	0.0352–0.0358
Biotemp®	0.0551–0.1001
Midel® eN	0.0171
Midel® 7131	0.0200–0.0468
BecFluid® 9002	0.0047–0.0054
Silicon oil Powersil® Fluid TR 50	0.0070

incompetence of dielectric liquid molecules to reorient when they are exposed to an alternating electric field, turning into the form of heat losses. PF is influenced by contaminating agents, for instance, moisture solvents and conductive particulates. For new dielectric insulating liquids, the dissipation factor follows more or less the polarity of the liquid (Tables 4.19 and 4.28). Ecologically friendly insulating liquids such as natural and synthetic esters indicate higher dielectric dissipation factor than nonpolar mineral oil liquids, particularly at higher temperature. For Table 4.28 the data are taken from [61].

4.4.4.1 Dielectric Dissipation Factor—Test Method

Dielectric properties are highly dependent on long-term degradation of the dielectric insulating liquids. The liquid absorbs chemical content of the electrical equipment and, due to operating temperature and vibration, the liquid ages over the time. One of the useful parameters to detect the aging degradation is the dissipation factor (tan δ) measurement. Over the operating time, moisture is distributed in the insulating material into the liquid, highly dependent on the operating temperature. The dissipation factor of the liquid can be analyzed by taking samples from the test object. Unused biological insulating liquids have inherently higher dissipation factors than mineral oil liquids.

Besides the tan δ, the additional parameter of specific resistance (resistivity) and relative permittivity of the liquid can be analyzed. It supports to distinguish between the two components moisture and ageing parameters.

Relative complex permittivity is the permittivity in a complex number representation under steady sinusoidal field conditions expressed as Eq. 4.36, where $\varepsilon_r^{'}$ and $\varepsilon_r^{''}$ have positive values.

$$\underline{\varepsilon_r} = \varepsilon_r^{'} - j\varepsilon_r^{''} = |\varepsilon_r|e^{-j\delta} \tag{4.36}$$

The complex permittivity $\underline{\varepsilon_r}$ is customarily quoted either in terms of $\varepsilon_r^{'}$ and $\varepsilon_r^{''}$, or in terms of ε_r and tan δ, $\varepsilon_r^{''}$ is termed loss index. The dielectric dissipation factor tan δ (loss tangent) is the numerical value of the ratio of the imaginary to the real part of the complex permittivity (Eq. 4.37).

$$\tan\delta = \frac{\varepsilon_r^{''}}{\varepsilon_r^{'}} \tag{4.37}$$

Thus, the dielectric dissipation factor tan δ of an insulating material is the tangent of the angle δ by which the phase difference φ between the applied voltage and the resulting current deviates from π/2 rad, when the solid insulating material is exclusively used as dielectric in a capacitive test specimen (capacitor). The dielectric dissipation factor can also be expressed by an equivalent circuit diagram using an ideal capacitor with a resistor in series or parallel connection (Eq. 4.38).

$$\tan\delta = \omega C_S \cdot R_S = \frac{1}{\omega C_P \cdot R_P} \tag{4.38}$$

Test procedure of dissipation factor: A sample of 50 ml is enough to perform the two tests of dissipation factor and resistivity. The liquid sample is poured into a standard test cell (Fig. 4.17) like IEC 60247 [114] under ambient laboratory condition. A power supply AC source of 2000 V in the frequency range between 48 and 62 Hz. The real and imaginary current is measured at the constant voltage and temperature 25 and 90 °C. The tan δ is calculated at the rated frequency and temperature.

The values can be converted for frequencies in the range from 48 to 62 Hz (Eq. 4.39) described in [138].

$$\tan\delta[f(50\,\text{Hz})] = \frac{f(\text{Hz})}{50} * \tan\delta[f(\text{Hz})] \tag{4.39}$$

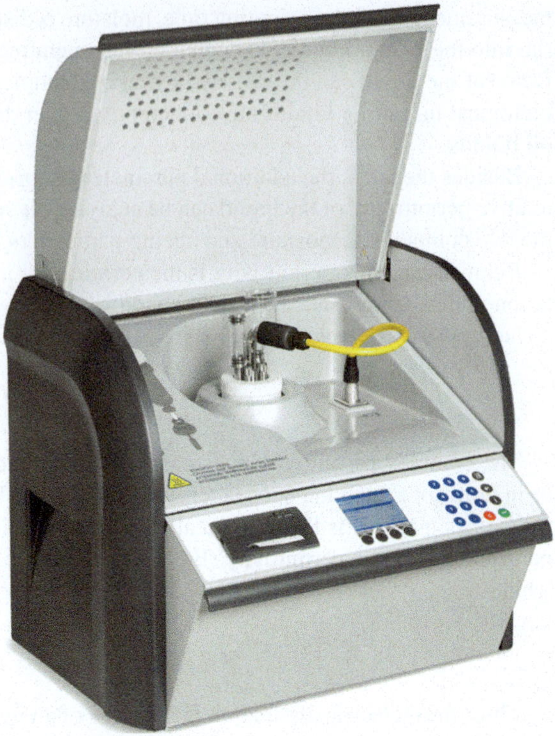

Fig. 4.17 Dissipation factor and resistivity measuring instrument (BAUR ©)

4.4.5 Volume Resistivity

Volume resistivity relates to the current flowing inside insulation material under the action of an electrical field. The resistivity of a liquid is a measure of its electrical insulating properties under conditions comparable to those of the test. High resistivity reflects low content of free ions and ion-forming particles, and normally indicates a low concentration of conductive contaminants [116]. Unused biological insulating liquids have inherently lower (one to two magnitude) volume resistivity than mineral oil liquids (Table 4.19).

4.4.5.1 Volume Resistivity—Test Method

Test procedure: The sample is charged with a constant DC positive and negative electrical field of 500 V. The negative and positive current is measured after 1 min time of field stabilization. The temperature shall be measured within an accuracy of ± 1 °C. The volume resistivity factor is calculated in the term of Ωm (Eq. 4.40) [114].

$$P = K * \frac{U}{I} \tag{4.40}$$

P Resistivity (Ωm)
U Reading of the test voltage (V)
I Reading of the current (A)
K Test cell constant (m). The cell constant K is calculated from the capacitance according to Eq. 4.41.

$$K = 0.036 * \pi * C \tag{4.41}$$

C Capacitance (pF) of the empty cell.

4.4.6 Partial Discharge Behavior

Partial discharges are defined as: Localized electrical discharges that only partially bridges the insulation between conductors and which can or cannot occur adjacent to a conductor. Partial discharges are in general a consequence of local electrical stress concentrations in the insulation or on the surface of the insulation. Generally, such discharges appear as pulses having a duration of much less than 1 μs [139].

From a physical point of view, self-sustaining electron avalanches may happen only in gaseous dielectrics. Consequently, typical discharge types occurring in ambient air, such as glow, streamer, and leader discharges, may also happen in gaseous inclusions due to imperfections in solid and liquid dielectrics. The pulse charge of glow discharges is in the order of a few pC. Streamer discharges may create pulse charges ranging from about 10 pC up to some 100 pC. A transition from streamer to leader discharges may occur if the pulse charge exceeds a few 1000 pC [140].

In most cases, insulation faults are associated with partial discharges. Partial discharge (PD) plays a significant role in causing the aging and deterioration of liquid/pressboard insulation systems and may ultimately lead to power transformer failures. Airgaps in solid dielectrics can result from many causes. In the case of transformer liquid/pressboard insulation systems, for example, it can be caused by incompletely immersing liquid into pressboard. The higher viscosity of the biological insulating liquids must be considered here.

Biological insulating liquids manifest larger gas development on account of PD (hydrogen with hints of acetylene) as compared with mineral oil liquids at similar voltage value [38]. Because of the degradation of insulating liquid and pressboard, electronegative gases are released into the airgap gradually, such as CO, CO_2 and O_2. As biological insulating liquids contain oxygen in its chemical structure while mineral oil does not, they are more susceptible to produce such electronegative gases under high electrical stress. It was found that vegetable oil has a slightly lower PDIV

(Partial Discharge Inception Voltage) than mineral oil. Although the magnitude of airgap PD is relatively low, it can do serious harm to the solid insulation that cannot be self-recover.

Eberhardt et al. tested and compared in [141] the PD behavior of three kinds of insulating liquids (mineral oil, synthetic ester, natural ester) in combination with transformer board. They arranged the electrodes as needle to plate, with a distance of 10 and 40 mm. They used tips with a radius of 40 μm. Between the electrodes, they placed a specimen of pressboard (3 mm thick and 190 mm in diameter). They found a significant low average PD level at both distances for mineral oil. The average PD in all ester liquids is higher than that for mineral oil. Especially the tests with board impregnated in synthetic ester and the vessel filled with natural ester produced comparatively the highest PD values. The PD values of the test series with the synthetic ester are higher than the values for mineral oil but at a lower level than the ones of the natural ester.

4.4.6.1 Partial Discharge Inception Voltages (PDIV)—Test Method

The partial discharge insulation voltage (PDIV) test is of importance as a measure of the liquid's ability to withstand electric stress without failure. It serves to indicate the presence of contaminating agents such agents as water, dirt, moisture cellulosic fibers, or conducting particles in the liquid, one or more of which may be present in significant concentrations when low PDIV values are obtained. However, a high PDIV does not necessarily indicate the absence of all contaminants; it may indicate more sensitively than the dielectric breakdown the concentrations of contaminants that are present in the liquid between the electrodes.

Principle and Measurement Procedures: An AC power high voltage source in the range of 50–100 kV is applied to a sphere—needle arrangement mounted into a test cell with a volume of 0.4–0.7 l filled with the sample liquid (Fig. 4.18). Using a constant voltage rate of rise, the partial discharge activity is monitored according to IEC 60270 standard [139]. After reaching a sustained 100 pC partial discharge activity, the voltage supply is stopped and the result is memorized. PD sustainability means to monitor the PD activity over a minimum period of one minute with several hundred of PD counts. Due to statistical scatter, a minimum of 6 tests are necessary and the average PDIV is calculated to get a valid result [142].

4.4.7 Electrostatic Charging Tendency (ECT)

Research conducted on damaged large power transformers has concluded that the insulating liquid's electrical charging tendency (ECT) is another factor that can contribute to the degradation of the insulation. The charging tendencies in the transformer are attributed to the physicochemical processes that are taking place at the separation surface between the insulating liquid and the solid components of the

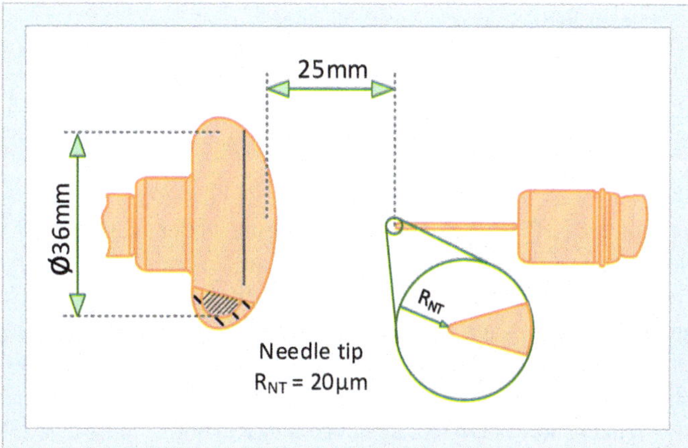

Fig. 4.18 Electrode arrangement for partial discharge inception voltage measurement (PDIV)

equipment (both the metallic components and the cellulose-based materials). The electrostatic charging tendency is not a material-specific value but represents the property of an insulating material pair.

When a solid is in contact with a liquid, the initially neutral liquid–solid system is polarized because of the physicochemical process that are taking place at the solid/liquid interface. This electrical double layer is formed progressively in time, even from the moment when the two materials get in contact. Because of the enormous heat which can be generated by the windings and core, large power transformers are cooled by the forced circulation of the dielectric insulating liquid. However, the liquid circulation can result in a separation of charge at the paper/liquid interface and in the appearance of a double layer. Then an electro kinetic phenomenon, known as flow electrification phenomenon, is generated. Depending on the liquid, this charge generation can be more or less important. The accumulation of these charges can lead to the initiation of partial discharges and even to the breakdown of the transformer. The transformer insulating liquid flow electrification is characterized by the value of the electrification currents; the higher is the electrification current, the higher its flow electrification tendency gets. Generally, the dielectric insulating liquid is charged positively and the solid insulating material negatively [125]. Only some atypical insulating liquids (as silicone oils) have induced a reserve behavior [143].

The formation and time evolution of an electrical double layer depends on the physicochemical properties of the insulating liquid (aging, impurities content) and pressboard (roughness, porosity), as well as on the liquid temperature.

Vihacencu et al. compared in an experimental "spinning disk system", where the flow of the insulating liquid is simulated by the rotation speed of the disk, two mineral oils with a vegetable oil [144]. With ρ_{SS}, the surface charge density in steady state is expressed in Table 4.29.

It has been determined that the vegetable oil has a up to 200 times higher surface charge density located at the liquid/paper interface compared to mineral oil.

Table 4.29 Surface charge density of different oil types

Oil type	Mineral oil 1			Mineral oil 2			Vegetable oil		
Rotation speed (rot/min)	200	400	600	200	400	600	200	400	600
ρ_{SS}(nC/m^2 s)	0.254	0.573	0.950	2.5	6.3	14.3	127	159	222

Podesser found in [145] that electrostatic charging tendency increase from mineral oil to synthetic ester and finally is at highest in natural ester (Fig. 4.19). Which of the two ester liquids has the greater tendency to charge seems to be related to the moisture level? In the dry state, natural ester showed a higher charging tendency. In the wet and very wet states, synthetic ester had a higher charging tendency (Fig. 4.20). Podesser shows in [145] that with laminar flow, all three insulating liquids have a significantly weaker increase in leakage currents (Reynolds number < 2300), which is unfavorable for heat transfer.

The results of Perrier and Beroual [125] are similar and show that ester liquids (natural and synthetic) create a larger number of charges than mineral and silicone oils. The assumption that this phenomenon could be linked to the liquid viscosity is not valid, because silicone oil is also very viscous, but it generates very few charges. It seems to be connected to the molecular structure (polarity) of the ester liquids. In [82], Perrier et al. demonstrate that electrostatic charging tendency of insulating liquids is connected to its resistivity. The higher the resistivity, the lower the electrical charging tendency of the insulating liquid. Vihacencu et al. write in [144] that the polar molecules in insulating liquids (vegetable oils) give them lower volume

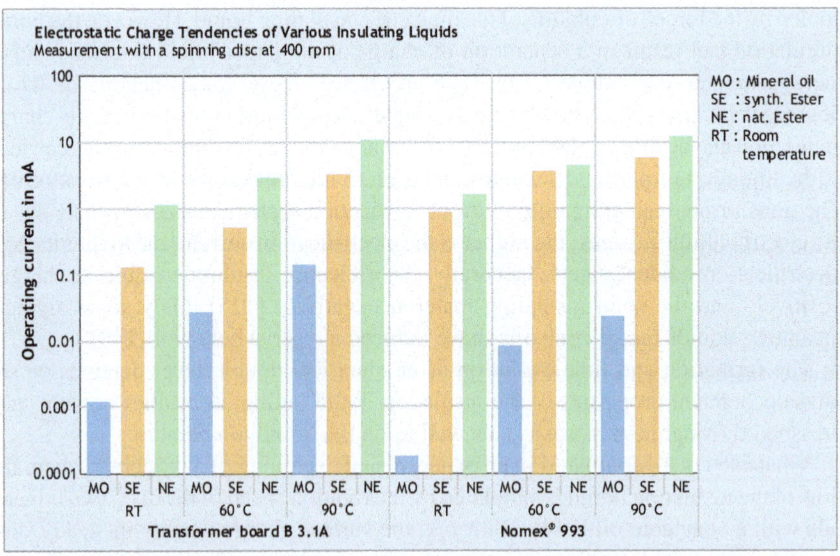

Fig. 4.19 Liquid influence on the electrostatic charge at different temperatures and solids

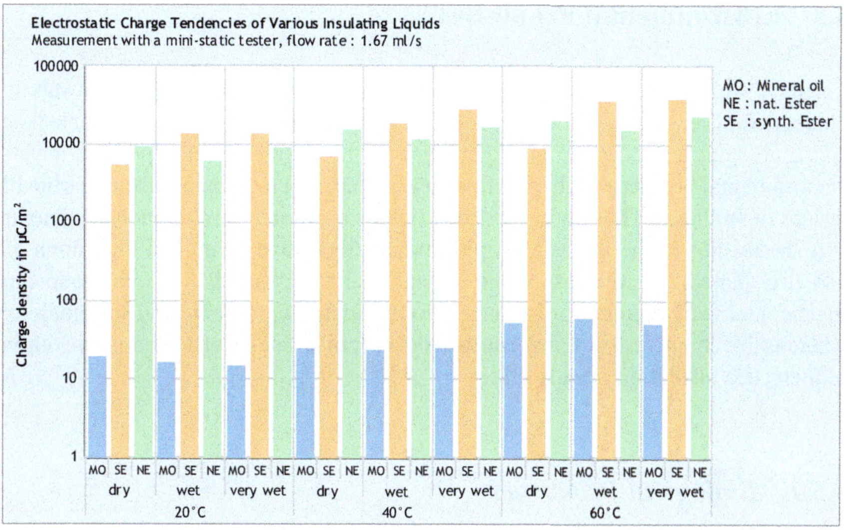

Fig. 4.20 Influence of liquids on the electrostatic charge at different temperatures and moistures

resistivity and higher values of relative permittivity and electrification tendency. They found an empirical relation between electrification current in steady state (ISS) and the real part of the relative permittivity (ε_r) measured at f = 1 MHz (Eq. 4.42).

$$I_{SS} = 7.73684 * 10^{-9} - 6.16032 * 10^{-9} * \varepsilon_r + 1.22168 * 10^{-9} * \varepsilon_r^2 \quad (4.42)$$

In [146], Zdanowski shows that electrical charging tendency is strongly dependent on the type of insulating liquid and pipe material. A conclusion of this study is the observation that a small amount of fresh or aged mineral oil (up to 10%) significantly reduces the electrical charging tendency.

The higher conductivity of esters (natural and synthetic esters) enables better evacuation of generated charges on the solid surface and thus limits the potential for the solid/liquid surface. Zelu et al. [143] concluded their study that even if the charge and current generation are largely higher with ester liquids than with mineral oils, one can note that the accumulated charge, being the most important indicator of electrification risks, is not so critical in ester liquids. Thus, in terms of hazard and according to the current knowledge, ester liquids do not present significantly concern.

4.5 Environmental Properties

The most significant concerns with insulating liquids may include their renewability, biodegradability, and their respect to environmental guidelines in the event of leakage and accidental fire. Mineral oil liquids are not biodegradable and possess adverse environmental profile, which specifies they are harmful to humanoid and marine life in case of spillages. Therefore, substitute liquids must obey environmental protocols and should not be perilous. Multiple investigators have examined environmental and fire-related effectiveness of biological insulating liquids. The environmental traits contain key factors, for instance, biodegradability, toxicity, and sustainability, whereas fire characteristics may constitute flash point, fire point and emission outline of dielectric insulating liquids [38].

4.5.1 Biological Properties

Biological insulating liquids must be derived from 100% renewable feedstock and are mostly edible oil based. Edible vegetable base oils have a relatively fast biodegradation rate and are naturally non-toxic. Biodegradation is a process in which natural materials deteriorate to smaller molecular weight matters by enzymes generated by the course of microorganisms. Organic substances experience deterioration in the existence of oxygen (aerobic) and in deficiency of oxygen (anaerobic). Rancidification is the first step in the biological degradation of biological insulating liquids. When oils are exposed to sunlight, photosynthesis tends to occur, which is disrupted, leading to release of the hydrogen ions that react with molecular oxygen to form loosely combined hydrogen peroxide. The unstable peroxide units with unsaturated bond of triglyceride form a glyceride peroxide which in turn splits into an aldehyde and forms the rancid compound that can be detected through sensory evidence.

Three pathways for rancidification are recognized: (1) Hydrolytic degradation refers to when triglycerides are hydrolyzed and free fatty acids are released. This reaction of lipid with water sometimes requires a catalyst, but results in the formation of free fatty acids. At higher temperature ≥ 100 °C, hydrolysis seems to be the dominant degradation process of vegetable oils. (2) Oxidative degradation by oxygen in the air. Via a free radical process, the double bonds of unsaturated biological insulating liquid can undergo cleavage, releasing volatile aldehydes and ketones. For low temperatures < 100 °C, oxidation is considered as the main path for vegetable oils degradation. Oxidation primarily occurs with unsaturated biological liquids. The process can be suppressed by the exclusion of oxygen or by the addition of antioxidants. (3) Microbial degradation refers to a process in which microorganisms, such as bacteria or molds, use their enzymes such as lipases to break down triglycerides.

In the OECD guideline for testing of chemicals, methods are described that permit the screening of chemicals for readily biodegradability in an aqueous medium [147]. The pass levels for readily biodegradability are 70% removal of DOC (dissolved

Table 4.30 Applicability of test methods

Test	Analytical method
CO_2 evolution	Respirometry: CO_2 evolution
MITI (Ministry of International Trade and Industry, Japan)	Respirometry: oxygen consumption
Closed Bottle	Respirometry: dissolved oxygen
Manometric respirometry	Oxygen consumption

organic carbon) and 60% of ThOD (theoretical oxygen demand) or $ThCO_2$ (theoretical carbon dioxide) production for respirometric methods. They are lower in the respirometric methods since, as some of the carbon from the test chemical is incorporated into new cells, the percentage of CO_2 produced is lower than the percentage of carbon being used. These pass values must be reached in a 10-day window within the 28-day period of the test. The 10-day window begins when the degree of biodegradation has reached 10% DOC, ThOD or $ThCO_2$ and must end before day 28 of the test. Chemicals which reach the pass levels after 28-day period are not deemed to be readily biodegradable. The EPA (United Environmental Protection Agency) guidelines OPPTS 835.3110 [148] are very similar to the OECD guideline [147]. Table 4.30 shows the test methods which can be used for poorly soluble liquids like biological insulating liquids.

Definition of **readily biodegradable**: An arbitrary classification of chemicals which have passed certain specified screening tests for ultimate biodegradability; these tests are so stringent that it is assumed that such compounds will rapidly and completely biodegrade in aquatic environments under aerobic conditions. While the base vegetable oil typically constitutes a minimum of 95% of the content of the insulating liquid, there are additives applied, generally for improved pour point and oxidation stability. Because the health and environmental properties of additives can impact the overall safety of the insulating liquid, it may be important to users to be able to determine the overall environmental and health properties. Figure 4.21 shows the biodegradability of the different liquids presented in [149, 149].

The curves show that the natural ester and synthetic ester are nearly completely degraded in less than 30 days, whereas the mineral oil is only degraded by about one third at this time. Silicone oil is practically non-biodegradable. The term "biodegradable" is not clearly defined; biodegradation takes place gradually in nature. In a first step, the "primary degradation", fragments are created that can still be harmful to the environment and, in terms of toxicity, can even be more harmful than the starting material. With complete biodegradation, the end products are harmless to the environment, namely carbon dioxide, water, and biomass. The CEC tests only make a statement about the primary biodegradation. The fact that the decomposition products that develop can be more damaging to the environment than the undegraded substances is not considered in this context. This must be taken into account when comparing the data listed in Table 4.31.

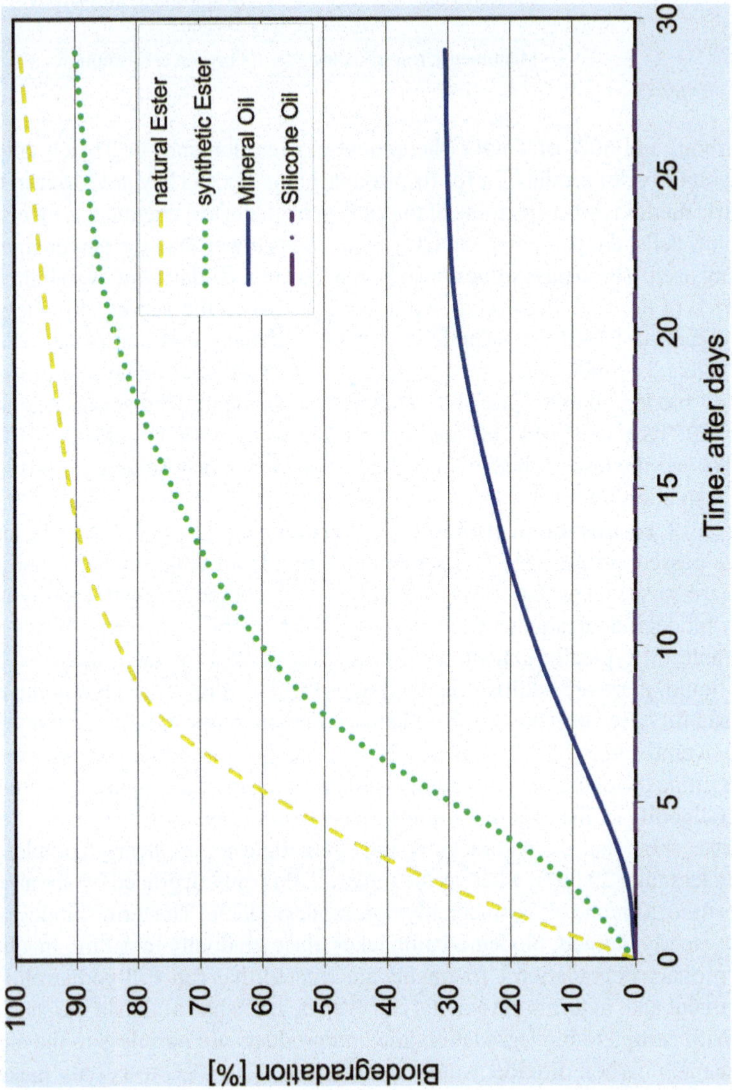

Fig. 4.21 Biodegradability of different liquids versus time duration

Table 4.31 Biodegradation

Insulating liquid	Test method	Biodegradability
Nynas Nytro® 4000X [5]	OECD 301[147]	Non-biodegradable
Nynas Nytro® BIO 300X [72]	OECD 301 F [147]	Readily biodegradable > 70%
Envirotemp™ 360 Synthetic ester [6]	OECD 301[147]	Readily biodegradable 89%
Midel 7131 [98]	OECD 301[147]	Fully/readily biodegradable
BecFluid® 9902 [99]	OECD 301[147]	Readily biodegradable 99%
Nycodiel® 1233 [9]	OECD 301 B [147]	90%
Nycodiel® 1244 [9]	OECD 301 B [147]	83%
Nycodiel® 1258 [9]	OECD 301 B [147]	84%
MATROL-BI® FDE01A [10]	OECD 301 B [147]	Readily biodegradable
Envirotemp® FR3™ Fluid [40]	EPA OPPTS 835.3110 [148]	Readily biodegradable
Midel eN 1204 [98]	OECD 301[147]	Fully/readily biodegradable
Midel eN 1215 [98]	OECD 301[147]	Fully/readily biodegradable
BIOTEMP® [74]	CEC L-33-A 21 day	100%
NeuGen Plus [21]	CEC-L-33-A-97	> 95%
Paryol Electra 7426 [22]	OECD 301 B, C or F	Readily biodegradable
MATROL-BI® FDV01A [23]	OECD 301 B [147]	Readily biodegradable
Powersil® Fluid TR 50 [25]	OECD 301[147]	Non-biodegradable

In [149] Stenborg describes that natural ester from sources such as soybeans, rapeseed or sunflowers have no harmful effects on fish toxicity [151] and that oral ingestion has no other effects other than diarrhea [152]. The biodegradability is referred to as easily biodegradable according to [153].

4.5.2 Ecological Properties

Typically, biological insulating liquids have been formulated to minimize health and environmental hazards. Although no known hazard is involved in the normal handling and use of biological insulating liquids, additives to the base vegetable oil may differ.

The biological insulating liquids provide ecological and environmental benefits at all stages of its life cycle when compared to traditional mineral oil-based insulating liquids. The raw materials used in traditional petroleum-based transformer oils come from a long chain of what can often be environmentally costly and damaging steps. From exploring and drilling to production, from refining to transportation and disposal, there is an environmental cost to the production of traditional transformer insulating oils. The spillage of mineral oil liquids can have an enormous environmental impact. Soil acts as an absorbent for biological insulating liquid and offers excellent condition for natural biodegradation. Accidents in the production and transport of petroleum and petroleum products, as very often evidenced by the media, have a devastating effect on fauna and flora. The cost of any spill goes beyond penalties and clean-up costs. It is the contamination of soil and waterways that poses the serious concern.

Biological insulating liquids are environmentally friendly as they are renewable, non-toxic, biodegradable and reduce the production of greenhouse gases. Fratelli Parodi reports in [154] that by using the natural ester PARYOL ELECTRA 7426 approximately 80% of greenhouse gas can be saved during the lifecycle of a transformer.

4.6 Interaction with Transformer Materials

Each transformer design and insulating liquid should be shown to be compatible under standard service conditions. Verification of compatibility shall occur to help assure that no excessive interaction or reaction occurs between materials in contact with the dielectric coolant. Standard test method ASTM D3455 [155] apart from Point 8.2.5, "Aged properties for the reference oil specimen" can be used for biological insulating liquids. The reference should be modified for biological insulating liquids.

4.6.1 Corrosive Sulfur Contamination

In the past, many of the transformer failures of transformers filled with mineral oil were attributable to corrosive sulfur contamination in the electrical insulating oils. These failures, reported in several countries around the world, have puzzled owners and manufactures alike, particularly since many of the problems have occurred in relatively new equipment. The failures were not confined to any certain transformer manufacturers or oil supplier. Even though the transformer liquids have passed industry specification and standard tests, the insulating liquids can contain thermally unstable sulfur-bearing compounds. Under transformer condition, these compounds can be converted to corrosive sulfur above all as the equipment operates under heavy load.

The existence of corrosive sulfur in transformer liquids is not new and was first reported many decades ago, but was thought to be eliminated through improved industry standards and detection methods. In the meantime, experts have developed a number of tests to identify and deal with transformers that have a high likelihood of failure due to corrosive sulfur [14, 156–158].

Due to the production process, as well as the sources of esters used, no corrosive sulfur problems in ester filled transformers are not known until now.

References

1. Nytro® 4000X. 2021. *Product Data Sheet*. Nynas.
2. IEC 62021-1. 2004. Insulating Liquids—Determination of acidity—Part 1: Automatic Potentiometric Titration.
3. IEC 62021-2. 2008. Insulating Liquids—Determination of Acidity—Part 2: Colourimetric Titration.
4. ISO 14596. 2007. Petroleum Products—Determination of Sulfur Content—Wavelength-Dispersive X-Ray Fluorescence Spectrometry.
5. NYTRO® BIO 300X. 2019. *Product Data Sheet*. Nynas.
6. ENVIROTEMP™ 360. 2018. *Synthetic Ester*. Cargill.
7. MIDEL® 7131. 2019. *Synthetic Ester Transformer Fluid*.
8. BecFluid® 9902. 2006. *Product Information*. ATLANTA Electric Insulation.
9. NYCODIEL® 1200 Series. 2017. *Synthetic Ester-Based Dielectric Fluids*. NYCO.
10. MATROL-BI® FDE01A. 2020. *Technical Data Sheet—Synthetic Ester Dielectric Fluid, Version 2.0*. NOVAMONT.
11. EPA 9045C. 1995. *Soil and Waste pH, Revision 3*.
12. ENVIROTEMP® FR3™ Fluid. 2001. Bulletin 00092, Product Information.
13. ASTM D974-14e2. 2014. Standard Test Method for Acid and Base Number by Color-Indicator Titration.
14. ASTM D1275-15. Standard Test Method for Corrosive Sulfur in Electrical Insulating Liquids.
15. IEC 62697-1. 2012. Test Methods for Quantitative Determination of Corrosive Sulfur Compounds in Unused and Used Insulating Liquids—Part 1: Test Method for Quantitative Determination of Dibenzyldisulfide (DBDS).
16. MIDEL eN 1204. 2019. Natural Ester Transformer Fluid (Rapeseed).
17. IEC 62021-3. 2014. Insulating Liquids—Determination of Acidity—Part 3: Test Methods for Non-mineral Insulating Oils.
18. MIDEL eN 1215. 2019. Natural Ester Transformer Fluid (Soya).
19. BIOTEMP. 2010. ABB Sensible Solution, The Superior Biodegradable, High Fire Point Dielectric Insulating Fluid. *Hitachi Energy*
20. IEC 62535. 2009. Insulating Liquids—Test Method for Detection of Potentially Corrosive Sulphur in Used and Unused Insulating Oil.
21. NeuGen Plus. 2021. Environmentally sustainable dielectric fluid. *Technical Data Sheet*.
22. Paryol Electra 7426®. 2014. Product Guide, A&A Fratelli Parodi Spa.
23. MATROL-BI® FDV01A. 2020. *Technical Data Sheet—Natural Ester Dielectric Fluid, Version 2.0*. NOVAMONT.
24. Evangelista, J. M. G. Jr., et al. 2017. Development of a New Bio-based Insulating Fluid from Jatropha Curcas Oil for Power Transformers. *Advances in Chemical Engineering and Science* 7: 235–255.
25. Powersil® Fluid TR 50. 2021. *Technical Data*. Wacker.
26. IEC 60836. 2015. Specifications for Unused Silicone Insulating Liquids for Electrotechnical Purposes.
27. Ingebringsten, S., et al. 2004. *Solubility of Carboxylic Acids in Paper (Kraft)-Oil Insulation System*. Tronheim: SINTEF Energy Research AS.
28. Yan, Z.W., et al. 2020. Measuring Low Molecular Weight Acids in Mineral and Ester Transformer Liquids. In *The 8th International Conference on Condition Monitoring and Diagnosis (CMD 2020)*, Thailand.
29. Yan, Z.W., et al. 2018. Extraction of Low Molecular Weight Acids from Transformer Liquids Using Water Extraction Technique. In *12th IEEE International Conference on the Properties and Applications of Dielectric Materials*, Xi'an, China.
30. Pashkovskaya, A.A., et al. 2018. Mechanism of Long-Chain Free Fatty Acid Protonation at the Membrane-Water Interface. *Biophysical Journal* 114: 2142–2151.
31. Kanicky, J.R., and D.O. Shah. 2002. Effect of Degree, Type, and Position of Unsaturation on the pK_a of Long-Chain Fatty Acids. *Journal of Colloid and Interface Science* 256: 201–207.

32. ASTM D664-18e2. 2018. Standard Test Method for Acid Number of Petroleum Products by Potentiometric Titration.
33. ASTM D1534-95. 2017. Standard Test Method for Approximate Acidity in Electrical Insulating Liquids by Color-Indicator Titration.
34. Gunstone, F. 2004. *The Chemistry of Oils and Fats, Sources, Composition, Properties and Uses*. Blackwell Publishing Ltd. ISBN 1-4051-1626-9.
35. ASTM D6871-17. 2017. Standard Specification for Natural (Vegetable Oil) Ester Fluids Used in Electrical Apparatus.
36. IEC 61125. 2018. Insulating Liquids—Test Methods for Oxidation Stability—Test Method for Evaluating the Oxidation Stability of Insulating Liquids in the Delivered State.
37. Darwin, A. 2011. The Use of Natural Esters in Transformers. In *MATPOST Conference*.
38. ASTM D2440-13. 2013. Standard Test Method for Oxidation Stability of Mineral Insulating Oil.
39. Rafiq, M., et al. 2020. Sustainable, Renewable and Environmental-Friendly Insulation Systems for High Voltage Applications. *Review, MDPI Molecules*. https://doi.org/10.3390/molecules25173901.
40. Envirotemp FR3 Fluid. 2019. *Formulated for Performance*. Cargill.
41. Gomna, A., et al. 2019. Review of Vegetable Oils Behavior at High Temperature for Solar Plants: Stability, Properties and Current Applications. *ScienceDirect, Solar Energy Materials and Solar Cells* 200. https://doi.org/10.1016/j.solmat.2019.109956.
42. Lashbrook, M., Coker, A. 2020. The Role of Ester Fluids in Fire Safe, Greener, Better Performing Power Networks. *Transformer Technology* 9.
43. ASTM D2300-08. 2017. Standard Test Method for Gassing of Electrical Insulating Liquids Under Electrical Stress and Ionization (Modified Pirelli Method).
44. IEEE C57.147. 2018. IEEE Guide for Acceptance and Maintenance of Natural Ester Liquids in Transformers.
45. ASTM D7150-13. 2020. Standard Test Method for the Determination of Gassing Characteristics of Insulating Liquids Under Thermal Stress.
46. Martin, D., et al. 2010. Preliminary results for dissolved gas levels in a vegetable oil-filled power transformer. *IEEE Electrical Insulation Magazine*. https://doi.org/10.1109/MEI.2010.5585007
47. Martin, R. 2014. Fluid Maintenance and Dissolved Gas Analysis for Ester-Based Transformer Liquids. In *5th Workshop, Oil/Paper and Gas-Insulated Systems in Components of Electrical Power Supply*, Graz.
48. Deng, X., et al. 2018. Experimental Research on Gassing Characteristics of Transformer Oils Under Low Thermal Decomposition. In *12th IEEE International Conference on the Properties and Applications of Dielectric Materials*, Xi'an, China.
49. Müller, A., et al. 2012. Solubility Study of Different Gases in Mineral and Est-based Transformer Oils. In *IEEE International Conference on Condition Monitoring and Diagnosis*, Bali, Indonesia.
50. ISO 12185. 1996. Crude Petroleum and Petroleum Products—Determination of Density—Oscillating U-Tube Method.
51. ISO 3675. 1998. Crude Petroleum and Liquid Petroleum Products—Laboratory Determination of Density—Hydrometer Method.
52. ASTM D1298-12b. 2017. Standard Test Method for Density, Relative Density (Specific Gravity), or API Gravity of Crude Petroleum and Liquid Petroleum Products by Hydrometer Method.
53. Olawuni, A., and I. Adejumobi. 2014. Breakdown Voltage Characteristics of Castor Oil as Alternative to Transformer Insulation Oil. *International Journal of Scientific Engineering and Research* 2(4) (2014). Paper ID: J2013229.
54. ASTM 1903-08. 2017. Standard Practice for Determining the Coefficient of Thermal Expansion of Electrical Insulating Liquids of Petroleum Origin, and Askarels.
55. Fasina, O., and Z. Colley. 2008. Viscosity and Specific Heat of Vegetable oils as Function of Temperature: 35 °C to 180 °C. *International Journal of Food Properties*. https://doi.org/10.1080/10942910701586273.

56. Diamante, L., and T. Lan. 2014. Absolute Viscosities of Vegetable Oils at Different Temperatures and Shear Rate Range of 64.5 to 4835 s^{-1}. *Hindawi Publishing Corporation, Journal of Food Processing*.
57. ISO 3104. 2020. *Petroleum Products—Transparent and Opaque Liquids—Determination of Kinematic Viscosity and Calculation of Dynamic Viscosity*, 3rd ed.
58. ASTM D445-21. 2021. Standard Test Method for Kinematic Viscosity of Transparent and Opaque Liquids (and Calculation of Dynamic Viscosity).
59. Mariprasath, T., and V. Kirubakaran. 2015. Pongamia Pinnata as Alternate Liquid Dielectrics in Distribution Transformer: A Critical Study on the Property of Viscosity. *Advances in Energy and Power* 3(1). https://doi.org/10.13189/aep.2015.030101.
60. Kumara, S., et al. 2017. Comparison of Coconut/Sesame/Castor Oils and Their Blends for Transformer Insulation. *Conference Paper*. https://doi.org/10.1109/ICHNFS.2017.8300410.
61. Pagger, E. 2013. *Alternative Insulating Liquids Compared to the Classic Mineral Oil*. Doctoral Thesis, Graz University of Technology.
62. McShane, C., et al. 2006. *Natural Ester Dielectric Fluid Development, Transmission & Distribution Conference*, Dallas.
63. Möller, J., and Nasser, J. 2002. *Schmierstoffe im Betrieb*, 2nd ed. Berlin: Springer.
64. Dubbel. 1983. *Handbook of Mechanical Engineering*, 15th revised ed. Berlin: Springer.
65. IEC 62961. 2018. Insulating Liquids—Test Methods for the Determination of Interfacial Tension of Insulating Liquids—Determination with the Ring Method.
66. ASTM D971-20. 2020. Standard Test Method for Interfacial Tension of Insulating Liquids Against Water by the Ring Method.
67. Vasovic, V., et al. 2014. Equilibrium Charts for Moisture in Paper and Pressboard Insulations in Mineral and Natural Ester Transformer Oils. *DEIS Feature* 10(2).
68. Pagger, E., et al. 2018. Natural Ester FR3 Insulating Liquid—Very Paper Friendly. In *12th IEEE International Conference on the Properties and Applications of Dielectric Materials*, Xi'an, China.
69. IEC 61203. 1992. Synthetic Organic Esters for Electrical Purposes—Guide for Maintenance of Transformer Esters in Equipment.
70. IEC 60814. 1997. Insulating Liquids—Oil-Impregnated Paper and Pressboard—Determination of Water by Automatic Coulometric Karl Fischer Titration.
71. ASTM D1533-20. 2020. Standard Test Method for Water in Insulating Liquids by Coulometric Karl Fischer Titration.
72. Feischl, T. 2021. Introduction Presentation epp.
73. Bowers, S. 2009. *Ester Transformatoren*. Midel.
74. Biotemp®. 2012. Pucaro Sensible Solution, Die Hochwertige, Biologisch Abbaubare und Schwer Entflammbare Dielektrische Isolierflüssigkeit.
75. ISO 12937. 2017. Petroleum Products—Determination of Water—Coulometric Karl Fischer Titration Method.
76. Analytix Notes Hydranal®. 2007. *Product Line Overview*. SIGMA-ALDRICH.
77. Methrom. 2010. 860 KF Thermoprep, Handbook.
78. Borsi, H., and E. Gockenbach. 2005. Properties of Ester Liquid Midel 7131 as an Alternative Liquid to Mineral Oil for Transformers. In *IEEE International Conference on Dielectric Liquids*.
79. Fofana, I., et al. 2002. Challenge of Mixed Insulating Liquids for Use in High Voltage Transformers, Part 1: Investigation of Mixed Liquids, *IEEE Electrical Insulation Magazine* 18(3).
80. Gockenbach, E., and H. Borsi. 2002. Performance and New Application of Ester Liquids. In *IEEE 14th International Conference on Dielectric Liquids*.
81. IEC 60296. 2020. Fluids for Electrotechnical Applications—Mineral Insulating Oils for Electrical Equipment.
82. Perrier, C., et al. 2006. Improvement of Power Transformers by Using Mixtures of Mineral Oil with Synthetic Esters. *IEEE Transactions on Dielectrics and Electrical Insulation*.

83. Suwarno, A., and I.S. Darma. 2008. Dielectric Properties of Mixtures Between Mineral Oil and Natural Ester from Palm Oil. *WSEAS Transaction on Power Systems* 3(2).
84. IEC 61039. 2008. Classification of Insulating Liquids.
85. IEEE C57.12.00. 2015. IEEE Standard for General Requirements for Liquid-Immersed Distribution, Power, and Regulating Transformers.
86. Rodenberg, C. 2010. How Transformers Can Explode, Popular Mechanics.
87. CNN. June 4, 2010.
88. TheHindu. June 4, 2010.
89. Pagger, E., and M. Thelen. 2017. Reduction of the Environmental Hazard Using Natural Esters. In *Oberlausitzer Energy Symposium 2017*, Zittau, Germany, 25–26 October 2017.
90. IEC 61936-1. 2015. Power Installations Exceeding 1 kV a.c.—Part 1: Common Rules.
91. ISO 2719. 2016. Determination of Flash Point—Pensky-Martens Closed Cup Method.
92. ASTM D93-20. 2020. Standard Test Methods for Flash Point by Pensky-Martens Closed Cup Tester.
93. ISO 2592. 2017. Petroleum and Related Products—Determination of Flash and Fire Points—Cleveland Open Cup Method.
94. ASTM D92-18. 2018. Standard Test Method for Flash and Fire Points by Cleveland Open Cup Tester.
95. Shengwei, C., et al. 2019. Fire Resistance of Transformers Filled with Natural Ester Insulating Liquid. In *The 14th IET International Conference on AC and DC Power Transmission (ACDC 2018), The Journal of Engineering*, vol. 2019 Issue 16.
96. Hoffmann, J.-F. 2015. *Thermal Energy Storage System with Natural or Recycled Materials for Concentrating Solar Power Plant*. Doctoral Thesis, University of Perpignan Via Domitia.
97. ASTM D240-19. 2019. Standard Test Method for Heat of Combustion of Liquid Hydrocarbon Fuels by Bomb Calorimeter.
98. Midel. 2019. Selection Guide.
99. Jonat, T. 2008. *BecFluid 9902—Eine Neue Isolierflüssigkeit auf Esterbasis*. Berlin: Inductica Konferenz.
100. Paryol Electra 7426®. 2016. *TG5—Fire Safety*. Fratelli Parodi Spa.
101. Hellebuyck, D. 2013. *Functional Performance Criteria for Comparison of Less Flammable Transformer Oils with Respect to Fire and Explosion Risk, Report 5424*. Department of Fire and Safety Engineering and Systems Safety, Lund University.
102. Transformer Technology. 2021. Issue 11
103. IEC 60695-11-10:2013. 2013. Fire Hazard Testing—Part 11-10: Test Flames—50 W Horizontal and Vertical Flame Test Methods.
104. Morad, N.A., and F. Panau. 2000. Liquid Specific Heat Capacity Estimation for Fatty Acids, Triacylglycerols, and Vegetable Oils Based in Their Fatty Acid Composition. *Journal of the American Oil Chemists' Society*.
105. ASTM E1269-11. 2018. Standard Test Method for Determining Specific Heat Capacity by Differential Scanning Calorimetry.
106. ASTM D2717-95. 2009. Standard Test Method for Thermal Conductivity of Liquids.
107. Kurzweil, P., et al. 2021. Environmental Impact and Aging Properties of Natural and Synthetic Transformer Oils Under Electrical Stress Conditions. *Advanced Sustainable Systems*.
108. Rapp, K.J., et al. 1999. Behavior of Ester Dielectric Fluids Near the Pour Point. In *IEEE Conference on Electrical Insulation and Dielectric Phenomena*.
109. ISO 3016:1994. 2017. Petroleum Products—Determination of Pour Point (DIN ISO 3016).
110. ASTM D97-17b. 2017. Standard Test Method for Pour Point of Petroleum Products.
111. Paryol Electra 7426®, TG7 –COLD ENERGIZATION, 2019.
112. Technical Data Sheet for Powersil®. 2009. Fluid TR 50.
113. IEC 60156. 2018. Insulating Liquids—Determination of the Breakdown Voltage at Power Frequency—Test Method.
114. IEC 60247. 2004. Insulating Liquids—Measurement of Relative Permittivity, Dielectric Dissipation Factor (tan δ) and D.C. Resistivity.

115. ASTM D924-15. 2015. Standard Test Method for Dissipation Factor (or Power Factor) and Relative Permittivity (Dielectric Constant) of Electrical Insulating Liquids.
116. ASTM D1169-19. 2019. Standard Test Method for Specific Resistance (Resistivity) of Electrical Insulating Liquids.
117. Hacker, V., and C. Sumereder. 2020. *Electrical Engineering*. De Gruyter.
118. Küchler, A. 2009. *High Voltage Engineering, Fundamentals—Technology—Applications*. Heidelberg: Springer.
119. Bartnikas, R. 1994. Engineering Dielectrics, Electrical Insulating Liquids. *ASTM* III.
120. Shah, Z.H., and Q.A. Tahir. 2011. *Dielectric Properties of Vegetable Oils*. JSR Publications.
121. Judendorfer, T. 2012. *Oil-Cellulose Insulation Systems for HVDC Applications*. Doctoral Thesis, Graz University of Technology.
122. Valery, B., and G. Davydov. 2014. Natural Ester Dielectric Liquids for Transformers and Other HV Apparatus. *Transmission & Distribution*.
123. Pagger, E. 2014. The Polar Behavior of Different Insulating Liquids and Their Impacts. In *5th Workshop—Oil/Paper- and Gas-Insulated Systems in Components of Electrical Power Supply*, Graz, Austria.
124. ASTM D1218-12. 2016. Standard Test Method for Refractive Index and Refractive Dispersion of Hydrocarbon Liquids.
125. Perrier, C., and A. Beroual. 2009. Experimental Investigations on Insulating Liquids for Power Transformers: Mineral, Ester, and Silicone Oils. *IEEE Electrical Insulation Magazine* 25(6).
126. Rapp, K., et al. 2009. Lightening Impulse Testing of Natural Ester Fluid Gaps and Insulation Interfaces. *IEEE Transactions on Dielectrics and Electrical Insulation* 16 (6): 1595–1603.
127. Pukel, G., et al. 2006. Transformer Diagnostics: Common Used and New Methods. In *International Conference on Condition Monitoring and Diagnosis*, Changwong, South Korea.
128. Lick, W., et al. 2006. Improvement of the Test Method for Insulating Oils. In International Conference on Condition Monitoring and Diagnosis, Changwon, South Korea.
129. FR3™ ENVIROTEMP™. 2015. *Natural Ester Fluid, Technical Perspectives*. ©2015 Cargill.
130. Gasser, H.P., et al. 2011. Aging of Pressboard in Different Insulating Liquids. In *International Conference on Dielectric Liquids*, Trondheim.
131. Martin, R. 2018. The Advantages of Using Ester Transformer Fluids in Sub-Saharan Africa, Webinar.
132. CIGRE Working Group. 2010. Experiences in Service with New Insulating Liquids. In *CIGRE Working Group A2.35*. Technical Brochure 436, p. 13
133. ASTM 1816-12. 2019. Standard Test Method for Dielectric Breakdown Voltage of Insulating Liquids Using VDE Electrodes.
134. ASTM D877/D877M-19. 2019. Standard Test Method for Dielectric Breakdown Voltage of Insulating Liquids Using Disk Electrodes.
135. Li, H., et al. 2020. Influence of Molecule Structure on Lightning Impulse Breakdown of Ester Liquids. *MDPI Energies*. https://doi.org/10.3390/en14041061.
136. ASTM D3300-20. 2020. Standard Test Method for Dielectric Breakdown Voltage of Insulating Liquids Under Impulse Conditions.
137. Chhajer, D. 2014. Power Factor Insulation Diagnosis, Demystifying Standard Practices. In *50th Annual Minnesota Power Systems Conference*, Megger.
138. IEC 61099. 2011. Insulating Liquids—Specification for Unused Synthetic Organic Esters for Electrical Purposes.
139. IEC 60270:2000. 2000. High-Voltage Test Techniques—Partial Discharge Measurements.
140. Lemke, E., et al. 2008. *Guide for Electrical Partial Discharge Measurements in Compliance to IEC 60270*. CIGRE Technical Brochure, WG D1.33.
141. Eberhardt, R., et al. 2010. Partial Discharge Behavior of an Alternative Insulating Liquid Compared to Mineral Oil. In *IEEE International Power Modulator and High Voltage Conference (IPMHVC)*, Atlanta.
142. IEC TR 61294:1993. 1993. Insulating Liquids—Determination of the Partial Discharge Inception Voltage (PDIV)—Test Procedure.

143. Zelu, Y., et al. 2011. Study on flow electrification hazards with ester oils. In *IEEE International Conference on Dielectric Liquids, Trondheim*. https://doi.org/10.1109/ICDL.2011.6015404.
144. Vihacencu, M., et al. 2013. Experimental Study of Electrical Properties on Mineral and Vegetable Transformer Oils. *UPB Scientific Bulletin, Series C* 75(3).
145. Podesser, J.A. 2015. *Vergleich der elektrischen Aufladung alternativer und konventioneller flüssig/fester Isoliersysteme für Leistungstransformatoren*. Doctoral Thesis, Graz University of Technology.
146. Zdanowski, M. 2020. Electrostatic Charging Tendency Analysis Concerning Retrofilling Power Transformers with Envirotemp FR3 Natural Ester. *MDPI Energies*. https://doi.org/10.3390/2n13174420.
147. OECD 301. 1992. *OECD Guideline for Testing of Chemicals*. Ready Biodegradability.
148. EPA. 1998. OPPTS 835.3110 Ready Biodegradability, Fate, Transport, and Transformation Test Guidelines.
149. Stenborg, P. 2008. Erfahrungen mit einer Natürlichen-Ester-Isolier- und Kühlflüssigkeit in Hochspannungstransformatoren, Stuttgarter Hochspannungssymposium.
150. Repsol BioElectra. 2014. Non-Inhibited Dielectric Oil Based on Natural Esters. Berlin: CIEWE.
151. OECD. 2019. *Test No. 203: Fish, Acute Toxicity Test, OECD Guidelines for the Testing of Chemicals, Section 2*. Paris: OECD Publishing.
152. OECD. 2002. *Test No. 420: Acute Oral Toxicity—Fixed Dose Procedure, OECD Guidelines for the Testing of Chemicals, Section 4*. Paris: OECD Publishing.
153. IEC 62697-2. 2018. Test Methods for Quantitative Determination of Corrosive Sulfur Compounds in Unused and Used Insulating Liquids—Part 2: Test Method for Quantitative Determination of Total Corrosive Sulfur (TCS).
154. Paryol Electra 7426®. 2015. TG6–CO2 SAVING.
155. ASTM D3455-11. 2011. Standard Test Methods for Compatibility of Construction Material with Electrical Insulating Oil of Petroleum Origin.
156. IEC TR 62697-3. 2018. Test Methods for Quantitative Determination of Corrosive Sulfur Compounds in Unused and Used Insulating Liquids—Part 3: Test Method for Quantitative Determination of Elemental Sulfur.
157. ASTM D130-19. 2019. Standard Test Method for Corrosiveness to Copper from Petroleum Products by Copper Strip Test.
158. ISO 2160. 1998. Petroleum Products—Corrosiveness to Copper—Copper Strip Test.

Chapter 5
Application of New Insulating Liquid in High Voltage Equipment

This chapter deals with the construction and operation of transformers from the cradle to the grave. Differences caused using different insulating liquids are worked out and shown. Reliable transformers are the result of a proven design, qualified manufacturing process, and the right choice of materials for the desired application [1].

As infrastructure comes due for replacement and everything we own becomes electrified, this aging of equipment and additional load demand means the need for more transformers. Keeping an eye on the climate target, this must happen with reduced carbon dioxide emission.

A transformer is defined as a static piece of apparatus with two or more windings, which by electromagnetic induction, transforms a system of alternating voltage and current into another system of voltage and current usually of different values and at the same frequency for the purpose of transmitting electrical power.

The construction of a transformer comprises two active components—the ferromagnetic core and the windings. The passive part of a transformer is the cooling system, consisting of tank, pipes, conservator, cooling fins and the cooling liquid. Another important part is the insulating system, which is in the most cases insulating liquid and cellulose. These insulations have also another task. The insulating liquids are also for cooling and the solid insulation is responsible for the mechanical strength.

A transformer uses the core's magnetic properties and current in the primary winding—connected to the source of electricity—to induce a current in the secondary winding—connected to the output or load. Alternating current in the primary winding induces a magnetic flux in the core, which in turn induces a voltage in the secondary winding. A voltage step-down results from the exchange of voltage for current from a higher level to a lower one, and its magnitude is determined by the ratio of turns in the primary and secondary winding.

Transformer bushing is an insulating liner in an opening situated on the tank through which conductors pass that allows connection to the electrical grid.

E. P. Pagger et al., *Biological Insulating Liquids*,
https://doi.org/10.1007/978-3-031-22460-7_5

5.1 General

Insulating liquids are a major part of the electrical insulation system in many types of electrical equipment, such as transformers, circuit breakers, bushings, cables, and capacitors. In the electricity sector, this is acknowledged that the lifespan of the high voltage device is primarily based upon used insulation arrangement (usually liquid and solid insulation materials) status. Often, the insulation liquid provides additional functions. It is also the cooling medium in transformers and cables, the arc extinguishing phase in switchgear, or even the lubricant in tap changers and circulating pumps, or as a preventive coating film to avoid corrosion of metallic areas. Each application then demands an insulating liquid with somewhat different electrical, chemical, and physical characteristics. In transformers, for example, the insulating liquid must possess high specific heat and thermal conductivity along with low viscosity and pour point to provide effective heat transfer. Electrical stresses in general are not extremely high. However, these stresses are high in capacitors, and the insulating liquid must provide exceptional resistance to partial electrical discharges (high discharge inception voltage, high gas adsorption). Less heat is generated in normal operation of a capacitor, and the physical characteristics of the insulating liquid to heat transfer are of less concern. Still another set of characteristics is required in circuit breakers. The electrical arc associated with operation must die abruptly as the breaker is open or closed. The requirements that the several kinds of electrical equipment place on liquid insulation are often in conflict, and no single class of insulation liquids serves equally well in all applications.

The behavior of mineral oil as insulating liquid is well understood, and designers have established rules for the construction of transformers through research, as well as trial and error, over many years. In modern times, the design of power transformers has become more and more sophisticated, with both electrical and thermal computer modelling now widely used. This allows designers to push the designs to their limits, whilst being relatively confident that the transformer will pass the final test if the manufacturing process is without fault.

The transformer industry is guided by standards. These standards are the basic for design and manufacture the equipment. Many standards organizations exist globally, and the applicable standards can vary depending on where they are located. But the widely used are IEC and IEEE standards [2–8].

In general, we use insulating liquids, having a high dielectric constant for distribution and power transformers. Comparison of the dielectric constant values of biological insulating liquids with that of mineral oil suggests that the biological insulating liquids having the higher value of dielectric constant may be used in transformers as a dielectric coolant, preferably.

At higher voltage levels, it is not always possible to use a mineral oil design transformer with an alternative insulating liquid. The variance between the insulating liquids means that they are often not able to be used within a single common design. Some design changes may need to be made to accommodate the different chemical makeup of the alternative insulating liquids.

Insulating materials will deteriorate under normal operating conditions. The aging rate of any material is influenced by several external stresses like thermal, electrical, mechanical, and environmental stresses. The composition and molecular structure of insulating materials can be decomposed by all these stresses, leading to a situation in which the material eventually can no longer fulfill its insulation functions. The transformer reliability depends largely on the properties of the electric insulation system, since the highest percentage of transformer failures result from insulation degradation. Thus, aging of the insulation has been recognized as one of the main causes of transformer failures. The active part of a transformer is the decisive function component of any transformer, conceiving about 90% of a transformer value. Insulating liquid plays an important role in the transformer insulation system. The degradation of the insulating liquid can result in an increase in humidity, an enhancement of the possibility of partial discharges in the insulation system, and a decrease in the breakdown strength of the liquid-paper insulation. The predicted lifetime of a transformer is determined by the failure rates of the insulation system.

5.2 Interaction with Other Transformer Materials

The electrical equipment used in the transmission and distribution of electrical energy is expected to have a service life of years or decades. Aging of insulating liquids results in degradation of insulating properties depends on time, load, temperature, humidity, and oxygen amount. In addition to possessing suitable properties initially, the insulating liquid should maintain them throughout the long useful life of the equipment. Metals (copper, iron, aluminum) are used as elements of construction, as conductors and, often, as magnetic materials. Other nonconducting or insulating materials (cellulose, wood, plastics) are present in most cases. The insulating liquid used in each type of equipment must have demonstrated compatibility with the other materials present in that equipment. Therefore, understanding compatibility is necessary and ensures safe operation. The requisite properties of the particular insulating liquid used must neither rapidly degrade these materials nor be degraded rapidly by them.

One well-known issue in transformer operation is the insulating liquid leakage from gaskets. Some rubbers used for gaskets (butyl, nitrile, and neoprene) and some plastic materials (PVC) or paints are attacked by some insulating liquids. Natural esters are gentler to these materials than synthetic esters. It has been observed that nitrile rubber gaskets get brittle in synthetic ester but swell in natural ester, with the grade of degeneration depending on the individual rubber composition [9]. The less influence the insulating liquid has on the gasket materials, the less deformation and deterioration of the gasket will occur and insulating liquid leakage from gaskets will be avoided.

It is well known that rubber gaskets can leach particles into the insulating liquid due to the solvency nature of the insulating liquid. This will not only degrade the quality of the gasket, but it can also reduce the electrical insulating properties of the insulating liquid and impact the overall insulation performance of the transformer system. The quality of the gasket material plays a key role in achieving compatibility with biological insulating liquids. In the case of power transformers, where longevity and long-term performance are very important, it is necessary to use high-quality grades, especially for any elastomeric components.

5.2.1 Interaction Between Solid and Liquid Insulation

Cellulose, like paper and pressboard, is still the most common form of solid insulation used in transformers. Cellulose paper is perfectly qualified as a solid insulating material for high voltage applications. This is due to a very low electrical conductivity, appropriate electrical permittivity, and its chemical properties. Furthermore, it is a cheap and renewable material and can be easily manufactured in several ways (molding, wrapping, bending, sawing, grinding, milling, etc.). Pressboard is used to separate large insulating liquid gaps into smaller gaps with higher dielectric strength (barrier package). Generally, barrier systems consist of pressboard cylinders, clamping rings, spacers, handmade wet molded parts like snouts, angle rings, caps, etc. [10]. Cellulose is a polymer structure made up of many glucose units arranged in chains linked by oxygen atoms (Fig. 5.1).

These glycosidic bonds of the polysaccharides (acetal bonds) are stable in neutral and moderately alkaline environments, disregarding possible oxidation reactions. In an acid environment, however, the bonds are hydrolyzed. The speed increase with increasing hydrogen ion activity, and the molecular weight (the degree of polymerization) of the polysaccharides decreases [11].

Cellulose is the most common carbon compound in nature, with the gross composition $(C_6H_{10}O_5)_n$. The number of molecules (n) can reach values up to 3000 and the molecular weight is between 50,000 and 500,000 g/mol. Cellulose paper plays a vital role in the operation of transformers by insulating the conductor windings. The cellulose structure consists of a crystalline and an amorphous area, Fig. 5.2. The proportions vary depending on the origin and treatment of the fibers. The crystalline

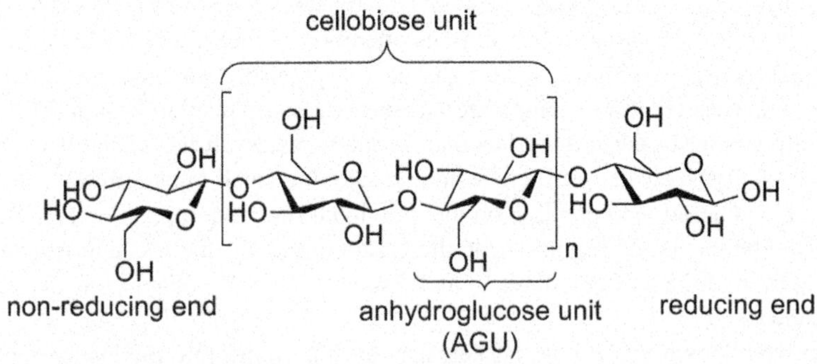

Fig. 5.1 Cellulose—molecular structure

area is the compacted area with a significantly higher degree of crosslinking. Several parallel molecular chains form microcrystalline very densely packed areas through intra- and intermolecular hydrogen bonds, which are connected by poorly ordered areas (amorphous areas). The amorphous regions of the cellulosic fibrils are more readily cleaved than the crystalline ones, and the hemicellulose is also susceptible to chemical attack. These elementary fibrils combine to form microfibrils, which in turn form fibrils and ultimately fibers. The average number of glucose units in the polymeric chains, called the degree of polymerization (DP), dictates how mechanically strong the paper is. The longer the chains, the stronger the paper.

The dielectric strength of pure cellulose is very low, as it contains large amounts of air and moisture. Therefore, insulating liquids are used as the second part in this insulation system to create superior insulation systems. The cellulose in transformers with liquid-paper insulation is subject to aging-related, sometimes irreversible, deterioration during operation. In contrast to insulating liquid, the properties of which can be significantly improved by regenerating or replacing the insulating liquid. For cellulose, the degradation of the mechanical properties is irreversible. The progressive deterioration of these strength values in connection with the decreasing winding pressure can lead to considerable winding damage and to immediate failures under high loads because of short circuits and inrush currents.

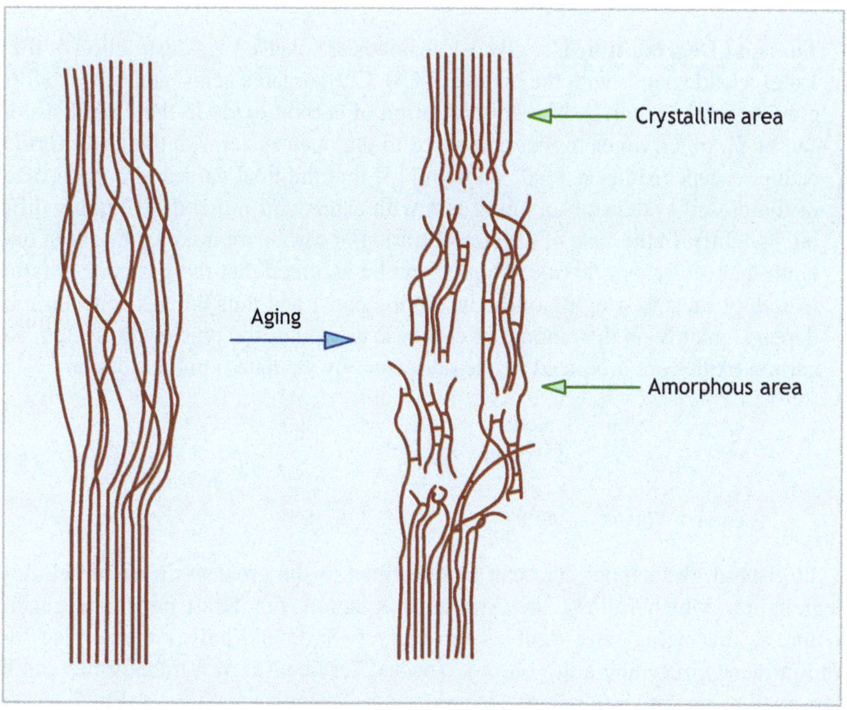

Fig. 5.2 Effect of aging on the structure of cellulose

A key indicator of the potential working lifetime of a transformer is the condition of the paper insulation. In a new transformer, the DP-value of the cellulose insulation is typically between 1000 and 1200. As a transformer ages, the cellulose begins to degrade and depolymerize forming chains of shorter length (lower DP-value). When the average DP-value reaches 200, the paper insulation is said to have reached the end of its working life. In contrast to paper samples in laboratory tests, the cellulose in real transformers does not age evenly, but—depending on the local temperatures of the windings—at sometimes significantly different speeds. The aging process is essentially one of depolymerization brought about by acid-hydrolysis, pyrolysis, and oxidation. It is commonly acknowledged that a high acidity in the insulating liquid will accelerate aging. The process is driven by hydrogen ions from dissociated acids. That means undissociated carboxylic acids, as they can be found in biological insulating liquids, do not depolymerize cellulose. It is the H^+- concentration that causes depolymerization, not the total acid concentration. It also means that water does not participate in the rate-controlling step. Water does, however, affect the H^+-concentration by causing carboxylic acids do dissociate, and in this way, it exerts an essential influence upon aging process. Water content increases during aging, which is due to hydrolysis processes, as this is an autocatalytic process in which water is consumed and generated from the breaking of the cellulose molecules.

As a first approximation, the degradation of cellulose can take place according to the three basic principles [12].

- **Thermal Degradation** The glycosidic bonds are cleaved via destruction of the β 1,4 glycoside bonds with the release of CO, CO_2, organic acids, and water to form glucose units (Eq. 5.1). The concentration of carbon oxide in the closed system can be up to ten times higher compared to the open system, as the gases formed cannot escape. Münster et al. show in [13] that the final values of carbon oxide in the closed system of an aging test with esters and mineral oil liquids differ only slightly. In the case of the concentration of carbon monoxide, it's about one-tenth that of carbon dioxide. Thus, it can be assumed that the gas concentration is mainly caused by aging of the insulating paper and thus the gas concentration depends mainly on the amount of cellulose as well as the type of the system and carbon oxides are produced to a lesser extent by insulate liquid oxidation.

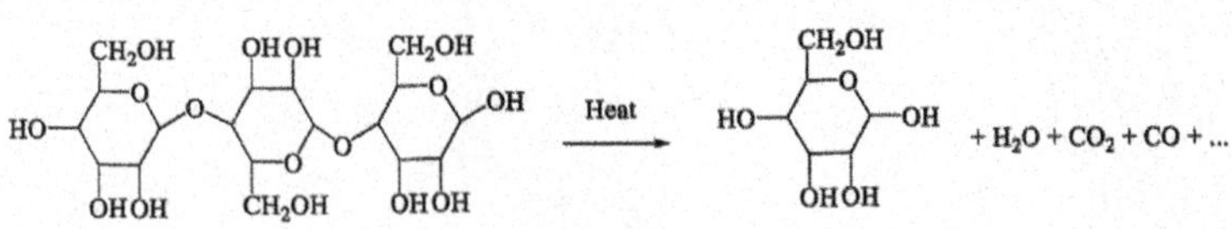

(5.1)

In general, the temperature can be described as the greatest driver of cellulose degradation, which follows the Arrhenius equation. For Kraft paper the general picture is that aging rates doubles for every 6–8 °C and halves the service life. The influence of temperature on the expected service life of a transformer can be described using the temperature dependence of Arrhenius as follows (Eq. 5.2):

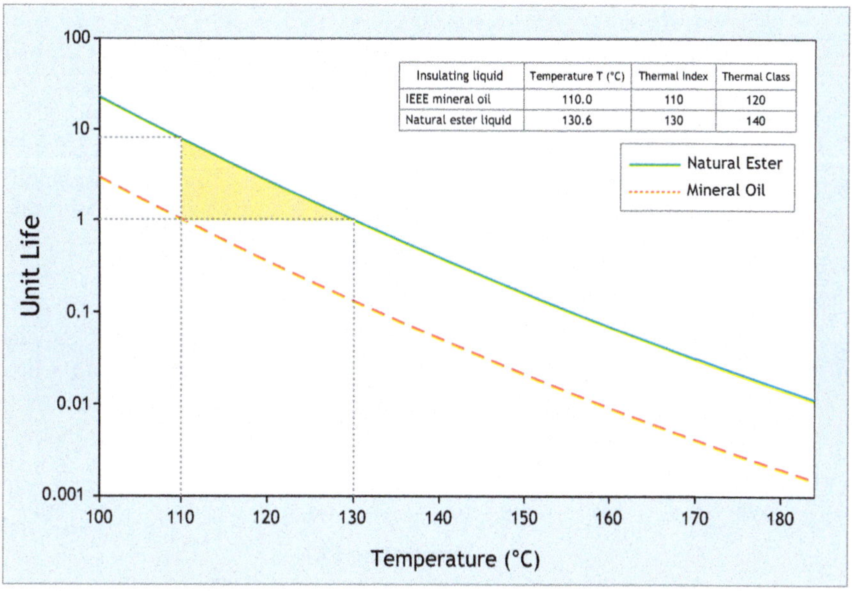

Fig. 5.3 Thermally upgraded paper unit life versus temperature

$$unit\,life(T) = A * e^{\frac{B}{(T+273)}} \tag{5.2}$$

T Temperature [°C]
A, B Constants, based on testing and reactions conditions

Figures 5.3 and 5.4 compare immersed paper life curves of biological insulating liquid to mineral oil [3] based on Eq. 5.2. Table 5.1 shows the aging constants for calculating the temperature limits.

For thermal upgraded Kraft paper (Fig. 5.3), there are the options.

a. Extend asset life at current 110 °C hotspot.
b. Increase load capability up to a hotspot of 130 °C with the same expectancy of solid insulation life as with mineral oil.
c. Extend solid insulation life and increase load capability with a hotspot x (110 °C < x < 130 °C).

For standard Kraft paper (Fig. 5.4), there are the options.

a. Extend asset life at current 95 °C hotspot.
b. Increase load capability up to a hotspot of 110 °C with the same expectancy of solid insulation life as with mineral oil.
c. Extend solid insulation life and increase load capability with a hotspot x (95 °C < x < 110 °C).

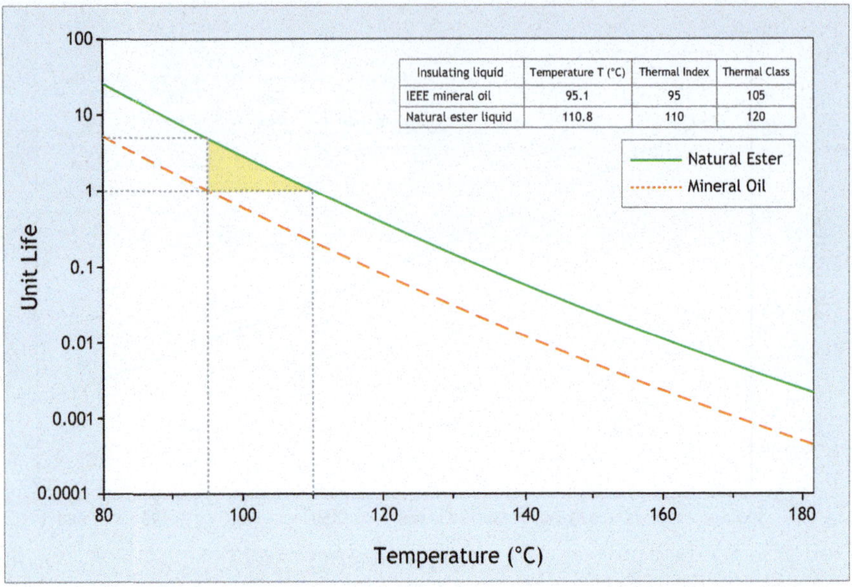

Fig. 5.4 Kraft paper unit life versus temperature

Table 5.1 Calculated constants and temperature indices

Liquid/paper	Constant B	Constant A (hour)	Temperature T (°C)	Thermal index	Thermal class
IEEE mineral oil/thermally upgraded Kraft paper	15,000	9.80×10^{-18}	110.0	110	120
Natural ester liquid/thermally upgraded Kraft paper	15,000	7.25×10^{-17}	130.6	130	140
IEEE mineral oil/cellulose-based paper	15,000	2.00×10^{-18}	95.1	95	105
Natural ester liquid/cellulose-based paper	15,000	1.06×10^{-17}	110.8	110	120

- **Oxidative Degradation** This paper degradation is caused by oxygen, which can enter through open breather system of the transformer or by faulty seals. Cellulose is highly susceptible to oxygen and oxidizing agents. In this case, the hydroxyl groups of the glucose units are oxidized via the aldehydes to the carboxylic acids with the formation of water (Eq. 5.3) and can accelerate the hydrolytic degradation of the cellulose. If the oxygen level in the insulating liquid is held below 2000 ppm, the rate of degradation of the full liquid-paper insulation system is reported to be five times lower than for a free breathing transformer [14]. The amount of water formed is related to the number of chain scissions of cellulose. Furthermore, water is an end-product of the oxidation of the paper as well as the insulating liquid. In an inert nitrogen gas atmosphere, only a slow reaction takes place, and the rates of aging are similarly proportional to the hydrogen ion concentration [11].

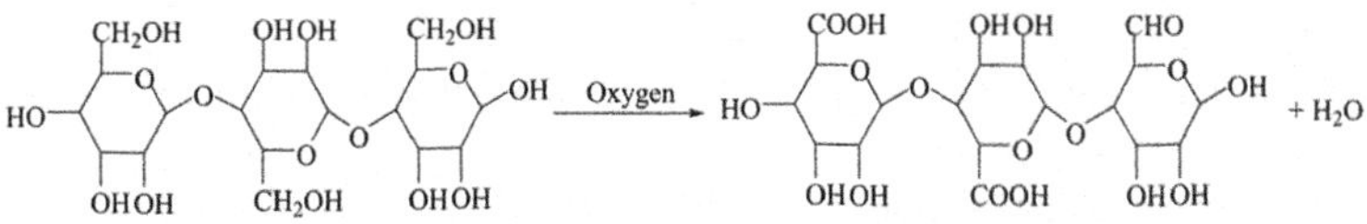

(5.3)

- **Hydration Degradation** In this case, the glycosidic bonds are split by the presence of hydrogen protons and water, releasing glucose molecules Eq. 5.4. The necessary water can stem from outside through the open breather system (exhausted silica gel cartridges) as well as via any leaking gaskets or from the insulating liquid, or as by-product of the de-polymerization process shown in Eqs. 5.1 and 5.3. It can be stated that the cellulose ages more slowly in the closed systems, as hydrolyses and oxidation are driving factors of aging. An increase in moisture content leads to a strong acceleration of the aging of paper [11].

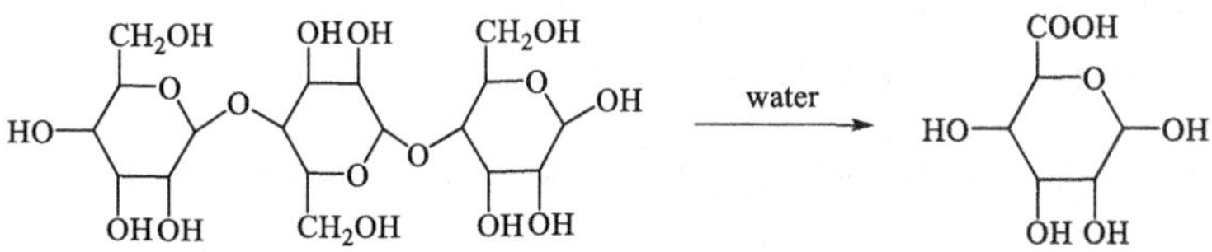

(5.4)

Since the aging is essentially a hydrolytic/oxidative process, the fact that the hemicellulose has a considerably higher rate of hydrolysis is important. The aging process is also catalyzed by acetic acid, which is produced during the cleavage of acetyl groups of the hemicellulose. It is clear that a high hemicellulose content gives a paper with poor aging stability [11].

Considering paper aging in general, in addition to the main components mentioned above, other components such as acetaldehyde, vanillin, methanol etc. will be generated, which have not yet been adequately considered as markers [15]. In theory, the ultimate degradation products of cellulose are water and carbon oxides. The decomposition and by-products that generate during cellulose breakdown get into the insulating liquid and concentrate there. Aging markers such as water, acids, CO_2 and CO can come from both paper aging and liquid aging, while furans are very cellulose-specific. Studies have suggested that paper will age more slowly in ester fluids, due to their interaction with water. Partially, this is due to the higher (some 20 to 50 times) water solubility, but the main contributor is hydrolysis of the ester, which consumes water, while producing long chain acids, which are less harmful to paper. It has also been proposed that in ester liquids, a process of transesterification takes place, which serves to reduce the paper aging rate further.

There have been a number of studies conducted looking at the comparative aging rate of paper in natural (biological insulating liquids) and synthetic ester liquids compared to mineral oil. In hydrolysis and aging of ester dielectrics, the production of low molecular weight, water soluble acids, is rare and limited. This is one of the reasons why the experimental work in this area shows that the integrity of cellulose is better preserved in these liquids. The paper with the synthetic ester ages faster, but this seems to slow down towards the end of the aging process [13]. The aging in the closed system only shows smaller differences between mineral oil and ester at the end of the aging process, so that it can generally be assumed that the increased water absorption of esters has a bigger influence at the end of the service life of the paper.

Münster et al. observed in [13] for ester liquids a very good correlation between DP (degree of polymerization) value and the carbon oxide concentration, when cellulose is aged in a closed system representing a hermetic transformer.

Since the direct determination of DP values of the paper insulation from winding areas of interest is not possible without destruction, an attempt is made to determine the DP value indirectly via the concentration of furanic compounds found in the insulating liquid. The goal is the investigation of the correlation between the aging markers and the DP value of the solid insulation to avoid direct sampling of the cellulose during operation.

When degradation of cellulose occurs one of the products that is formed is the family of compounds known as furans. Furans contain one oxygen atom in a five-membered ring (Fig. 5.5). Some of the physicochemical properties associated with this family of compounds include low boiling point, high flammability, a strong ether odor and miscibility with most organic solvents.

There are many different types of furan derivatives. The main derivatives examinate in transformer insulating liquids are.

- 5-hydroxymethyl-2-furfuraldehyde (5-HMF) (Fig. 5.6)
- 2-furfuralaldehyde (2-FAL) (Fig. 5.7)
- 2-acetyl furan (2-ACF) (Fig. 5.8)
- 5-methyl-2-furfuralaldehyde (5-MEF) (Fig. 5.9)
- 2-furfuryl alcohol (2-FOL) (Fig. 5.10)

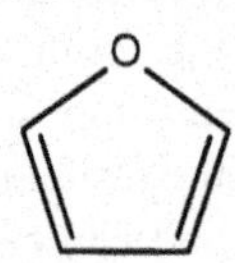

Fig. 5.5 Furan

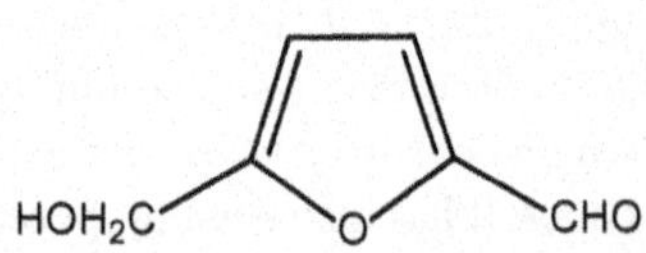

Fig. 5.6 5-hydroxymethyl-2-furfuraldehyde

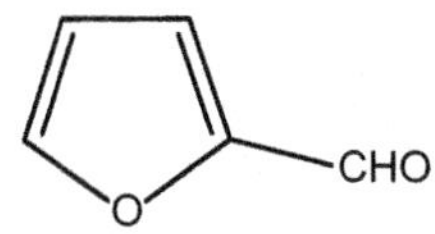

Fig. 5.7 2-furfuraldehyde

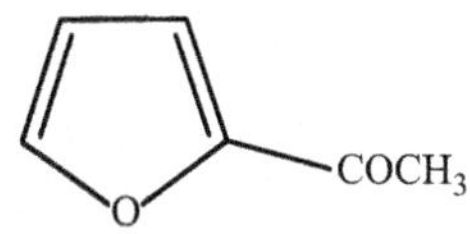

Fig. 5.8 2-acetyl furan

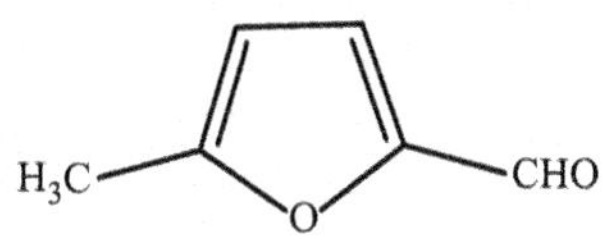

Fig. 5.9 5-methyl-2-furfuraldehyde

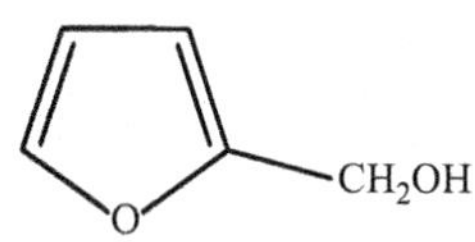

Fig. 5.10 Furfuryl alcohol

Changes in furanic compounds profiles are produced by different causes Table 5.2 [16].

The presence of furanic compounds is not generally considered to significantly influence the aging of insulating liquid or paper. However, the measurement of furanic compound content, especially 2-FAL, has found use in transformer diagnostics. A correlation between the degree of polymerization of the paper and 2-FAL (or total furanic compound) content of the insulating liquid has been observed. There are several mathematical approaches (Table 5.3) for this in the literature [17–20] and [21] which, however, lead to quite different calculation results. The 2-FAL concentration in Table 5.3 is expressed in mg/kg (ppm). The Fig. 5.11 shows the curves calculated using the formulas in the Table 5.3 in the range from 0.1 to 4 ppm. The curves scatter over a wide range.

Table 5.2 Causes for generation of furanic compounds

Compound	Observed cause
2-FAL	Overheating and normal aging
5-M2F	High temperatures
2-ACF	Rare, no definite cause
5-HMF	Oxidation
2-FOL	High moisture

Table 5.3 Calculation of the DP value from the 2-FAL concentration

Autor	Formula
Chendong	$DP = (1.51 - \log_{10} 2FAL)/0.0035$
Scholnik	$DP = (1.17 - \log_{10} 2FAL)/0.00288$
Pahlavanpour	$DP = 800/((0.186 * 2FAL) + 1)$
DePablo	$DP = 1850/(2.3 + 2FAL)$
Stebbin	$DP = (4.51 - \log_{10}(2FAL * 1000 * 0.88))/0.0035$
Heisler and Banzer	$DP = 325 * \left(\frac{19}{13} - \log_{10} 2FAL\right)$ $100 \leq \mathrm{DP} \leq 900$
Martin (1)	$DP = \frac{2.25 - \log_{10} 2FAL}{0.0046}$ For a liquid/cellulose ratio of 4
Martin (2)	$DP = \frac{2.06 - \log_{10} 2FAL}{0.0046}$ For a liquid/cellulose ratio of 6.2
Burton	$DP = \frac{2.5 - \log_{10} 2FAL}{0.005}$
Vuarchex	$DP = \frac{2.60 - \log_{10} 2FAL}{0.0049}$

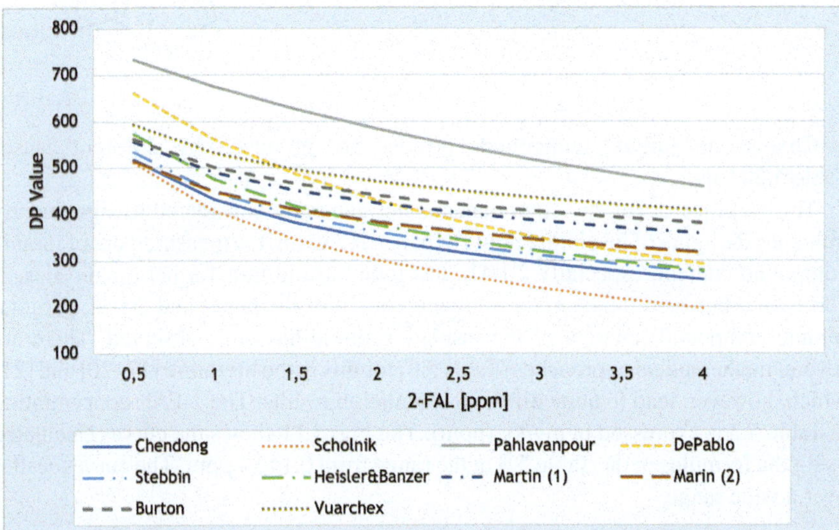

Fig. 5.11 DP value versus 2-FAL concentration

To measure furanic compounds from an insulation liquid sample is relatively simple, but the interpretation is complex, as in an aging process more than one mechanism is involved. Table 5.4 shows roughly different states of insulating paper based on the concentration of 2-FAL in the insulating liquid.

Table 5.4 Correlation between 2-FAL concentration and the state of a transformer [18]

Condition	Concentration [ppm]
Good	2-FAL < 0.5
Acceptable	0.5 ≤ 2-FAL ≤ 1
Need caution	1 < 2-FAL ≤ 1.5
Poor	1.5 < 2-FAL ≤ 2
Very poor	2-FAL > 2

There are several mathematic models using DP values to estimate transformer's life expectancy in the literature [18]. In [22], the solid insulation life expectancy in years is expressed as a function of temperature T and the parameters E and A (Eq. 5.5). Arrhenius model for loss of life is constructed on the theory that the only condition for aging is in addition to the DP values the temperature and it depends on the hottest spot in a transformer.

$$Expected\ Life(years) = \frac{\frac{1}{DP_t} - \frac{1}{DP_0}}{A * 24 * 365} * e^{\frac{E_A}{R*T}} \tag{5.5}$$

DP_t DP value after aging period
DP_0 Initial degree of polymerization
A Constant depending on the chemical environment
E_A Activation energy [kJ/mol]
R Molar gas constant [8.314 J/(mol*K)]
T Absolute temperature in Kelvin

Lundgaard found in his experiments [23] an activation energy of 125 kJ/mol when water was added and for oxygen 96 kJ/mol, respectively. In [22] data for the activation energy E_A and the environmental factor A are given. They scatter in a large range—for E_A from 97 to 117 kJ/mol and for A from 10^6 to 10^{12}. Lundgaard indicates in [23] a value for A with 10^8 to 10^9 for an activation index of 111 kJ/mol.

The A-factor depends on the existence of organic acids and water, which can dissociate these acids. Low molecular weight soluble acids that are formed by paper aging and to some degree also by insulating liquid aging are more active than the larger hydrophobic acids, which manly stem from the mineral oil aging. The fact that acid catalyzed hydrolysis generates organic acids and at the same time is governed by their presence makes the process auto acceleratory. It has been shown that acids will react with the cellulose in an esterification process. Therefore, not all produced acids will be active in the hydrolytic process.

In [22], results of studies on accelerated aging of cellulose in different insulating liquids are given. The temperature used for the experiments has ranged from 90 to 170 °C and most of the work has been carried out under sealed conditions, since biological insulating liquids are not suitable for breathing applications. Even though the aging studies have been conducted at these elevated temperatures, the mechanism that reduce the aging rate like water migration, hydrolysis, and transesterification also

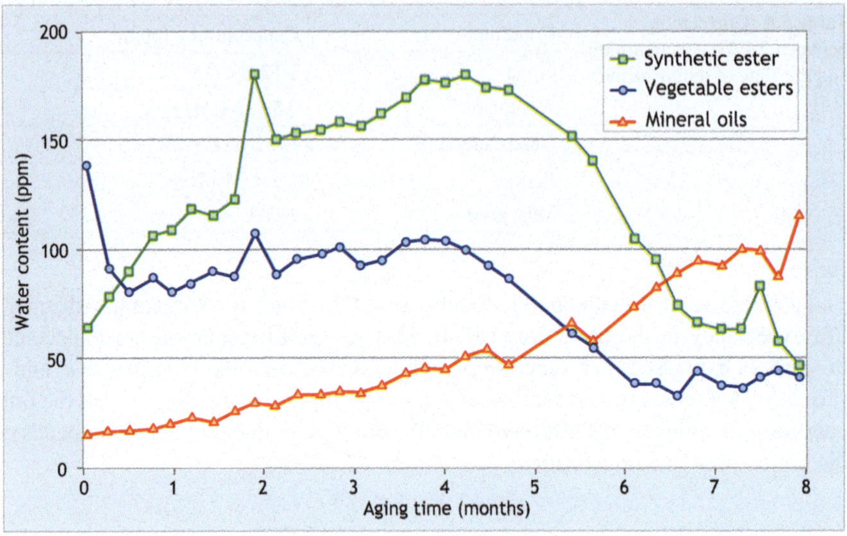

Fig. 5.12 Water content of insulating liquid, aged at 135 °C [24]

starts to occur at normal operating temperatures. The studies demonstrate that over a range of temperatures; the cellulose aging rate is slower in biological insulating liquids than in mineral oil. A key finding of the studies is that the DP value to level off is at higher value in biological insulating liquids than in mineral oil. This can partly be explained by a reduced content of free water in the cellulose due to hydrolyzation (Fig. 5.12) [24]. Also, at higher temperatures, the high-water solubility of biological insulating liquids will result in drying out of the cellulose. An argument to operate the transformer at higher temperatures.

The use of several mathematical models for the estimation of lifetime of power transformers has shown that these methods are not suitable to predict the condition of a transformer accurately [18]. It is needed to study the tendency of physicochemical and dielectric properties of insulating liquid during the transformer's operation. The condition of a transformer cannot be determined considering only the most recent measures of insulating liquid's properties. This is because, as the insulating liquid and paper age, the measured variables and the degradation products do not follow a continued tendency over time, fluctuations occur. Additionally, variables such as the working environment or load factor should be considered because these variables have an influence on insulating aging.

5.2.1.1 Water Partition Phenomenon

Partition (or equilibrium) curves describe how water will distribute between a liquid and a solid at a specific temperature. There is a large difference between mineral oil and ester filled devices—see Fig. 5.13 from [22]. The water solubility of the natural ester at 40 °C is about as high as that of mineral oil at 120 °C. Mineral oil, as a nonpolar substance, has a low affinity for water, however, its water solubility increases markedly with temperature. Other liquids such as esters and biological insulating liquids dissolve more water than mineral oil and have an increase of solubility with temperature. They can absorb up to fifty times more water than mineral oil and, for this reason, there is a different partitioning curve between liquid and cellulose. The higher water solubility of the esters enable that more water diffuses from the cellulose into the insulating liquid, thus reducing hydrolyses processes in the cellulose. How much water in the liquid can be absorbed depends on the temperature and the kind of liquid. However, using these curves for aged insulation systems can give large errors in the moisture content. Aged insulation liquids show a significant increase in the water absorbing capacity and the equilibrium curves are strongly shifted to higher water in the insulating liquid for the same water content in cellulose. There are clear differences between the open and closed systems. In the open system, the water content is higher due to the higher rate of chain scissions, as the DP value in this system is lower.

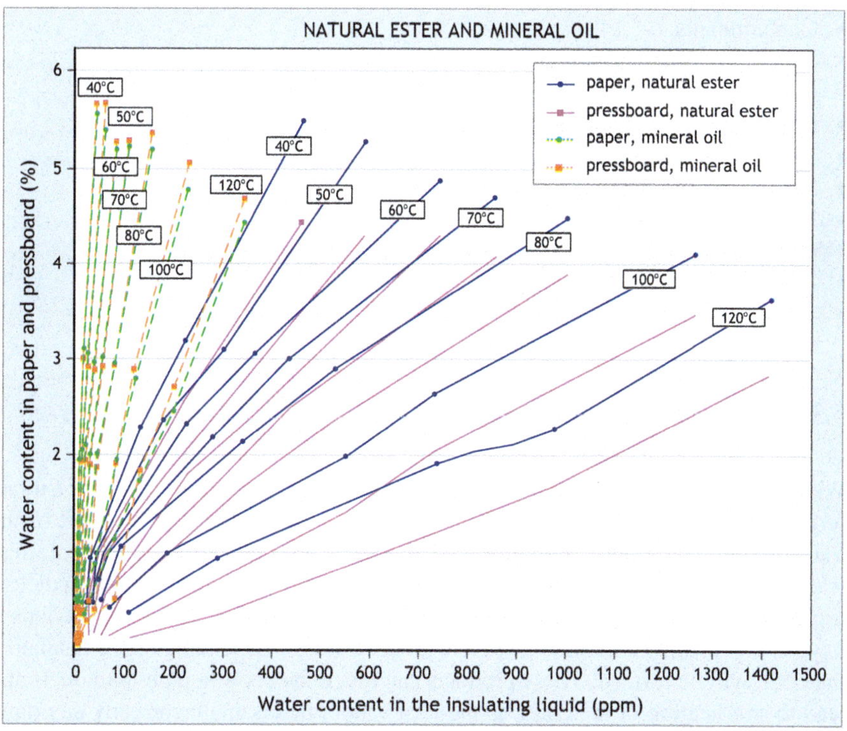

Fig. 5.13 Perrier-Lukic equilibrium curves for paper and pressboard in mineral oil and natural ester

5.2.2 *Electrical Charging Tendency*

The static electrification phenomena or electrostatic charging tendency (ECT) and therewith connected problems have been reported in Japan in the 1970s for the first time [25]. Obviously, due to the movement of one insulating material (streaming insulating liquid) over a second one (cellulosic insulation), the insulation system builds up charges. It was discovered that the similarity of all failed transformers was a forced oil cooling system. Due to the flowing of an insulating liquid (with a low electrical conductivity) past an insulating solid, charge separation occurs at the interface. Roughly speaking, each transformer with a liquid-cellulose insulation system and forced liquid circulation is therefore subjected to static electrification in one way or another due to the relative movement of two insulating materials. Judendorfer reported [25] that the following parameter can be relevant for static electrification:

- Insulating liquid temperature
- Insulating liquid moisture content
- Insulating liquid flow rate
- Turbulence of insulating liquid flow
- Contaminants
- Surface condition of solid insulation
- Charging tendency of insulating liquid
- Energization/Electrical field strength
- Dielectric strength of moving insulating liquid
- Migration of moisture
- Particulate matter
- Charge injection
- Pumps
- Orifice effects.

5.3 Degradation Mechanism of Insulation Systems

With approximately 38%, the windings failures are the most frequent root of a transformer problem, which explains why the condition assessment of this part of the transformer is particularly important [13]. Electrical, mechanical, and chemical stress lead to the aging of insulation systems. Breakdown of the chemical bounds in dielectric liquids and cellulose can generate soluble gases and colloidal suspension. If oxygen is present, soluble and insoluble products (slag), depending on the insulation material, will be formed. These products can block the cooling pipes and ducts and lead to overheating of the coils, as the heat cannot be dissipate correctly anymore. Slag formation is hardly seen using biological insulating liquids. In [26], Yuliastuti describes an aging test with an aging time up to 1436 h with the result that new mineral oil changed the color from light yellow brown to dark and some sludges were formed in the liquid. The break down voltage of mineral oil decreased by 5.9%

and that of natural ester by 4.7% during this period. The tan δ results show that the value of natural ester liquid increased approximately 82.9% from the initial value until the end of aging, whereas the tan δ value of mineral oil keeps very stable at low level. For the natural ester liquid sample, the color did not assertively change or form any sludge tracery.

5.4 Transformer Design

At voltage level up to 66 kV, it is usually possible to use a mineral oil designed transformer with biological insulating liquids. Some design changes need to be made to accommodate the different chemical makeup of the biological insulating liquids. In fact, electrical, thermophysical and chemical properties of insulating liquids will influence ducts and heat exchanger sizes, pumping consumption, vessels pressure rating or pipes and all materials which are in contact with the liquid.

Yürekten writes [10] that solid insulation components have a considerable effect in order to reduce the cost of transformers. An optimized design of the insulation system can make it possible to reduce the weight and overall size of transformers, without compromising the quality. That means using smaller active parts, smaller tanks, less insulating liquid, and smaller magnetic core.

Designing a transformer is a balancing act between optimizing the design constraints such as short circuit strength, total losses, temperature rise, and noise level with the ability to minimize the overall transformer cost [27]. In liquid filled transformers, dielectric liquids are used to cool the windings and provide optimal performance. In self cooled transformers, the cycle of the liquid is governed naturally by convection. Natural convection can also be assisted by a series of fans directing air against the radiators, increasing heat transfer and subsequent rate of cooling in the windings. In large power transformers, it is also possible to have a level of forced liquid circulation where a pump assists the circulation of the liquid. This generally provides a lower top liquid temperature and more uniform temperatures within the windings. For specifying the correct pump size, it should be made aware of the biological insulating liquid viscosity and the required pumping rate, suction lift, and discharge head. The following factors should be considered:

- Because the viscosity of biological insulating liquids is generally higher than conventional mineral oil, care should be used in selecting a pump with the horsepower and capacity required. First, determine the maximum flow rate required and then select a pump and motor that can handle this flow rate at the lowest temperature (highest viscosity) that could be encountered. Using the flow model to determine the electrical charge tendency. Podesser found that the pressure losses for ester liquids are significantly higher than for mineral oil, especially at 20 °C, which can be attributed to the higher viscosity of these liquids [28].

- The most recommended pump for biological insulating liquids is the positive displacement gear pump. A standard iron pump with either mechanical seal or stuffing box is also satisfactory.

The partially different properties of the biological insulating liquids compared to mineral oil must be considered when designing the transformer devices. To counteract the higher temperature rise because of the viscosity, it is possible to have larger cooling channels in the windings and between the windings and barriers. The higher viscosity of the biological insulating liquids must be considered when impregnating the cellulose by increasing the impregnation temperature, extending the impregnation time, and possibly by lowering the boiler pressure. The preferred method of filling transformers is under vacuum conditions. Transformer manufactures have found it advisable to allow biological insulating liquid immersed transformers to wait longer than normally allowed for mineral oil after filling and breaking vacuum before energizing or high voltage testing. A conservative option is to wait until the transformer has cooled to room temperature [29]. The additional wait time is recommended because biological insulating liquids generally take significantly longer to impregnate cellulose insulation than mineral oil under the same conditions (see Sect. 6.3.1). Transformers with heavy thickness of pressboard insulation require standing times that are adequate to allow the required impregnation. The impregnation rate of biological insulating liquids is a function of liquid temperature and the thickness of the cellulose material to be saturated. Recommended minimum standing times vary depending on the type of pressboard, thickness, initial temperature, ambient temperature, voltage class, etc. If such guidance is unavailable, a minimum of a 3 to 4 times longer impregnation time is recommended [30].

Additional vacuum processing of the biological insulating liquid is recommended to sufficiently degas the liquid prior to filling the transformer tank, to help avoid excessive foaming. Keep an eye on transformer tank vacuum limits. Commercial dehydration and degassing units are available that can process biological insulating liquids to acceptable levels of dissolved water and dissolved air. Proper processing temperature helps ensure sufficient degasification and dehydration of the biological insulating liquid prior to introduction into the transformer. After the biological insulating liquid is processed through the degasifier and particulate filter, it should be introduced directly into the transformer under vacuum.

Existing storage tanks that have been used for mineral insulating liquid can be used for biological insulating liquids if the following conditions are met:

- Transfer pumps and lines are of adequate capacity to pump the more viscous liquid. If the tank and transfer system are situated so that the liquid may have to be moved while it is cold, use of electric- or steam-line tracing and tank-heating apparatus may be necessary.
- The tank should be thoroughly drained and flushed with 60–80 °C biological insulating liquid before being filled to help avoid contamination.
- It is important that pumps and lines are properly grounded during liquid transfer to prevent build-up of a static electric charge.

Medium and large power transformer tanks may be filled with biological insulating liquids in the plant for testing, then drained and filled with dry gas prior to shipping. Dry nitrogen is technically preferred to help avoid the potential for thin film oxidation to occur between the time of the drainage at the factory and the field filling.

The viscosity of dielectric coolants within the range of normal operating temperatures is important because it can impact both the cooling and performance of some other transformer components, such as load tap changers, which are immersed with the same insulating liquid.

As oxidation processes dramatically increase the viscosity for the natural esters in open systems, an application in open breathing transformers should be avoided. To ensure and maintain optimum performance of natural esters, exposure to oxygen and moisture must be minimized. Compared with standard transformers, the hermetic power transformer with a vacuum type of tap changer has not only the advantage of less aging, but it also requires less maintenance during the total lifetime [31]. The most effective way to avoid and minimize aging is to design the transformer hermetically sealed and prevent contact with atmosphere (oxygen and moisture). This transformer may not have neither an oil conservator nor dehydrating breathers, but radiators with integrated functionalities of cooling and oil expansion at rising temperature [32]. The hermetic design with expanding radiators results in a significantly reduced aging, and this process is then largely limited to thermal aging. A sealed system based either on a nitrogen cushion or a rubber bag in the expansion tank can minimize the ingress of aging substances like water and oxygen. When using an expansion tank with a rubber bag (Fig. 5.14), it is advantageous if the air side is connected to a dehydrating breather unit (Fig. 5.15). This means that dehumidified air is always present on the air side of the rubber bag, which limits the diffusion of moisture through the rubber bag into the insulating liquid.

How an automatized dehydrator works:

- In normal operation, the air flowing to the insulating liquid conservator is routed across the desiccant (silica gel) and thus dried.
- The degree of moisture of the air flowing to the expansion tank is permanently monitored. The regeneration of the desiccant is activated when the moisture content reaches the defined limed value depending on the behavior of the equipment.
- Heating up the silica gel desiccant during an exhalation process.

Because of the possibility to bind water chemically and physically, natural- and synthetic ester can dry out the cellulose. This means a large increase in transformer insulation life. Most transformer materials (copper, steel, aluminum, cellulose, etc.) are compatible with ester liquids. Care must be taken for the gaskets. Elastomers, including NBR types with higher nitrile content, silicone or fluoropolymer, are recommended. Gaskets with higher temperature demands warrant the use of silicone or fluoropolymer (Viton) compositions [33, 34]. Fluorosilicone rubber is a good alternative gasket material which can be used with natural ester liquids without any adverse effects [35].

Fig. 5.14 Expansion tank with rubber below

Care must be taken when using mobile parts, like switch gear, in the insulating liquid. As biological insulating liquids have a higher pour point compared to mineral oil, it must be sure that the temperature is high enough that the liquid will not become solid. This can be done by running the transformer at no-load. In the event of a longer shutdown and extremely cold temperatures, suitable actions should be taken to bring the temperature to the temperature level specified by the manufacturer of the insulating liquid. Moore et al. [36] and Cargill [37] have reported that the same cold startup procedures that are used for power transformers filled with mineral oil can be used when they are filled with natural ester liquids with similar physical characteristics. They have no dielectric issues observed during energization and startup.

5.4.1 Risk Analysis

Risk is the **effect** of **uncertainty** on **objectives** [38]. **Effect** is the deviation from what is expected—positive and/or negative. **Uncertainty** is the lack of information properly assesses an event and its consequences or probability. **Objectives**

Fig. 5.15 Dehydrating breather

are different aspects (finance, health and safety, environment) at different levels (strategic, organization-wide, project, product, and process).

Since the late 1990s the inspection and maintenance approaches in industry have been globally moving from prescriptive, time-based towards risk-based ones. This trend has been clearly driven by the wish to increase the on-stream time, to reduce unscheduled downtime due to breakdown maintenance and/or reduce equipment condition, which would ultimately cause a shutdown or have an undesirable impact on safety. In the present scenario power utilities prefer condition-based maintenance rather than time-based maintenance for power transformers.

Fire safety is a key concern for users of insulating liquids. The protection goals to the minimization of fire hazard arising from the use of electrical insulating liquids are [39]:

- Electrotechnical equipment and systems
- People, building structures and their contents
- Environment.

The practical objective shall be to prevent ignition, but if ignition occurs, to control the fire with respect to heat, toxic smoke, and soot, preferably within the enclosure of the electrotechnical equipment. How this objective is to be accomplished will differ from case to case with different strategies according to the desired level of risk acceptance. Three aspects are important in the decision-making process:

The Economical Aspect: When a mineral oil filled transformer failure results in a fire, the transformer will often be damaged to a degree where repair is not economic. Traditionally, the aim is therefore not to save the transformer if a transformer fire occurs but rather to minimize:

- The replacement cost of the substation
- Damage to neighboring installations through heat radiation and corrosive smoke deposits
- The cost of outage time
- The loss of reputation.

Life Safety Aspect: The safety of the staff and for the population must always be given.

Environmental Aspect: Environmental consideration is a driving engine of the use of alternative insulating liquids. Biological insulating liquids and synthetic esters are biodegradable in the case of a liquid spill. Smoke produced during the combustion of these liquids is less toxic in nature.

Every maintenance action means and causes costs. In modern maintenance, people are moving away from time-based to condition-based maintenance [40]. A maintenance concept is optimal if it guarantees the optimal performance of the assets and enables them to be used for as long as possible at the lowest possible costs while, complying with all safety and environmental aspects.

A risk matrix (Fig. 5.16) helps.

Fig. 5.16 Risk matrix

- Assessing probability (PoF) and consequence (CoF) of failure.
- Defining risk by combining PoF and CoF.
- To show the acceptability and tolerability limits (colors in the matrix—**red**: not tolerable, risk reducing measures required; **yellow**: tolerable but not acceptable, use as low as reasonably possible principle and consider further investigation; **green**: acceptable).
- Represent a color class with a set of strategies, actions, and measures.
- To allow a risk comparison with a useful ranking.

For example, the probability of spillage of a transformer filled with mineral oil and biological insulating liquids is nearly the same. But as the biological insulating liquids are biodegradable, the consequence (P_1) is less compared to mineral oil (P_2) (Fig. 5.17).

Or the consequence of a fire is almost the same no matter if it stems from mineral oil or from biological insulating liquids. But the probability that a transformer filled with biological insulating liquids catches fire is because of the high flash- and fire point practically zero (P_1). In the case of mineral oil filled transformers, the probability is much higher (P_2), which is proven by the statistic, too (Fig. 5.18).

The probability of failure depends on the methods for the PoF-curve. In practice, the bathtub curve is the most commonly used today.

Many transformers fail because their insulation system is no longer able to withstand stresses created during events such as lightning, switching impulse, overload, secondary short circuit, line fault, etc. An event such as those outlined above causes the insulation system to experience localized stress. When the insulation system is

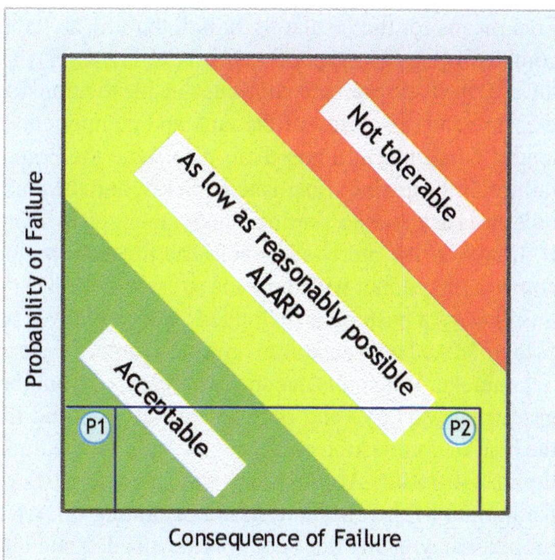

Fig. 5.17 Risk matrix—spillage of insulating liquid

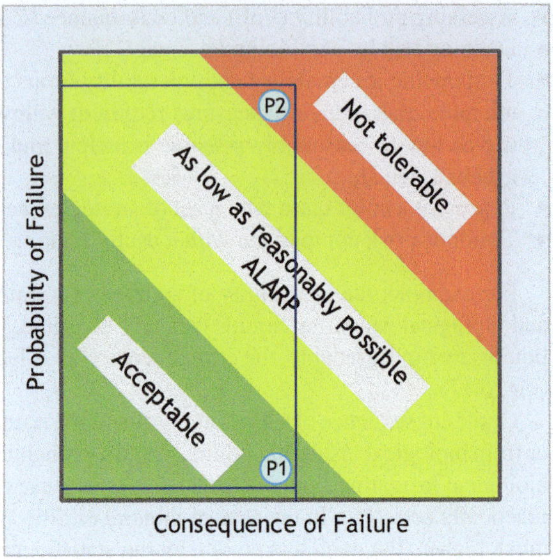

Fig. 5.18 Risk matrix—pool fire

unable to withstand that stress, a breach (typically somewhere in the paper) occurs. From this breach, an arc generates and continues until it is extinguished as either the source is removed, or the distance becomes so large that the dielectric liquid quenches it. How great the fault is depending on how long the arc survives. During the life of the arc, much of the energy associated with the arc is consuming/destroying the materials surrounding the arc. The remaining energy is heating the surrounding materials. This means for the insulating liquid, that the arc cracks their molecules, generating combustible gases (H_2, CH_4; C_2H_6, C_2H_4, C_2H_2) at a substantial rate. As long as the arc survives, pressure continues inside the transformer tank. The rapid buildup of pressure and the ability of the tank and components to withstand the physical challenges being placed upon them determine the potential impact of the catastrophic failure. If the arc is extinguished quickly, and the tank and components withstand the pressure such that no venting occurs, the transformer maybe simply stops operating. If, however, the arc is sustained; the pressure builds to a point where the tank or components cannot withstand the stress, the weakest part of the tank will be compromised (e.g., a bushing is dislodged, or a weld gives way, etc.) and volatile gases will escape. Mixed with air and an ignition source, the gases can explode, causing substantial damage. When this occurs, heat released from the initial burning of combustible gases may vaporize and burn an insulating liquid that is close to its flash point. If the heat of combustion continues to vaporize the liquid, a sustained fire on top of the liquid will result. At this point, the characteristics of the insulating liquid become paramount [41]. Petroleum-based insulating oil, while exhibiting reliable dielectric properties, typically does not provide an adequate margin of fire safety during transformer failure. Insulating liquids with a fire point of approximately 170 °C or less (Sect. 4.3, Table 4.15) requires significantly less heat to reach its fire point compared to less-flammable liquids like biological insulating liquids. The scenarios of fire can be broken down into 3 base cases: pool fire, spray fire, vapor/gas cloud explosion or any combination. Transformers containing insulating liquid can be exposed to an external fire, too.

Pool fire experience with mineral oil-filled transformers has shown that, if the transformer tank is ruptured by a catastrophic failure caused by a high energy internal arc, the insulating liquid can be ejected as a spray. This spray burns intensely for a short time and can itself cause damage, but, in most recorded accidents, a considerable contribution to total fire damage was caused by the high heat release rate from the resulting burning pool of insulating liquid. For this reason, the possibility of a pool fire must be a matter for particular consideration [39]. The fire behavior of a pool fire is generally much less efficient than that of a spray fire. Thus, a pool fire would be expected to produce larger amounts of smoke than a spray fire, while a spray fire would be expected to have a larger instantaneous heat release rate than a pool fire.

Vegetable oils have a comparatively high flash- and fire point. With a fire point >300 °C, vegetable oils are classified as K liquids according to IEC 61039 [42]. Equipment filled with biological insulating liquids has therefore a lower risk potential. This makes it easier to use liquid filled devices (transformers) in sensitive regions, such as in densely populated metropolitan areas. In critical areas, biological insulating liquids are often the most appropriate insulating liquid because of their high fire safety credibility. Safety devices such as fire protection walls, collection pan and fire sprinkling system can be dispensed or need to be built less expensively. The distances to adjacent buildings or to neighbor electrical devices can be reduced [43, 44].

In case of spillage biological insulating liquids don't reach ground water as quickly as mineral oil, as they are more viscose. Cargill recommends using bioremediation to remediate ground spills of Envirotemp FR3 natural ester [45]. To accelerate the process, Cargill advocates adding biomass consuming micro-organisms to the site by spreading active yeast over a spill site and adding water to activate the micro-organisms contained in the yeast. The micro-organisms will consume Envirotemp FR3 fluid by producing water and carbon dioxide, thereby effectively removing it from the environment. The costs are not comparable when mineral oil gets into the ground. Then the contaminated soil must be dredged, treated, and replaced with a new one.

5.4.2 Temperature Monitoring

Transformer solid insulation life depends on insulating liquid and temperature. Above all, the hot spot temperature is an important parameter. Biological insulating liquids allow a higher operating temperature of the transformer. This temperature can be simulated and calculated, but the best way is to measure it directly. In this way, a dynamic load control of the transformer is possible. Fiber optic monitors are designed to monitor fiber optic hot spot temperature sensors installed inside high voltage power transformers. Immunity to electrical interference and the high dielectric constant procured by fiber optic sensors allows direct contact with high voltage components. It is the only technology that monitors the true winding hot spot temperature in real-time [46]. The sensors should be installed to be in contact with the winding's hottest spot. If the correct place is selected, the sensor must be placed carefully between the windings (Fig. 5.19). Care must be taken of making the cable cross the edge of the winding with a shallow angle to have a large bending radius.

The core of any fiber optic cable is made of glass. For this reason, it should be handled with care and should not be viewed like a standard conventional electrical cable. If it is pinched, twisted, bent sharply, or crushed, the glass core will break, and the light attenuation will occur at this point. Table 5.5 compares fiber optic cable with different core diameter.

The use of fiber optics sensors to directly measure temperature of hot spots is already part of the standard IEC 60067-7 [47].

Fig. 5.19 Installation of temperature sensor [46]

Table 5.5 Mechanical properties of fiber optic cable

Parameter	Core diameter 62.5 μm	Core diameter 200 μm
Transmission loss per km	< 3.5 dB	< 8 dB
Bending radius for long term mechanical reliability	> 17 mm	> 28 mm
Bending radius for short term mechanical reliability	> 10 mm	> 17 mm

5.4.3 Biological Liquids—Accompanied Tests

To ensure complete documentation, the quality of the insulating liquid should be determined from delivery to commissioning. Upon receipt, biological insulating liquids meeting or exceeding the limits presented in Table 5.6 are considered to be acceptable.

Table 5.6 Acceptable values for receipt of shipments of unused biological insulating liquids IEEE C57.147 [29], IEC 62770 [6]

Properties	Test Method		Requirement	
	ASTM	ISO/IEC	IEEE C57147	IEC 62770
Flash point [°C]	ASTM D92 [48]	ISO 2719 [49]	≥ 275	≥ 275
Fire point [°C]	ASTM D92 [48]	ISO 2592 [50]	≥ 300	≥ 300
Kinematic viscosity [mm^2/s]	ASTM D445 [51]	ISO 3104 [52]		
0 °C			≤ 500	–
40 °C			≤ 50	≤ 50
100 °C			≤ 15	≤ 15
Pour point [°C]	ASTM D97 [53] ASTM D5949 [54] ASTM D5950 [55]	ISO 3016 [56]	≤ −10	≤ −10
Color	ASTM 1500 [57]		L1.0	–
Relative density [kg/m^3]	ASTM 1298 [58]	ISO 3675 [59] ISO 12185 [60]	≤ 0.96	≤ 1.00
Corrosive sulfur	ASTM D1275 [61]	IEC 62535 [62]	Non corrosive	Non corrosive
DBDS		IEC 62697-1 [63]	–	Below detection limit
Total additives [%]		IEC 60666 [64]	–	≤5
Neutralization number [mg KOH/g]	ASTM D974 [65]	IEC 62021-3 [66]	≤ 0.06	≤ 0.06
Water content at 20 °C [mg/kg]	ASTM D1533 [67]	IEC 60814 [68]	≤ 200	≤ 200
Dielectric breakdown [kV]	ASTM D1816 [69]	IEC 60156 [70]		
1 mm gap			≥ 20	–
2 mm gap			≥ 35	–
2.5 mm gap			–	≥ 35

(continued)

Table 5.6 (continued)

Properties	Test Method		Requirement	
	ASTM	ISO/IEC	IEEE C57147	IEC 62770
Dissipation factor [%]	ASTM D924	IEC 60247		
25 °C	[71]	[72]	≤ 0.2	–
90 °C			–	≤ 0.05
100 °C			≤ 4.0	–
Impulse voltage at 25 °C, 25.4 mm gap [kV]	ASTM D3300 [73]		≥ 130	–
Biodegradability		OECD 301 [74] OPPTS 835.3110 [75]	Readily biodegradable	Readily biodegradable
Oxidation stability (48 h., 120 °C) IEC 61,125 [76]				
Neutralization number [mg KOH/g]		IEC 62021-3 [66]	–	≤ 0.6
Kinematic viscosity at 40 °C, increase over the initial value [mm^2/s]		ISO 3104 [52]	–	≤ 30%
Dissipation factor [%]		IEC 60247 [72]	–	≤ 0.5

In sampling insulating liquid that is contained in apparatus, use care to obtain a representative sample. After the filling is completed and the standing time is also completed, tests on the biological insulating liquid should be made before energization of the transformer. A substantial proportion of electrical equipment is supplied to the final user already filled with biological insulating liquids. In such cases, as the natural ester has already come into contact with insulating and other materials, it can no longer be considered as unused biological insulating liquid as defined in IEC 62770 [6] and IEEE C57.147 [29]. Therefore, its properties shall be regarded separately (Tables 5.7 and 5.8).

Table 5.7 Test limits for unused biological insulating liquids received in new equipment, prior to energization IEEE C57.147 [29]

Test and ASTM method	Value for voltage class		
	≤ 69 kV	> 69 kV < 230 kV	≥ 230 kV
Dielectric strength-minimum [kV] ASTM D1816 [69]			
1 mm gap	25	30	35
2,5 mm gap	45	55	60
Dissipation factor at 25 °C, maximum, ASTM D924 [71]	0.5	0.5	0.5
Color, maximum, ASTM 1500 [57]	L1.0	L1.0	L1.0
Visual examination ASTM D1524 [77]	Bright and clear	Bright and clear	Bright and clear
Neutralization number (acidity), maximum, [mg KOH/g] ASTM D974 [65]	0.06	0.06	0.06
Water content at 20 °C [mg/kg], maximum ASTM D1533 [67]	300	150	100
Fire point [°C], minimum, ASTM D92 [48]	300	300	300
Kinematic viscosity at 40 C, maximum [mm^2/s] ASTM D445 [51]	50	50	50
Total dissolved gas [%], maximum, ASTM D3612 [78]	–	–	0.5

Table 5.8 Recommended limits for biological insulating liquids properties after filling in new electrical transformers and reactors prior energization IEC 62,975 [79]

Test and IEC method	Value for voltage class		
	≤ 72.5 kV	> 72.5 kV ≤170 kV	> 170 kV
Dielectric strength [kV] IEC 60156 [70]			
2,5 mm gap	≥ 55	≥ 60	≥ 60
Dissipation factor at 90 °C, IEC 60247 [72]	≤ 0.07		
Color, ISO 2049 [80] ASTM D1544 [81]	≤ 2.0		
Appearance	Clear, free from sediment matter		
Neutralization number (acidity), [mg KOH/g] IEC 62021-3 [66]	≤ 0.08		
Water content [mg/kg], IEC 60814 [68]	≤ 200	≤ 150	≤ 100
Fire point [°C], ISO 2592 [50]	≥ 300		
Density at 20 °C [g/ml], ISO 3675 [59] ISO 12185 [60]	≤ 1		
Kinematic viscosity at 40 C, [mm^2/s] ISO 3104 [52]	≤ 50		
Total dissolved gas [%], ASTM D3612 [78]	≤ 1.5		

Table 5.9 Suggested limits for continued use of in-service biological insulating liquids [29]

Test and ASTM method	Value for voltage class		
	≤ 69 kV	> 69 kV < 230 kV	≥ 230 kV
Dielectric strength-minimum [kV] ASTM D1816 [69]			
1 mm gap	23	28	30
2 mm gap	40	47	50
Water content at 20 °C [mg/kg], maximum ASTM D1533 [67]	450	350	200
Neutralization number (acidity), maximum, [mg KOH/g] ASTM D974 [66]	0.3	0.3	0.3
Fire point [°C], minimum, ASTM D92 [48]	300	300	300
Kinematic viscosity [mm^2/s] increase from initial value [%]	≤ 10	≤ 10	≤ 10

IEEE acceptable limits for in-service biological insulating liquids [29] and [82] are shown in Table 5.9 and for IEC 62975 in Table 5.10. The values for kinematic viscosity in Table 5.9 were corrected and adapted. IEC 62975 makes no differences in transformer classes but distinguishes between good, fair, and poor. However, a fire point falling below 300 °C should not impact the functionality of the transformer.

Table 5.10 Recommended limits for in service biological insulating liquids in transformers [79]

Property	Good	Fair	Poor
Dielectric strength [kV] IEC 60156 [70] 2,5 mm gap	> 60	50–60	< 50
Dissipation factor at 90 °C, IEC 60247 [72]	< 0.15	0.15–0.3	> 0.3
Oxidation inhibitor content [%] of original value	> 70	30–70	< 30
Color and appearance ISO 2049 [80]	Clear and without visible contamination	Darker than as new and/or pale and/or appearance of turbidly	
Neutralization number (acidity), [mg KOH/g] IEC 62021-3 [66]	< 0.3	0.3–0.5	> 0.5
Water content [mg/kg], IEC 60814 [68]	≤ 100	100–300	> 300
Fire point [°C], ISO 2592 [50]	≥ 300	< 300	
Flash point [°C], ISO 2592 [50]	≥ 250	< 250	

(continued)

Table 5.10 (continued)

Property	Good	Fair	Poor
Interfacial tension [mN/m] IEC 62961 [83]	> 20	14–20	< 14
Kinematic viscosity at 40 C, [mm^2/s] ISO 3104 [52], increase over original value [%]	< 10	10–15	> 15

5.4.4 *Oxidation Stability*

All practical measures should be taken to avoid continuous, long-term exposure to unlimited air exchange, particularly at operating temperatures. Oxidation essentially occurs only at the surface of the biological insulating liquid exposed to air, so thin film exposure is of greater concern than liquid in the transformer tank. It is recommended to minimize the time and temperature that surfaces with films of biological insulating liquids are exposed to air.

Different types of biological insulating liquids formulations can have different recommended air exposure time limits depending on the type of base vegetable oils and the type and amount of oxidation inhibitors.

Biological insulation liquids oxidize differently than mineral oil. Oxidation causes polymerization of the liquids, forming larger molecules that remain in solution, the rate of which is highly temperature dependent. Conversely, mineral oil oxidation forms reactive short chain acids and carbonaceous sludge, which can precipitate out of the oil. However, if the surface of the bulk liquid is exposed to continual exchange of air in the head space over several years, the viscosity of biological insulating liquids can measurably increase, resulting in lowered cooling capability of the liquid. This may necessitate some type of corrective action. Such polymerization of biological insulating liquids in the transformer tank should not impact the dielectric strength of the insulation system.

5.4.5 *Accelerated Aging*

In electric power transformers, the insulating liquids can be submitted to extreme stress conditions like electrical stress and high temperature. It is known that the thermal stress combined with oxygen degrade the triglycerides and produces free fatty acids, which increase the insulating liquid acid number. In this context, Evangelista et al. have done accelerating aging tests to evaluate the stability of jatropha curcas oil without antioxidants, and with 1% and 2% of tert-butylhydroquinone (TBHQ) [84]. This antioxidant is considered, in general, more effective than butylated hydroxy anisole (BHA) and butylated hydroxytoluene (BHT), other commonly antioxidants used in vegetable oils (see Sect. 2.7.1). TBHQ is moderately soluble in oils and fats, and to not complexes with iron or copper.

The results of the fatty acids profiles are presented in Table 5.11. Jatropha curcas oil shows that after the aging without antioxidant, the percentages of oil in the sample and polyunsaturated fatty acids in the oil have both decreased considerably. Obviously, a large part of the triglycerides was destroyed, which could be reduced by adding antioxidants.

The aged oil without antioxidants presents a higher acid number and viscosity than the oil before aging (Table 5.12). It is known that unsaturated vegetable oils are more prone to oxidation. This oxidation increases the free acids amount because the triglyceride molecule cleavage, leads to hydrogenation of unsaturated vegetable oils, and polymerization of triglycerides. Therefore, the increase in acidity and viscosity observed for the oil without antioxidants demonstrates that this oil was considerably oxidized during its accelerated aging. This result is expected, and it confirms the necessity of adding antioxidants when using as insulating dielectric insulating liquid.

The work from Evangelista et al. also shows the influence of the concentration (using 1% and 2% per weight TBHQ) of antioxidant to biological liquids. It's a common range using antioxidant in vegetable oils. As shown in Tables 5.11 and 5.12, the addition of 1% TBHQ was not able to prevent oil oxidation and even can cause a negative effect. On the other hand, the addition of 2% TBQH was able to prevent partly oxidation of the jatropha curcas oil during its accelerated aging. As shown in Table 5.11, reduction in the percentage of oil in the sample and polyunsaturated fatty acids in the oil were considerably smaller for the oil with 2% of TBHQ than that for the oil without antioxidants. In addition, acid number and viscosity determined for

Table 5.11 Fatty acid profile of jatropha curcas oil tested in accelerated aging

Percentage of fatty acid	Before aging	After aging (140 °C, 96 h)		
Kind of acid	Without antioxidant	Without antioxidant	TBHQ 1%	TBHQ 2%
Myristic (C14:0)	0.1	0.4	0.8	1.0
Palmitic (C16:0)	14.7	20.8	22.5	18.5
Stearic (C18:0)	5.9	7.4	9.0	7.0
Total of saturated fatty acid	**20.7**	**28.6**	**32.3**	**26.5**
Palmitoleic (C16:1)	1.2	1.4	1.6	1.4
Oleic (C18:1)	37.1	43.8	44.9	40.1
Total of monounsaturated fatty acids	**38.3**	**45.2**	**46.5**	**41.5**
Linoleic (C18:2)	39.2	22.0	17.8	29.2
Total of polyunsaturated acids	**39.2**	**22.0**	**17.8**	**29.2**
Percentage of oil in the sample	**100**	**69.7**	**72.4**	**94.3**

Table 5.12 Viscosity and acid number of oils tested in accelerated aging

	Before aging		After aging (140 °C, 96 h)	
	Acid number [mg KOH/g]	Viscosity at 40 °C [$mm^{2/}s$]	Acid number [mg KOH/g]	Viscosity at 40 °C [$mm^{2/}s$]
Without antioxidant	0.089	42.7	1.529	182.8
1% TBHQ	0.136	40.2	2.002	453.4
2% TBHQ	0.152	39.9	0.779	82.2

the oil with 2% of TBHQ after aging were both smaller than that found for the oil without antioxidants, as present in Table 5.12.

It is the responsibility of the manufacturer of the insulating liquids to inform the end user about the type of additives used and their required concentration. Only if the type and the required concentration level of the additives used are known, they can be checked during in service to guarantee a safe operation of the device.

5.4.6 *Transport of Lost Energy (Heat Transfer)*

Heat transfer is one of the essential functions that must be provided by the insulating liquid because of the high thermal stresses by electrical losses usually found in power transformers. The insulating liquid is used to carry the heat generated in the transformer near the coils, the core, and circuits to the cooling surfaces out of the device. Heat transfer by the dielectric insulating liquids reduces the temperature in transformer windings under both normal and emergency overload conditions. This, in turn, prolongs the life of the transformer by minimizing the rate of degradation of the cellulosic paper and pressboard used as the primary electrical insulation in the winding. Heat transfer is influenced by both thermal conductivity and by natural or forced convection. The convection represents all the properties that lead to heat transfer by liquid displacement (viscosity, specific heat, thermal expansion coefficient), whereas the thermal conductivity is realized within a liquid displacement. The heat transfer can be evaluated by the following relationship (Eq. 5.6) [85].

$$P = \frac{C_p * \lambda * \alpha}{\nu} \quad (5.6)$$

In the equation (Eq. 5.6) P means the cooling criteria, C_p is the specific heat, λ is the thermal conductivity, α is the thermal expansion coefficient, and ν is the kinematic viscosity. The knowledge of these properties of the dielectric insulating liquids used in a transformer, at the various operating temperatures, is essential for transformer design engineers to decide:

a. The type of liquid circulation (natural or forced convection, with the addition of liquid circulation pumps).
b. The size and configuration of the cooling ducts that will provide the adequate liquid flow to dissipate the heat generated.
c. The thermal time constants of the equipment, under normal and overloading conditions and in case of accidental forced-pumping failures [86].

P is a comparison criterion enabling the analysis of different liquids. Figure 5.20 shows the parameter *P* versus temperature of dielectric insulating liquids, of which the necessary data were completely available. The higher the *P* value, the better the heat transfer. The bio-based dielectric insulating Nynas Bio 300X has the best cooling behavior, followed by the mineral oil Nynas 4000X and the ester liquids. Within the ester liquids, the difference is not very great. By comparing *P* for the different characteristics appearing in equation (Eq. 5.6), it is observed that the kinematic viscosity is the most influential parameter for the transfer of heat.

The partially increased kinematic viscosity of the ester compared to mineral oil must be considered in the design of the devices (transformers). However, in the upper temperature range, where the cooling properties become more important, the differences between biological insulating liquids and mineral oil are no longer so great. It is crucial that the Reynolds number (Eq. 5.7) under the operating conditions of the transformer should be in the turbulent range (> 2300).

$$N_{Re} = \frac{\rho * v * d}{\eta} = \frac{v * d}{\nu} \tag{5.7}$$

ρ Density [kg/m^3]
v Velocity [m/s]

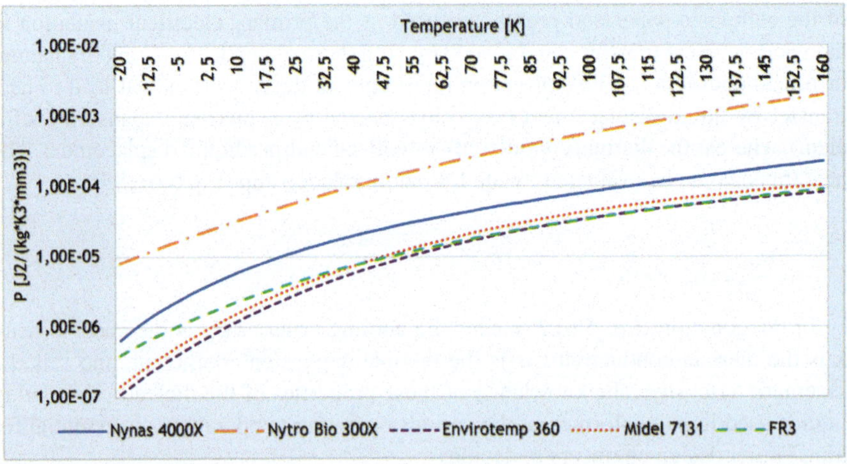

Fig. 5.20 Heat transfer [P]

d Characteristic linear dimension [m]
η Dynamic viscosity [kg/m*s]
ν Kinematic viscosity [m^2/s].

Maneerot et al. studied the thermal characteristics of natural ester (FR3) immersed transformer with mineral oil and palm oil immersed transformer [87]. They designed and constructed three single-phase transformers with a ratio of 400/1000 V and a load of 10 kVA. After physiochemical checks of the insulating liquids and impregnated solid insulation, transformers were operated at 100% of designed rated with inductor load for four months by which the temperature, current and voltage were recorded every hour. The results for the thermal behavior show that liquids with a higher viscosity have a higher increase in temperature (Fig. 5.21). But the difference is not very much and is nearly negligible for this kind of transformer.

With a suitable adaptation of the ducts of the inner cooling circuit in the active part, plus an adaptation of the pipes and the cooling fins in the outer area, the problem can be managed well for most of the transformers. Usually, the specific heat capacity and the thermal conductivity which defines the ability of a material to conduct the heat from a point to another point of its bulk, are for vegetable oils higher than that of mineral oil, which means that higher energy losses can be dissipated with the same volume. The higher the thermal conductivity, the more uniform the temperature of the insulating liquid will be in the transformer. It means that heat will be conducted away from the coils easier and thus would help to prevent hot spot. The higher temperature performance of the biological insulating liquids can be used to run the transformer at a higher temperature level by using less insulating liquid and construction materials. This higher exergy can be very helpful if the waste energy is used for further process heat like district heating, etc.

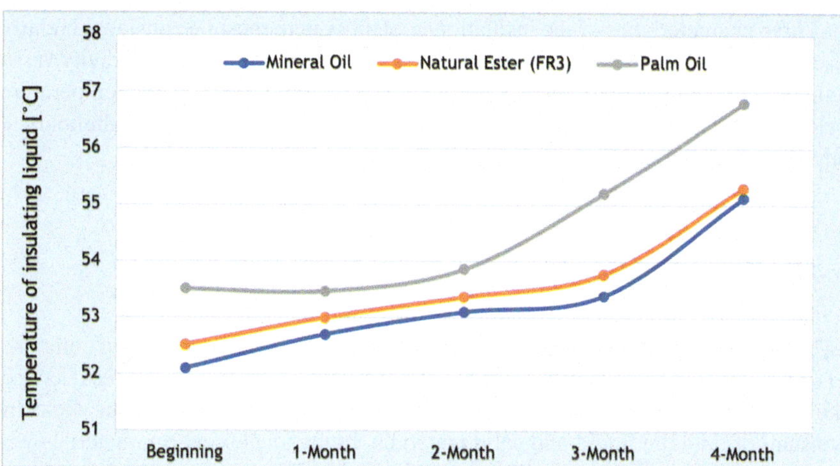

Fig. 5.21 Thermal profile of the insulating liquid during the trial period

5.4.7 Cold Temperature Behavior

Inside of a transformer, the temperature of the fluid depends on ambient temperature, volume of liquid, time at ambient temperature, rate of cooling, and the electrical load. As the liquid temperature decreases, the relative viscosity increases, potentially affecting cooling. The pour point of the dielectric insulating liquids differs in a wide range. The pour point of petroleum-based liquids and synthetic esters is approximately −60 °C and from biological insulating liquids −20 °C, respectively. However, pour point is simply a test that compares relative cold flow properties of liquids, but does not determine the applicability of insulating liquid in a transformer application, and is not indicative of performance below its pour point. The pour point can be useful for liquid type identification and for determining the type of equipment in which it can be used.

Poly methyl methacrylate lowers the pour point of the insulating liquid (see Sect. 2.7.2). Usually, the concentration is less than 1%. Poly methyl methacrylate works by suppressing the development of large crystals during solidification of the insulating liquid, suspending nucleation, modifying the size and formation of crystals, co-crystallization, or polymorphism. Ghani et al. show in [35] a significant improvement in pour point behavior of the vegetable oils by adding 1% of poly methyl methacrylate (Fig. 5.22) and an insignificant effect on the vegetable oil viscosity (Fig. 5.23).

HOSO	High oleic sunflower oil
SO	Sunflower oil
CO	Castor oil
SYO	Soybean oil
RO	Rapeseed oil.

Three characteristics of the insulation/coolant system must be considered relative to cold start. These are dielectric strength versus temperature, specific gravity versus temperature, and the thermal characteristics of the liquid. For start-up temperatures below −20 °C, the transformer should be held at no-load for at least eight hours for any insulating liquid [88].

5.4.8 Electrical Stress

AC fields depend on the permittivity distribution. In an insulation system composed of liquid and solid insulating materials, the distribution of electrical stress between the solid and liquid dielectrics is governed by the relative values of the dielectric constant of both the liquid and solid materials. In the ideal scenario, materials with the same dielectric constant will be used for both solid and liquid insulation since, this provides an even distribution of stress across the whole system. The electrical stress is inversely proportional to the dielectric constant value, hence the material

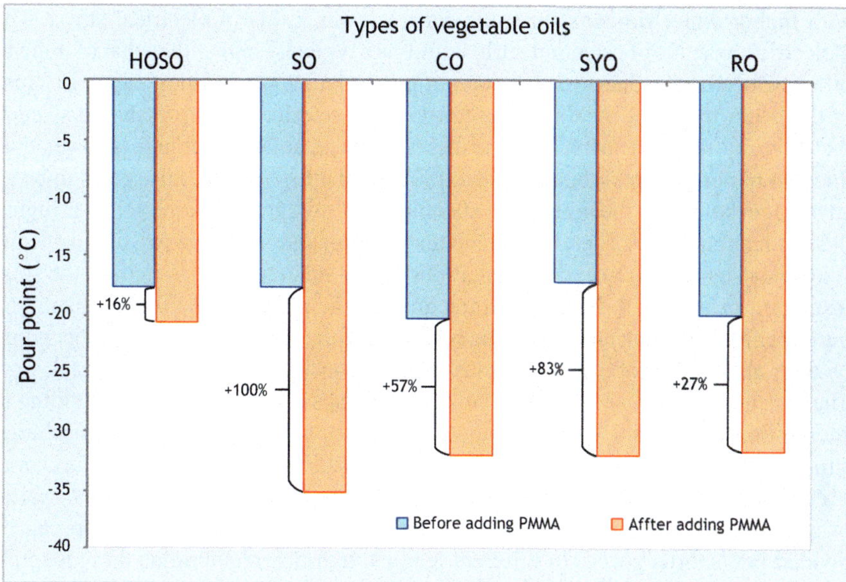

Fig. 5.22 Improvement in pour point behavior of vegetable oils

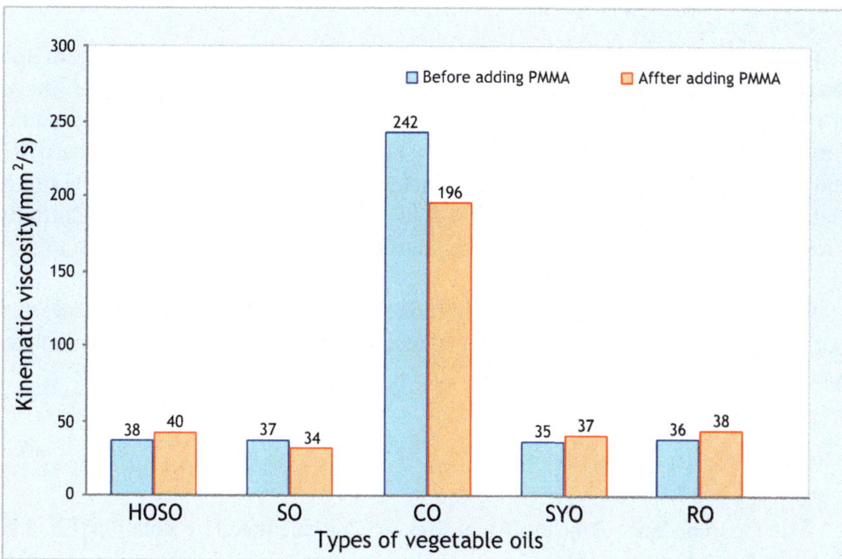

Fig. 5.23 Modification of the viscosity of vegetable oils by adding 1% poly methyl methacrylate

with higher dielectric constant value carries lower levels of electrical stress. The dielectric constants of ester dielectric liquids are typically higher than that of mineral oil and closer to the dielectric constant of pressboard and insulation paper. Because of the better match of the dielectric constants in an ester liquid/pressboard system, the disparity in the distribution of stresses between solid and liquid is lower than that in a mineral oil/pressboard system [89]. On the first glance, this seems to be an advantage, but it also means that the absolute field strength inside the solids is higher than with mineral oils. Therefore, inadequate impregnation or little cavities inside the solids have a more distinct detrimental impact on the dielectric strength of the solid insulating material [9]. Note that impregnating the paper with different insulating liquids slightly changes the dielectric constants of the impregnated paper, depending on their absorption capacity. This must also be considered when looking at design. The dielectric strength of wide insulating liquid gaps can be enhanced by dividing in shorter distances. Transformer manufactures can attain smaller volume of insulation structure in this way. If the gaps are subdivided, large statistical scatter of discharge occurrence will be reduced to the scatter of limited gaps. Because of this, scatter of complete system is smaller. In non-uniform field, insulating liquid gap can be divided into shorter spaces of different lengths. In order to determine these lengths, designers have to use electrostatic analyze programs and special design curves. At the same time the barriers must guarantee mechanical stability of the system. The relative permittivity of dielectric insulating liquids is not affected by accelerated aging processes [26].

In [90], Pagger immersed pressboard with a thickness of 2 mm with different insulating liquids. Before testing, the pressboard was dried under vacuum (1300 Pa) at 105 °C for 24 h. The intake of insulating liquid into a pressboard is for all liquids very similar. With a dielectric constant of 4.1 (50 Hz, 20 °C) for a dry pressboard [91] and the ones of the insulating liquids (Sect. 4.4.1, Table 4.19), a theoretical mixed dielectric constant is calculated (Table 5.13). Due to the difference in permittivity, stress in ester liquids is up to 7% lower and stress in solid insulation is up to 30% higher than in mineral oil.

With the simplified version of Lashbrook [92], the distribution of electrical stress can be calculated as follows. The impedance of the insulation system is calculated according to the equation (Eq. 5.8).

$$Z = \frac{d_p}{\varepsilon_p} + \frac{d_l}{\varepsilon_l} \tag{5.8}$$

With the impedance Z the electrical stress can be calculated by equation (Eq. 5.9).

$$E_n = \frac{U}{\varepsilon_n * Z} \tag{5.9}$$

The Finite Element Method (FEM) has been used by King Mongkut's Institute of Technology Ladkrabang to solve the electric field problem. The schematic model and constant parameters used in the simulation are summarized in Table 5.14.

Table 5.13 Composition of liquid immersed pressboard

Insulating liquid	Pressboard [%]	Liquid [%]	ε_r (50 Hz, 20 °C) Liquid	ε_r (50 Hz, 20 °C) Immersed pressboard	Liquid stress [kV/mm]	Paper stress [kV/mm]	Stress difference liquid-paper [kV/mm]
Nynas 4000x	75.9	24.1	2.18	3.64	11.70	7.01	4.69
FR3	67.7	32.3	3.15	3.79	10.88	9.04	1.84
Biotemp	70.0	30.0	3.11	3.80	10.92	8.94	1.98
Midel eN	69.5	30.5	3.12	3.79	10.91	8.98	1.93
Midel 7131	66.8	33.2	3.18	3.79	10.86	9.11	1.75
BecFluid	72.4	27.6	3.24	3.86	10.86	9.11	1.75
Silicone oil	71.2	28.8	2.68	3.69	11.24	8.16	3.08

Z Impedance [Ω]
d_p Thickness of pressboard [mm]
d_l Insulating liquid gap [mm]
ε_p Permittivity of impregnated pressboard [F/m]
ε_l Permittivity of insulating liquid [F/m]
ε_n Permittivity of the multi-layer system [F/m]
U Applied voltage [kV]
E_n Electrical stress in the paper-liquid system [kV/mm].

Table 5.13 and Figs. 5.24, 5.25, 5.26, 5.27, 5.28, 5.29 and 5.30 show the difference between insulating liquid and immersed paper under electrical stress. For this calculation, a 2 mm thick pressboard, a 5 mm liquid gap and an applied voltage of 72.5 kV were used. The difference in electrical stress between pressboard and ester liquids is lower compared to mineral oil and silicone oil, indicating a more even distribution (Fig. 5.31). Based on the point, it is important for designer to consider the permittivity difference between mineral oil and biological insulating liquids bringing benefits in certain areas of the insulation structure but require changes to reduce electrical stress in others. Biological insulating liquids performance is poorer for the field around sharp electrodes and the lightning impulse breakdown voltage of biological insulating liquids can be up to 20% lower compared to mineral oil [93].

Pay attention—the differences in the liquid's electrical withstand properties between ester and mineral oil are not to be neglected. Specifically, in the event of fast transient impulse stresses, esters tend to exhibit a significantly lower withstand capability in comparison to mineral oil [94]. However, if this is understood, then it is possible to re-design to reduce stress in key areas, usually at the winding ends and around static rings. But transformers using vegetable oil are also better equipped

Table 5.14 The schematic model and parameters used in the simulations

Schematic model	Constant parameters	
	Material	Relative permittivity
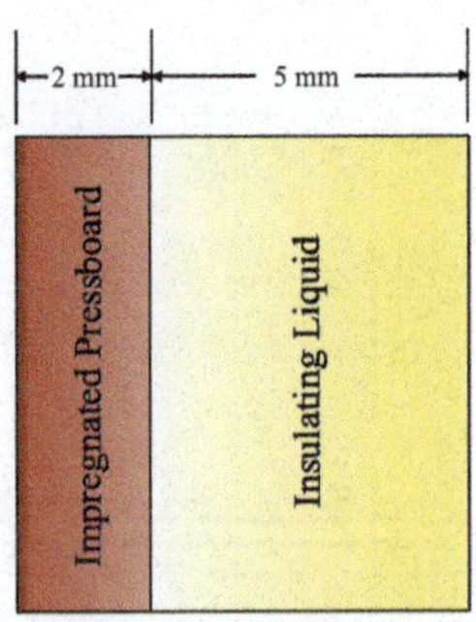 Multi-layer insulation model between impregnated pressboard and insulating liquid	Impregnated pressboard	ε_p
	Insulating liquid	ε_f

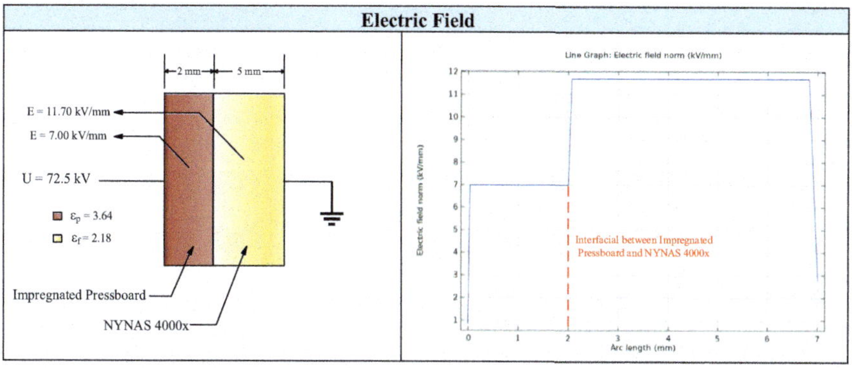

Fig. 5.24 Multilayer insulation model between impregnated pressboard and mineral oil NYNAS 4000X

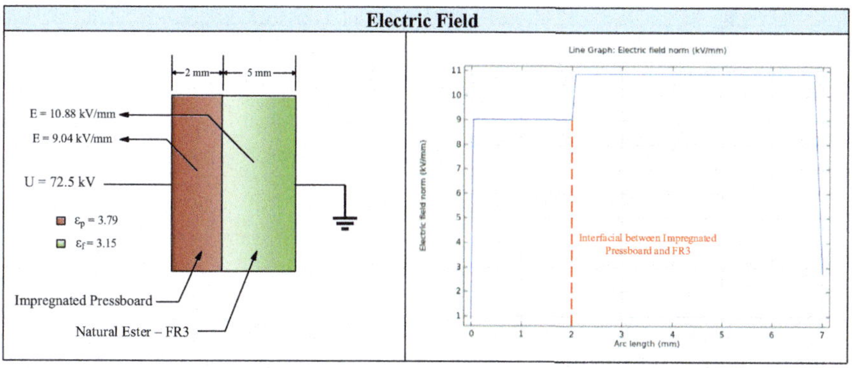

Fig. 5.25 Multilayer insulation model between impregnated pressboard and natural ester FR3

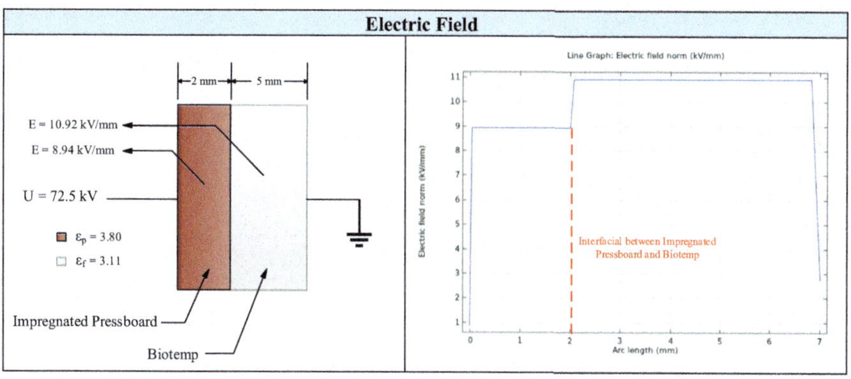

Fig. 5.26 Multilayer insulation model between impregnated pressboard and natural ester Biotemp

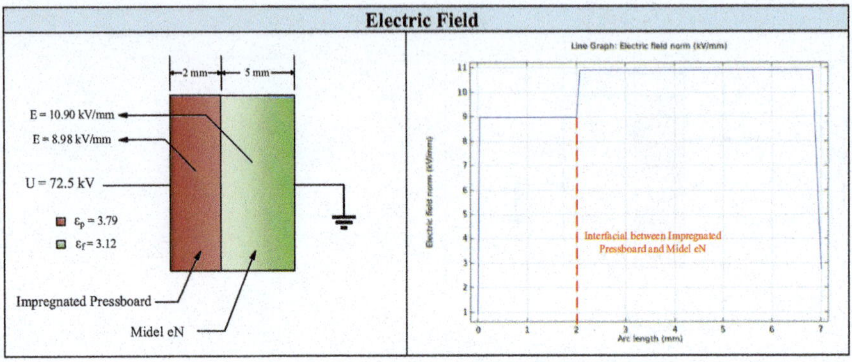

Fig. 5.27 Multilayer insulation model between impregnated pressboard and natural ester Midel eN

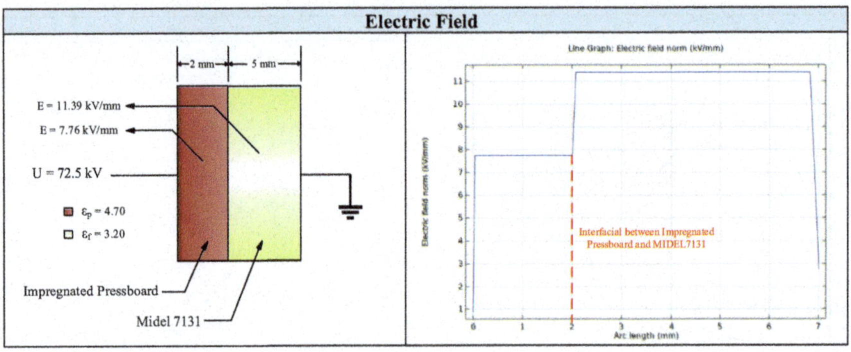

Fig. 5.28 Multilayer insulation model between impregnated pressboard and synthetic ester Midel 7131

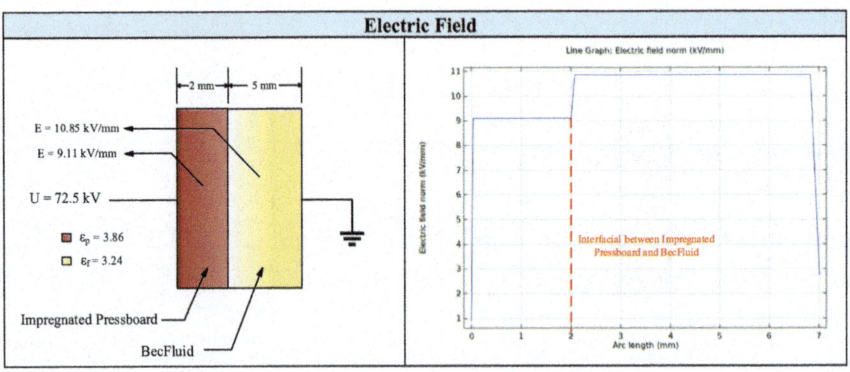

Fig. 5.29 Multilayer insulation model between impregnated pressboard and synthetic ester BecFluid

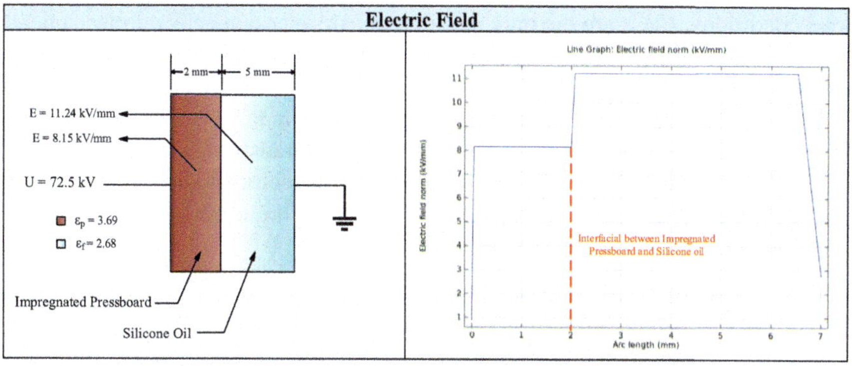

Fig. 5.30 Multilayer insulation model between impregnated pressboard and Silicone oil

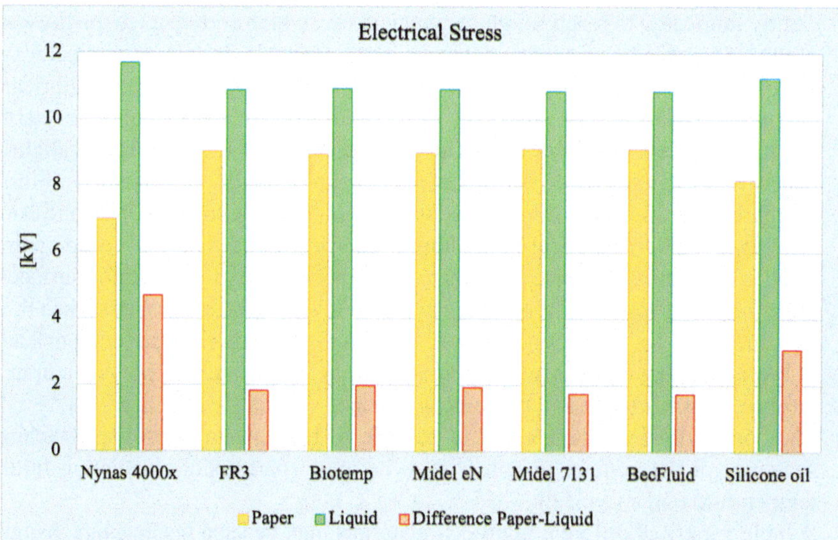

Fig. 5.31 Electrical stress comparison

than mineral oil transformers to absorb problems in the grid, things like capacity overloads. They fail less [95].

5.4.9 *Switching Technology*

Regulated power transformers are equipped with load tap changers (LTC) to change their ratio and, subsequently, adapt the transformer output voltage to the respective conditions. This enables the power supply network to be kept stable under changing load conditions. LTCs are the only part in a transformer that actively mechanically operates. Tap changers are complex electro-mechanical devices that also should adapt to high voltage conditions. This combination makes them unique components in energy supply technology. The frequency of failure in tap changers is higher than failures in transformers. While insulating liquids in a transformer have dual tasks of cooling and, in combination with the solid insulation electrically insulating windings and bushings against high voltage, a suitable insulating liquid for tap changers should also fulfill the following recommendations [29]:

- The switching arcs should be cooled and quenched by the surrounding liquid. The cooling behavior of ester fluids is slightly different from that of mineral oil. This affects the temperature rise of contacts and resistors and may lead to a slight reduction in the maximum allowable through-current to reliably avoid overheating of the contacts [96].
- All mechanically moving parts (gears, selector contacts, etc.) should be sufficiently lubricated to reach a high mechanical life, which correlates to the lifespan of the transformer. Lubricating behavior is an extremely important parameter for tap-changers. Vegetable oils present good potential as high temperature lubricant due to their structure and properties [97]. The triglyceride structure of vegetable oils provides desirable qualities in a lubricant comparable to mineral oil or slightly better. Frotscher reported [96] that mechanical endurance tests over 1.5 million switching operations with high molecular weight hydrocarbon (HMWH) and ester fluids have shown excellent lubricating properties. Long polar fatty acid chains provide high strength lubricant films that interact strongly with metallic surfaces, reducing both friction and wear. Vegetable oils make the best lubricants when they have high levels of monounsaturated acids. High saturated fatty acids content leads to low cold behavior of insulating liquids, while high content of polyunsaturated fatty acids like linoleic and linolenic are prone to oxidative stability.
- The spring driven switches should be able to help ensure a proper switching sequence of the contact system within the entire permissible insulating liquid temperature range (typically −25 °C to +125 °C).
- A large variety of different high-tech materials may be used inside a tap changer to achieve high electrical and mechanical functionality and a long working life; all of those should be compatible with the insulating liquid used.

Thus, the dielectric, mechanical, chemical, and thermal properties of the insulating liquids should be evaluated.

For example, the higher viscosity can limit tap changer operation in cold insulating liquids. Biological insulating liquids require prevention of continuous contact with air (sealed tank design) to avoid oxidation and subsequent increase in the viscosity of the insulating liquid due to polymerization of the ester molecules. A tap changer has a limited amount of spring force to operate the diverter mechanism. If the insulating liquid is too viscous, the switching operation begins but cannot be completed, a failure could be a result. Viscosity also affects the arc quenching behavior and can lead to a reduced switching capacity.

Biological insulating liquids also show different breakdown behavior in long gaps in highly non-uniform electrode configuration such as needle to plate and needle to sphere. This is due to a different streamer propagation mechanism that exists at voltages higher than the partial discharge inception level. Fast streamer can develop at significantly lower voltages compared to mineral oil. This effect is particularly visible in the case of inhomogeneous electrode arrangements [96]. They have a long stopping length, so there is a significant possibility they can bridge long insulating liquid gaps and cause a breakdown. This can occur under impulse voltage at inhomogeneous electrode configurations with uncoated electrodes, as commonly applied on tap changers. The geometric shape of tap changer electrodes and contacts is the result of a compromise between mechanical function, mechanical endurance, number of taps, required load current, and electrical insulation. The established compromises for mineral oil cannot be assumed to be optimal for other types of insulating liquids. For LTCs working in insulating liquid surroundings parameters like liquid temperature difference between the main tank and LTC compartment, DGA, motor drive torque, motor drive current index, operating time, contact wear or vibro-acoustic signature can be a good indicator to detect any abnormalities.

The arc extinction of conventional tap changers takes place in the tap changer insulating liquid. This results in oil contamination and shorter service intervals. Therefor a new switching technology has to be adopted to combine the lifetime advantages of the hermetic transformer design with the switching performance. This new technology uses vacuum interrupters, which fully encapsulate the switching arc. Because of the extremely low internal pressure in vacuum interrupters, only a small gap of a few millimeters is required to achieve a high dielectric strength. The short arcing time generates less energy, which, together with the high rate of metal vapor recombination, minimizes the contact wear. Therefore, vacuum interrupters significantly improve the switching capability in respect of numbers of operations and values of currents, which have to be interrupted.

The tap changer is used to change the turn ratio between windings in the transformer. This ratio determines the voltage ratio between the windings and is essential for the stabilization of network voltage under variable load conditions. An "On load tap changer (OLTC)" normally has a regulation range of $\pm 20\%$ of the rated line voltage. Regulation is performed in roughly 9 to 35 steps and operates 10 to 20 times a day in normal grid applications. Due to the fact that during the switching operations no oil contamination occurs, and constant service condition are maintained over the whole lifetime, the service interval could be extended from 50,000 to 100,000 up to 300,000 operations independent of the kind of insulating liquid.

When calculating and comparing the cost of maintenance in the overall life-cycle costs for conventional and new transformers, it becomes evident that hermetically sealed transformers—always combined with vacuum based OLTC—offer considerable savings in addition to reduced paper and oil aging, write Fink et al. [32].

5.4.10 Dielectric Response Measurement

The dielectric response of a transformer is influenced by the conductivities of the materials (liquid and solid insulation). As the conductivities of insulating liquids vary in a great range (being much higher for e.g., esters than for mineral oil liquids), one cannot use the characteristics and interpretation rules for mineral oil insulated transformers for units insulated with other insulating liquids. However, the methods are in principle generally applicable once based on the appropriate understanding [22].

5.4.11 Sustainable Peak Load Transformers

A transformation of the power supply to renewable sources of energy, accompanied by a substantial increase in energy efficiency, is the appropriate strategy for sustainable peak load transformers filled with biological insulating liquids. Considering all properties, biological insulating liquids can increase the transformer overloading capability, its performance, its useful life (about 40% higher), and can decrease its rate of failure. Its use requires nor or just minor modifications and it is compatible with the existing electric power infrastructure (see Sect. 6.6.1). A reduction of the insulating liquid volume and, in consequence, a smaller transformer size can be achieved as a result of optimized insulation system.

5.5 Condition Monitoring and Diagnosis

Liquid immersed transformers are one of the most expensive and essential equipment in power systems. During operation, liquid immersed transformers are subject to various stresses, like mechanical, chemical, thermal, environmental, and electrical stresses. Eddy currents and stray losses may induce very high temperatures (700–1000 °C) in transformer cores. High energy arcing in the high voltage windings and towards the transformer tank, core, and low voltage windings, with arc temperatures above 6000 °C are observed as well as lower energy arcing like sparks in the insulation liquid or tracking along the paper insulation. Carbon particles are formed because of arcing and discharges in the insulating liquid phase; these are particularly evident in circuit breakers and tap-changers, where insulating liquid has an arc-quenching function. High molecular weight compounds are formed when a gas phase containing insulating liquids vapors is subjected to ionization and corona discharges. All these stresses can lead to aging and deterioration of their insulation system. In addition, moisture from insulation degradation and coming from external environment can accelerate the aging process and reduce the dielectric strength. In today's increasingly large and instable power demand, if a power or large distribution transformer fails, it will likely cause interruption of power supply to the energy system and bring significant economic losses. Therefore, quickly and accurately diagnose the type of faults during transformer operation has become an important issue in promoting the grid process. The condition of the electrical device can be assessed via several chemical, physical, and electrical parameters. Parameters that essentially also apply to the assessment of new insulating liquids are described in Chap. 4. The inspection of biological insulating liquids should follow a similar procedure as those being used for mineral oil. As described in Chap. 2, the ester linkage is somewhat polar and this leads to attributes such as higher water solubility, higher tan δ and lower volume resistivity compared to mineral oil liquids. It also means that some parameters routinely monitored for mineral oil are less useful in assessing the condition of ester insulating liquids. For example, there is no proven link between interfacial tension value and fluid condition in ester liquids, despite this being a commonly used measurement for mineral oil. The lower variation of the interfacial tension in natural esters does not affect the insulating liquid performance, but it limits the use of interfacial tension as an indicator of insulating liquid degradation. The assessment of aging markers in insulating liquids to evaluate the condition of a transformer and, especially, the indirect assessment of the solid insulation, is depending on the type of sealing system, as well as on the insulating liquid.

p-Anisidine value, developed for edible oils, may also prove to be a useful indicator of the condition of aged biological insulating liquids used in transformers. The method [98] determines principally the amount of aldehydes (2-alkenals and 2,4-dienals) in biological insulating liquids. The aldehydes are formed as byproducts during oxidation of the vegetable oil base.

5.5.1 Measurement Accuracy

Measurement performance is defined by dynamics like measurement range, response time, accuracy, and stability. Another parameter related to measurement performance is the lowest detection limit, which is the lowest concentration that the measurement instrument can reliably distinguish from zero. Sensitivity is the relationship between the change in measurement output and the change in measured quantity. In practice, all measurements involve some imperfections or uncertainty. Accuracy is the agreement of the measured value with the true value. Mostly, the true value is not known precisely.

The specified accuracy may or may not include repeatability, which is an instrument's ability to provide a similar result when the measurement is repeated under constant conditions. Figure 5.32 shows the relationship between accuracy and repeatability.

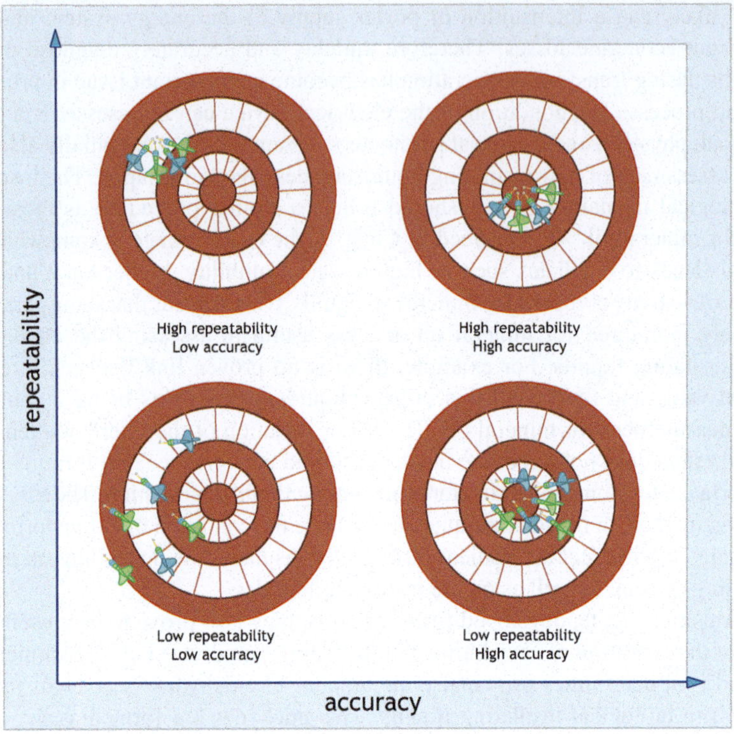

Fig. 5.32 Showing accuracy and repeatability schematically

5.5.2 *Insulting Liquid Sampling*

The biggest uncertainty source for the analytic quality is usually related to the quality of the insulating liquid sample. Sampling should be performed after as long as possible period of stable operation. Sampling performed right after a large change in load and/or temperature—except for special reason—should be avoided. To maximize the liquid homogeneity, it is recommended, sampling the transformer's liquid whilst it is on load, which facilitates the markers being in equilibrium between the two phases (solid insulation/liquid). It's important to know how it was collected—under which surrounding circumstances (weather, temperature, load, etc.). Storage and transport can have an influence on the quality of the insulating liquid. A significant amount of hydrogen can escape from the insulating liquid and humidity and ambient air gases like oxygen and carbon dioxide can contaminate the sample, resulting in inaccurate laboratory analyzes. Details for dissolved gas analyzes sampling are described in Sect. 5.11.1.

5.5.3 Color

Color is a rapid and easily performed quality control technique, which in some cases may indicate contamination of the insulating liquid and can reflect the extent of the oxidation very qualitatively. The visual change is more pronounced in the open system than in the closed system, due to increased oxidation processes. While an increase in color number during service is an indicator of liquid deterioration or contamination in mineral oil, this may not be the case for biological insulating liquids. Biological insulating liquid manufacturers may add clear colorants for identification purposes. If insulating liquids have their own color, this must be considered. In the case of the natural ester FR3, for example, the added green colorant fades as the insulating liquid ages.

5.5.3.1 Color—Test Method

Color is determined by ASTM Test Method D1500 [57] or ISO 2049 [80] by comparing the liquid sample with colored glass disks ranging in value from 0.5 to 8.0. Visual examination of oil, according to ASTM Test Method D1524 [77], is also useful to determine whether a sample should be sent to a laboratory for further investigation.

5.5.4 Acid Number

Acids will represent a very wide range of acid strength. However, it has demonstrated that the strongest and most polar acids, such as low molecular weight carboxylic acids, as they are much more reactive, are those that will have the greatest influence on paper degradation. In mineral oil, the formation of acidic components is commonly associated with oxidation. Acids are formed at a relatively late stage in the oxidation process. In biological insulating liquids, acidic components arise from diverse processes and may not be associated with adverse effects. The acidic components are produced mainly from hydrolysis, pyrolysis, and oxidation of biological insulating liquid. The presence of dissolved water in the liquid facilitates the hydrolysis reaction. It is common for the acid number that it significantly increases during the first months of transformer operation, attributable to the hydrolysis reaction with the initial moisture from the insulating paper (Eq. 5.10).

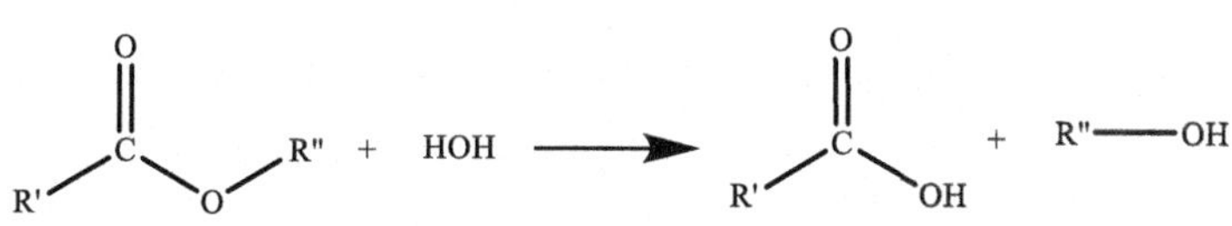

(5.10)

Pyrolysis of the ester bonds also yields fatty acids. The heat that causes the release of the fatty acids also causes some of them to break down further. Consequently, an increase in acid number from this process directly corresponds to an increase in dissolved hydrocarbon and carbon oxide gases. As with hydrolysis, the fatty acids introduced by this process are long chain organic acids and are not considered being detrimental. Oxidation of the biological insulating liquids results in the production of shorter chain acids. But the oxidation process mainly attacks unsaturated parts of the carbon chain in the molecular structure, leading to an increase in viscosity. This process does not itself lead to acid production. Due to this, natural esters are not recommended for breathing equipment and, as such, the effect of oxidation will be minimal. The most easily oxidized sites in biological insulating liquids produce acids with chain lengths in the range of C_7 to C_{11} [29].

Synthetic esters of the type currently used tend to have a very high level of oxidation stability, comparative or better than high grade inhibited mineral oil. They can be utilized in breathing systems and oxidation will lead to similar acids being formed to those found in mineral oil. However, in these liquids, the more dominant process for acid production is likely to be hydrolysis. Hydrolysis of both natural and synthetic esters leads to a breaking of the ester linkage in the molecule, which in turn produces long chain fatty acids. These acids are almost universally insoluble in water and to not tend to be absorbed by the paper. It appears that long chain acids have little effect on paper aging. There is yet no information published on potential sludge formation in esters [22].

It is therefore clear that the concentration of the low molecular acids (LMA) is a key indicator of the degradation environment for transformer paper insulation. Knowledge of this degradation environment and its relationship with the degradation state of the paper insulation, is not only essential to the establishment of a method to control the insulation degradation, but also vital to the future prediction of the aging rate of the transformer paper insulation. Unfortunately, there is no standard method available to quantify LMA concentration until now.

Yan et al. showed [99], that the different water solubility of LMA and high molecular acid (HMA) can be used to determine the LMA in insulating liquids (mineral oil, synthetic ester). With the water separation method, nearly 100% of the LMA in both Gemini X and Midel 7131 were extracted. The presence of HMA does not affect the effectiveness of the water extraction-based measurement of LMA from either mineral oil or synthetic ester at low acidity concentration.

5.5.5 *Interfacial Tension*

The interfacial tension between electrical insulating liquids and water is an indirect measure of the surfactant content or the insulating liquid that migrates under charge attraction into the water at the interface. The surfactants which decrease the interfacial tension value are polar or ionic soluble-contamination or biological insulting liquid products. Water molecules are strongly attracted to one another and require a certain force to break their interfacial tension. Surfactant species are attracted by the polar charges on water molecules. As surfactants in the biological insulating liquid are attracted across the biological insulating liquid—water interface, they obstruct some of the water-to-water attractions that weaken the tensile forces of the water interface. The amount and type of surfactants determine the amount of weakening of these forces. This weakening is measured as a decrease in the interfacial tension value. Interfacial tension is measured in millinewton per meter (see Sect. 4.2.3.2).

Biological insulating liquids have inherently lower interfacial tension compared to mineral oil. For field-aged biological insulating liquids, further investigation should be done when there is more than a 40% decrease in the interfacial tension value from initial transformer samples taken prior to energization [29].

5.5.6 *Pour Point*

The pour point has little significance as far as contamination or deterioration is concerned.

5.5.7 *Viscosity*

The viscosity at the operating temperatures of electrical insulating liquids influences their heat transfer properties in natural and forced (pumped) convective flow and, consequently, the temperature-rise of operating transformers containing them. Biological insulating liquids typically have higher viscosity than mineral oil. An increase in viscosity over time can indicate excessive polymerization of biological insulating liquid from oxidation, typically due to abnormal exposure to air and heat.

5.5.8 *Water Content*

Water may be present in an insulating liquid in several forms. The presence of free water can be detected by visual examination in the form of separated droplets or as a cloud (turbidity) dispersed throughout the liquid. This type of water invariably results in decreased dielectric strength. Water in solution cannot be detected visually and is normally determined by either physical or chemical means. The dielectric strength of the liquid decreases as the dissolved water increases, more significantly as the level approaches the saturation point (see Sect. 4.4.2.1).

5.5.9 Dissipation Factor

Field data of biological insulating liquids indicates a higher rate of increase in the dissipation factors under normal operating conditions relative to mineral oil.

5.5.10 Volume Resistivity

The volume resistivity is influenced by the polar nature of the ester linkage. All ester liquids have a DC volume resistivity which, is typically an order of magnitude lower than that of mineral oil liquids. This does not affect the liquid's ability to operate as an effective AC dielectric, but what users will notice is that measurements of insulation resistance on a transformer filled with ester liquid will be lower than would be expected for mineral oil.

5.5.11 Dissolved Gas Analyzes

Dissolved gas analysis is a diagnostic tool used in the maintenance of transformers. The analysis methods interpret specific dissolved gas levels (patterns of concentration) in the dielectric liquid to help understand the health of the transformer. Since several decades dissolved gas analyses (DGA) have been very common for the assessment and in the fault diagnosis of liquid filled electrical devices. Under electrical and thermal stresses, insulating liquid and liquid-impregnated insulating materials may deteriorate and produce gaseous disintegrated derivates that disband into the liquid. The quantity structure and degree of production of these disbanded gases assist as a worthy index of category and sternness of defects or irregularities occurring within the equipment [100].

Grid accidents can be dramatically reduced by analyzing the characteristic gases dissolved in the insulating liquid to monitor the oil-paper insulation characteristics and changes, accurately grasp and asses the liquid-paper insulation state in transformer to develop reasonable operation, maintenance, repair, or renewal plans. All (normally operating and faulted) transformers produce gases. The principal causes of excessive gas formation within a transformer are abnormal thermal and electrical stresses. Insulating liquid molecules subjected to electrical breakdown or high temperatures undergo various decomposition reactions, which lead to the formation of fission gases. Higher reaction temperature favors more dehydrogenation and the formation of unsaturated gases, as these have higher bond energies and are more stable at high temperatures. If the concentration of the dissolved gasses does not exceed certain limits or even the liquid solubility, the composition of the dissolved gasses can be used for analyzes if the relation of their content is evaluated.

High levels of gassing should always be investigated, especially for the gasses that indicate high temperature or high energy faults, such as acetylene. Acetylene (965 kJ/mol) is thus mainly associated with arcing (where temperatures can reach thousands of degrees), ethylene (605 kJ/mol) with high temperature hot spot (300–800 °C) and hydrogen (432 kJ/mol) with low temperature gas-phase discharges. Identification of the actual cause of a fault can be further complicated by the fact that several faults can be occurring simultaneously. Distinction between the gases formed initially and as a later consequence of a fault may also be a problem in some cases (for instance, a hot spot followed by short circuit arcing). The main advantage of the DGA technique is that it detects the gases in the oil phase, providing the earliest possible detection of the incipient fault.

Discovery of fault gases to low concentrations can serve as a sensitive early warning of equipment failure. Of the existing methods of diagnosing incipient faults, DGA is the most widely used for transformer's maintenance by utilities and for fault diagnoses by researchers. Events, such as mechanical, electrical, and thermal faults of liquid immersed transformers can lead to degradation of the insulating system by producing specific fission gases. The keys to understand the health of the transformers are:

- Understanding which gases are generated under specific conditions, and
- Determining the rate at which these gases are being generated by the transformer.

The causes of fission gas generation include corona, partial discharge, low energy sparks, arcing, cellulose, and insulation liquid overheating. In this case, fission gases such as carbon monoxide (CO), carbon dioxide (CO_2), hydrogen (H_2), methane (CH_4), ethane (C_2H_6), ethylene (C_2H_4), ethine (C_2H_2), propane (C_3H_8) and propene (C_3H_6) are dissolved in the insulation liquid in different proportions. Although the insulation liquid can contain oxygen (O_2) and nitrogen (N_2), these gases and with them, some amount of CO_2 can enter the transformer from outside and not directly related to the degradation process of the insulation system. Depending on the degradation process, oxygen can even be consumed because of the chemical reaction.

In principle, DGA can be used for all kinds of insulating liquids. DGA is one of the best ways to detect internal hidden problems and faults quickly and efficiently close to the real time view without shutting down the device. The detection of certain gases generated and dissolved in the liquid of power transformers in service is often the first available indication of a malfunction that, if uncorrected, can eventually cause a transformer failure. Because it detects incipient faults, DGA can help prevent further damage.

The existing measurement and calculation methods of the content of gases dissolved in insulating liquid as well as the relevant standards of fault diagnoses are mostly developed for mineral oil. While the gases produced in biological insulating liquids and cellulose insulated systems are the same as those produced in mineral oil/cellulose systems, the circumstances and quantities in which they are produced are sometimes because of their chemical structures different. This confirms that the diagnosis methods developed for mineral oil are not fully appropriate for the diagnosis of electrical and thermal faults in biological insulating liquids and need some modification. Martin et al. [101] analyzed dissolved gas levels of a normally operating power transformers filled with Envirotemp FR3 and indicate that ethane and hydrogen are significantly elevated compared to the same transformers filled with mineral oil. There are three principal differences caused by the chemical structure of these liquids, as follows [29]:

- Ethane is produced from the oxidation of linolenic acid, a significant compound in some biological insulating liquids (see Sect. 2.3.2). For example, soybean and rapeseed based biological insulating liquids have sufficient linolenic acid content routinely produces measurable amounts of ethane. After the initial operation of a transformer under normal conditions, the ethane level can rise to a few hundred ppm. This generation is considered "stray gassing" and a fault condition. Ethane generation tends to level off after a few weeks to a few months, depending on the operation temperatures. Also, exposure to sunlight or fluorescent light can increase the amount of ethane in the insulating liquid. UV exposure should be avoided, particularly for containers of sampling prior to testing. The increase in the other hydrocarbon gases to follow the onset or development of fault conditions should be watched.
- Carbon oxides can be produced from the pyrolysis of biological insulating liquids in amounts and proportions that can mask or confuse the carbon oxide production from the pyrolysis of cellulose. There are clear differences between the open and closed systems. In the open system, gases that have already been formed can be released into the environment, since the aim is always to achieve equilibrium with the surrounding medium. In the closed system, the gases formed cannot escape, which is why the gas concentration is two to seven times higher than in the open system. In conjunction with paper aging natural esters show a higher concentration of carbon oxides compared to synthetic ester. The CO_2/CO ratio is markedly higher for ester liquids compared to mineral oil liquids, which must be considered when using esters in transformers [13].
- A different hydrocarbon gas profile is produced from pyrolysis using biological insulating liquids. The same hydrocarbon gases are produced from heating, but the proportions are different from those produced from mineral oil liquids. Studies of Martin et al. [101] show a constant but slightly elevated level of hydrogen in normally operating FR3 filled transformers, and this may sometimes be incorrectly interpreted as indicating a partial discharge fault. Xiang et al. came to the same results, but also for camelia oil, for temperatures below 300 °C [102].

Testing (ASTM D3284 [103]) and field experience indicates that under the same magnitude of electrical overstress, biological insulating liquids typically produce somewhat less volume of the gasses compared to mineral oil. Xiang et al. showed in their studies [102] that under electrical stress simulation in all cases a significant amount of H_2 and C_2H_2 are generated and that for mineral oil, Envirotemp FR3, and camellia oil, the main fault gases of a breakdown fault are the same. Their results show that C_2H_2 is the main dissolved gas of electrical breakdown for mineral oil,

Envirotemp FR3, and camellia oil. Under partial discharge condition the contents of fault gases are mainly dependent on discharge energy, and it's hard to distinguish a partial discharge pattern through DGA. The samples have one thing in common: C_2H_2 does not largely exist, which shows that C_2H_2 is only the main dissolved gas of high-energy discharge.

However, for the same thermal overstress, some biological insulating liquids produce significantly more volume of the gases than others, depending on the type of base liquid. There are differences in gas solubility coefficients between the various biological insulating liquids and mineral oil and their respective values should be used for data interpretation.

In principle, there are two possibilities for DGA:

- Collecting samples and analyzing in the laboratory (laboratory DGA)
- Online monitoring.

In dissolved gas analyzes whether in laboratory or online, gases must be extracted from the insulating liquid to take measurement. The more complete the extraction, the less uncertainty there will be related to concentration calculation. There are three extraction methods:

- Vacuum method (Toepler pump, partial degassing)
- Stripping method
- Head space method.

With a mercury filled Toepler pump (Fig. 5.33), an efficiency of almost 100% is possible whereas the headspace method (HS) can achieve no more than 30% (Fig. 5.34).

Gas samples from relays (Buchholz) should be taken from the equipment as soon as possible after gas accumulation has been signaled. Changes in the composition of the gases caused by the selective reabsorption of components can occur when free gases remain in contact with oil.

5.5.11.1 Dissolved-Gas-Analyzes—Sampling Methods

The result of transformer examination is influenced by the quality of the insulating liquid sample that is used for DGA. Sampling of insulating liquids is described in IEC 60475 [104]. This standard is applicable to insulating liquids whose viscosity at the sampling temperature is less than 1500 mm^2/s. It can be used for mineral oil and non-mineral oil (such as synthetic esters, natural esters—biological insulating fluids—and silicone oils). More information about sampling of gases of free and dissolved gases is described in IEC 60567 [105].

Separate sampling devices must be provided exclusively for each type of liquid. For non-mineral oil (biological insulating fluids), the seals and hoses must be compatible with the liquid. For these insulating fluids, the seals should not be made of NBR (nitrile butadiene rubber) or silicone rubber, but of PTFE (polytetrafluorethylene). Care must be taken in drawing samples to prevent contamination and to ensure a

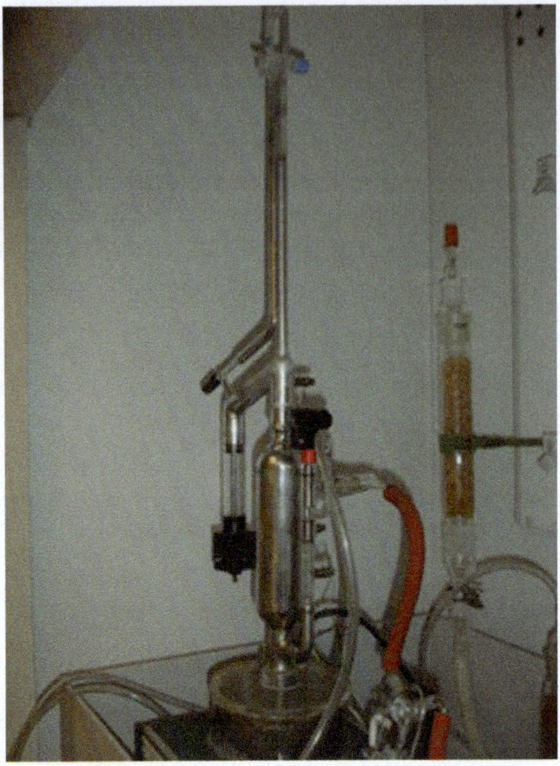

Fig. 5.33 Mercury—Toepler pump

truly representative sample. It is also important that the quantity and composition of dissolved gases remain unchanged during transport to the laboratory.

Sample containers may be glass syringes (Fig. 5.35), stainless steel cylinders (Fig. 5.36), or metal cans (Fig. 5.37) having flexible sides to tolerate pressure variation (to allow for volume changes caused by expansion or contraction of the insulating liquid).

- Glass Syringes (Fig. 5.35), of a suitable size terminated with a Luer lock fitting to which is attached a three-way stopcock, should be used. Syringes having precision ground barrels and pistons are preferred.
- Stainless Steel Sampling Cylinders (Fig. 5.36), equipped with valves on each end, may be used for sampling. The oil sampling tube is held vertically and the cocks on the sampling cylinder are opened. Now, the sampling valve on the equipment is opened carefully, so that insulating liquid flows through the sample tube to waste. The sample tube is completely filled and about two liters of insulating liquid are allowed to flow waste until no air bubbles are visible in the escaping insulating liquid. The insulating liquid flow is then closed by shutting off first the outlet cock, then the inlet cock and finally sampling valve on the transformer (Fig. 5.38). These

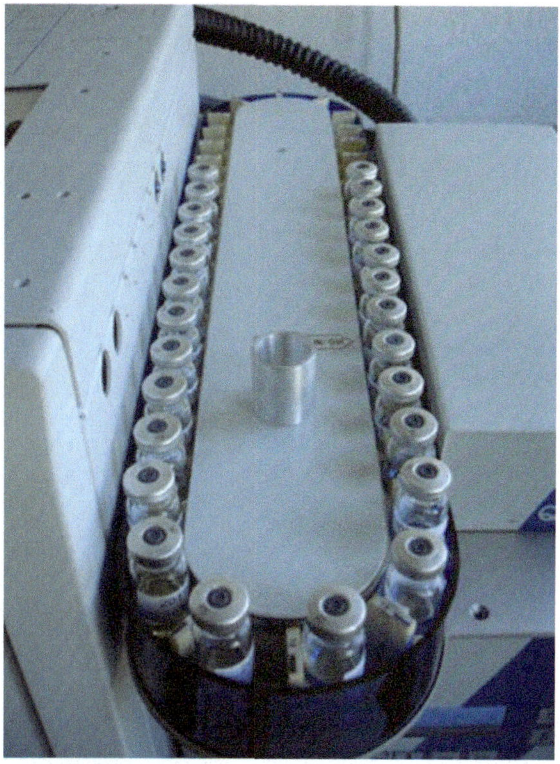

Fig. 5.34 Headspace vials in an autosampler

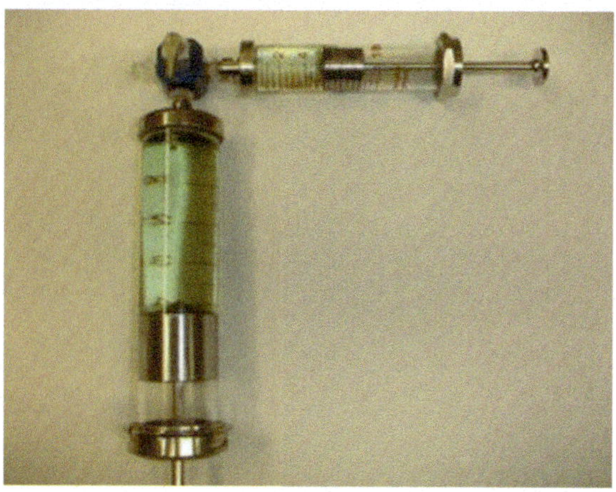

Fig. 5.35 Glass syringes

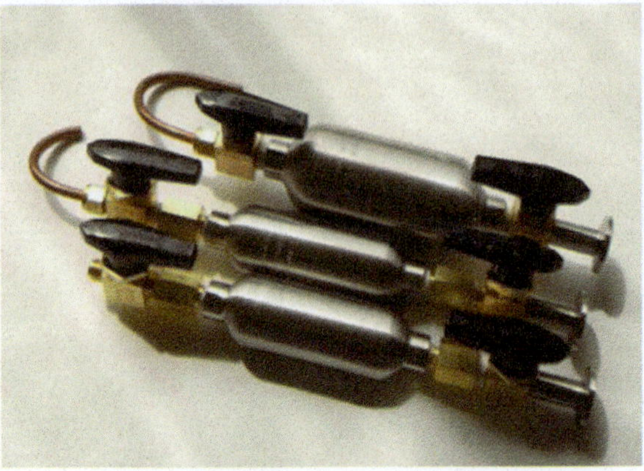

Fig. 5.36 Stainless steel cylinders

Fig. 5.37 Aluminum can

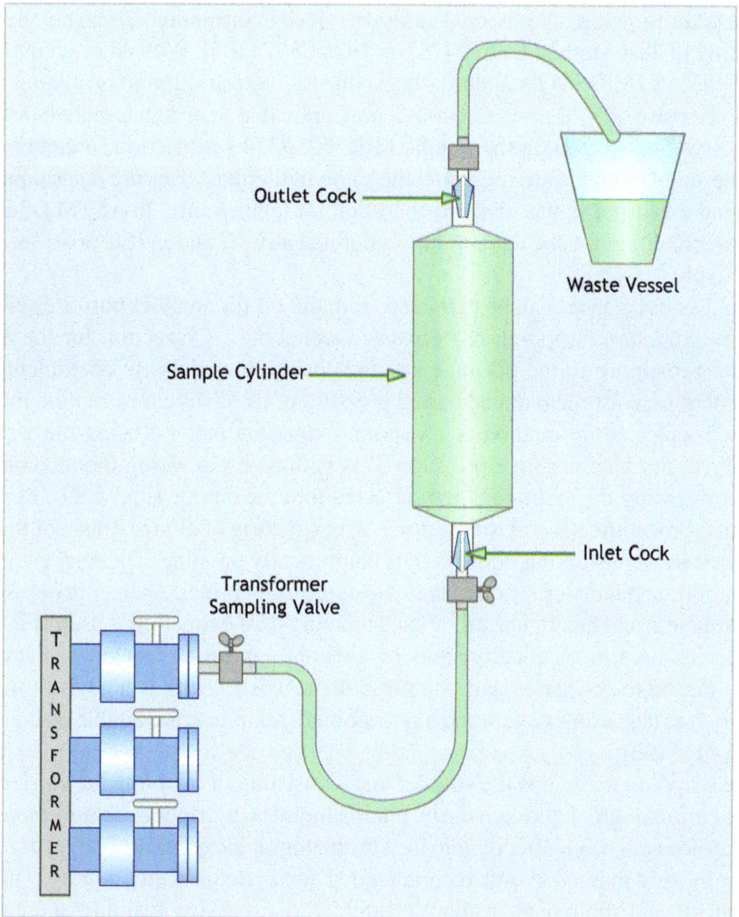

Fig. 5.38 Filling sample cylinder

cylinders have the disadvantage of not allowing visual inspection of the interior of the cylinder and they are not flexible in case of pressure fluctuations.

- Flexible-Sided Metal Cans (Fig. 5.37), having screw caps may be used, too.

5.5.11.2 Dissolved-Gas-Analyzes—Test Methods

Laboratory DGA is affected by many factors, like the quality of the oil sample, the quality of the used equipment, the reliability of the labor workers, etc. Further essential influencing factors are the gas extraction method, the gas type, and the concentration level.

The determination of individual gases dissolved in insulating oil can be carried out according to Test Method D3612 [78] or IEC 60567 [105]. With an exact interpretation of the ASTM D3612 standard, this is due to the restriction to insulating liquids with a viscosity of $\leq$20 mm^2/s at 40 °C, not applicable to most alternative insulating liquids (see Sect. 4.2.2). In the standard IEC 60567 this restriction cannot be found, even the same methods are used, and the same individual gases are determined. The two standards differ in the standard condition for temperature. In ASTM D3612, the standard condition for the temperature is defined as 0 °C and in IEC 60567 as 20 °C, respectively.

The dissolved gases can be extracted from the oil for analysis purposes either by vacuum extraction, stripping or headspace technology. Correction for the amount of gases remaining in the oil must be made using the solubility coefficient of the component gases at the reduced partial pressure of the different extraction methods.

The Toepler pump method is a vacuum extraction method using the weight of mercury to produce negative pressure. This is done by lowering the mercury level and compressing the additional gas extracted into the burette (Fig. 5.39). Repeating this same procedure several times (up to 20) extraction of almost 100% of the gases contained in the insulating liquid is thus theoretically possible. The total gas content is an important quality criterion of the state of the transformer. As mercury presents an environmental and health hazard, often a mercury-free extraction is used (Fig. 5.40). This device uses mechanical pistons or a diaphragm pump instead of mercury to extract the dissolved gases and compress them. Dissolved gas analyzers with full mercury free degassing system and a syringe auto-sampler are available on the market (Fig. 5.41) [106].

The stripping method is the simpler one, consisting of bubbling the carrier gas of the gas chromatograph through the insulating liquid on a stripper column containing a high surface area bead, directly into the chromatographic column. Testing of silicone liquids by this method is not recommended for systems which are also used for mineral oil and biological insulating liquids, as excessive foaming should cause contamination of columns after the stripper.

In the vacuum method (Toepler pump, extraction method), transfer to the gas chromatograph is done by a gas syringe or switching valve via gas loop from the burette of the degassing device.

If using headspace methods of DGA, it is also important to understand that gasses dissolve into biological insulating liquids in different proportions than in mineral oil. Solubility coefficients give an indication of how much gas can be dissolved into a liquid at normal atmospheric pressure. The headspace method brings an insulating liquid sample in contact with a gas phase (headspace) in a closed vessel (vial) purged with argon (Fig. 5.42).

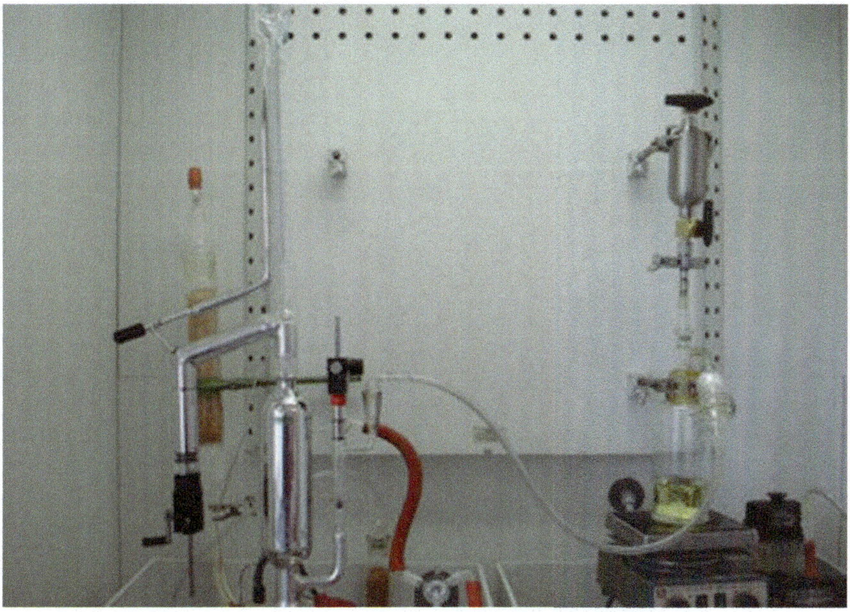

Fig. 5.39 Toepler pump—Degassing process

The headspace vials are filled as follows (Fig. 5.43):

- The syringe must be filled with insulating liquid in the correct way. The piston should be in the right position for measuring the volume.
- To transfer the insulating liquid into the vial, the syringe is pierced through the septum in a horizontal position. Under no circumstances should the syringe be held by the plunger so that insulating liquid is not squeezed out of the syringe before the injection.
- To fill the vial, turn the syringe vertically upwards and press the insulating liquid into the vial. Because of the argon that is already present, a counterpressure is created. Then pull the vial off upwards. The filled insulating liquid additionally seals the septum.
- The samples should be transported and stored with the septum facing down.

The dissolved gases contained in the insulating liquid are then equilibrated in the two phases in contact under controlled conditions (in accordance with Henry's law). At equilibrium, the relationship between the remaining concentration of a gas in the insulating liquid (C_L), its concentration in the headspace (C_G), and its initial concentration in the insulation liquid (C_L^o) may be deduced by mass equivalence (Eq. 5.11) [105]. In the case of the headspace method, the extraction efficiency is low, and the concentrations of the gases in the extracted gas phase are typically 100 times lower than in the other methods. At low gas-in-oil concentration levels,

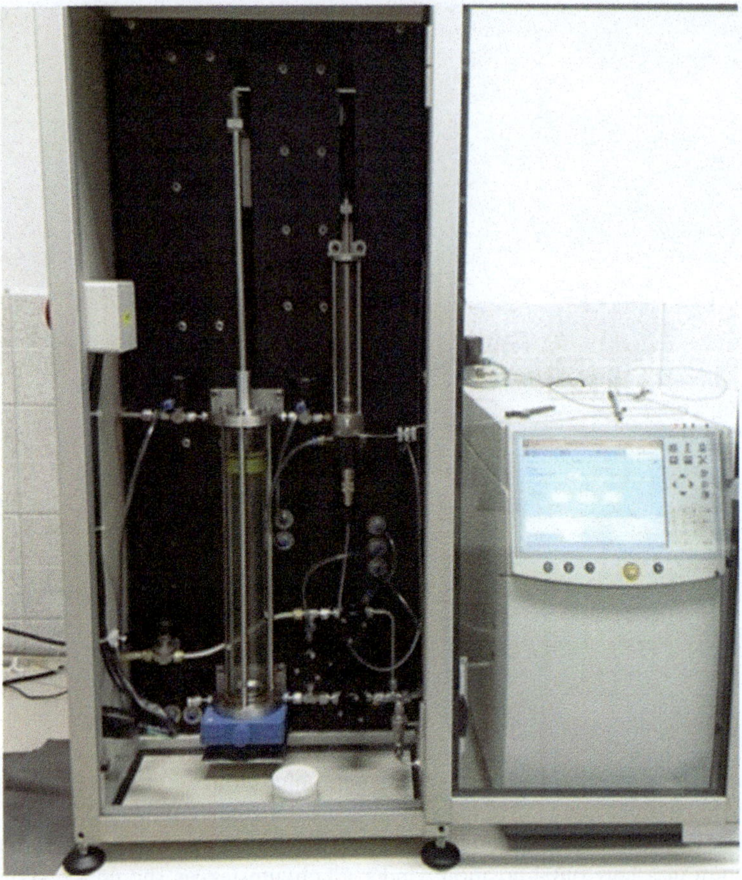

Fig. 5.40 Mercury free extraction pump (Picture provided by Milan Vidmar Electric Power Research Institute-Ljubljana)

detection limits are in the nanoliter per liter range and therefore low detection limits in the gas phase of the headspace are necessary. This requires the use of sensitive equipment and procedures.

$$C_L^o * V_L = C_L * V_L + C_G * V_G \quad (5.11)$$

V_L Volume of the insulating liquid sample [ml].
V_G Volume of the headspace [ml].

There is a direct proportionality between the concentration of this gas in the two phases in equilibrium, as follow Eq. 5.12.

$$C_L = K * C_G \quad (5.12)$$

Fig. 5.41 TOP TOGA GC

Fig. 5.42 With argon purged and filled vials

K Partition coefficient (Tables 5.15 and 5.16).

There are some differences in the solubility of gasses with natural and synthetic esters, most notably acetylene and carbon dioxide. This means that if these gasses are produced during a fault, a larger quantity will be dissolved into the insulating liquid.

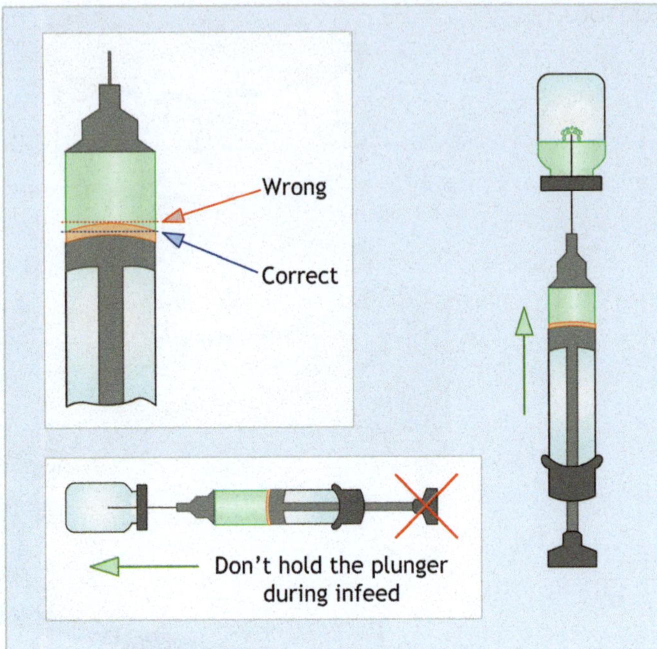

Fig. 5.43 Filling headspace vial

Table 5.16 shows the results of the solubility coefficient evaluated by the Cigré WG group [109].

By substituting C_L in Eq. 5.11 by its expression from Eq. 5.12, the mass equivalent relationship becomes Eq. 5.13 and based on this equation

$$C_L^o * V_L = K * C_G * V_L + C_G * V_G \tag{5.13}$$

C_L^o can be extracted (Eq. 5.14).

$$C_L^o = C_G * \left(K + \frac{V_G}{V_L} \right) \tag{5.14}$$

Equation 5.14 forms the basis of the headspace sampling method and shows that the initial concentration of a dissolved gas in insulating liquid be determined by analyzing an aliquot portion of the headspace when equilibrium is reached.

At equilibrium, the headspace is over pressurized with argon and then the content of a loop is filled by the depressurization of the headspace against the ambient atmospheric pressure. The gases contained in the loop are then introduced into a gas chromatograph.

A wide range of chromatographic conditions has been successfully employed. Both argon and helium have been used as carrier gases. In some cases, a separate

Table 5.15 Examples of partition coefficients of different insulating liquids

Gas	Mineral oil					Natural ester	Synthetic ester	Silicone oil	
	IEC 60567 (at 20 °C) [105]	IEC 60567 (at 25 °C) [105]	IEC 60567 (at 70 °C) [105]	ASTM D3612 (25 °C) [78]	IEEE C57.104 (25 °C) [107]	IEC 60567 (25 °C) [105]	IEC 60567 (25 °C) [105]	IEC 60,567 (25 °C) [105]	IEEE C57.146 (25 °C) [108]
H_2	0.040	0.0556	0.074	0.0558	0.0429	0.0470	0.0510	0.0924	0.057
N_2	0.070	0.0907	0.110	0.0968	0.0745	0.0728	0.0872	0.157	0.143
CO	0.100	0.132	0.120	0.133	0.102	0.111	0.127	0.204	0.096
O_2	0.130	0.172	0.170	0.179	0.138	0.134	0.152	0.266	0.175
CH_4	0.340	0.429	0.440	0.438	0.337	0.341	0.381	0.589	0.5148
CO_2	0.930	1.090	1.020	1.170	0.900	1.540	2.050	1.630	1.401
C_2H_2	1.000	1.240	0.930	1.220	0.938	2.680	4.380	2.040	[a]
C_2H_4	1.470	1.840	1.470	1.760	1.350	1.690	1.870	2.180	[a]
C_2H_6	2.180	2.820	2.090	2.590	1.990	2.160	2.190	3.100	[a]
C_3H_6	–	–	5.040	–	–	–	–	–	
C_3H_8	–	–	5.370	11.000	–	–	–	–	
C_4H_6	–	–	10.100	–	–	–	–	–	

[a] Values have not been determined experimentally

Table 5.16 Solubility/partition coefficients K of gases in mineral and non-mineral oils at 20 °C

Gas	Mineral oil	Natural ester	Synthetic ester	Silicon oil
H_2	0.0504	0.0471	0.0479	0.0869
N_2	0.091	0.074	0.091	0.155
CO	0.125	0.108	0.130	0.189
O_2	0.172	0.134	0.152	0.267
CH_4	0.423	0.341	0.378	0.580
CO_2	1.100	1.540	2.080	1,630
C_2H_2	1.250	2.580	4.260	2.040
C_2H_4	1.810	1.670	1.850	2.180
C_2H_6	2.880	2.140	2.200	3.100
C_3H_6	9.640	7.460	7.670	9.910

gas chromatograph (GC) or other device is used for the detection and quantification of hydrogen when helium is used as a carrier gas. If helium is used as a carrier gas with thermal conductivity detector, medium to high concentration of hydrogen may give a nonlinear response, due to the closed heat capacity values of helium and hydrogen. The limit of detection will be higher than with an argon carrier gas under similar conditions. If nitrogen is used as a carrier gas, nitrogen cannot be detected in the sample. With the use of an argon carrier gas, a catalytic converter containing powdered nickel located after the chromatographic columns is used to convert carbon monoxide and carbon dioxide to methane for detection with flame ionization detector for acceptable sensitivity. With helium as a carrier gas, a catalytic converter is not necessary but may be used to enhance sensitivity. A flame ionization detector, instead of a thermal conductivity detector, is often used to detect hydrocarbon gases due to its greater sensitivity for these components. The permanent gases H_2, O_2, and N_2 are detected using thermal conductivity detector (TCD). A wide range of injector, column, and detector temperatures can be used. Both isothermal and temperature programs can be used to provide adequate separation and sensitivity.

In the calculation, the obtained value of C_L^o can be corrected to the standard thermal and pressure condition (STP) as follows (Eq. 5.15).

$$C_L^o(STP) = C_L^o * \frac{P_a}{101.325} * \frac{273}{T_a} \tag{5.15}$$

P_a Laboratory ambient pressure [Pa]
T_a Laboratory ambient temperature [K].

5.5.11.3 Online Monitoring

Within the last years, different systems for online analyzes have been developed. These systems use diverse kinds of sensors like semiconductor sensors, infrared spectroscopy (Fig. 5.44) or photo acoustical effect.

All these methods use sensors that are different from the laboratory experienced version with degassing and gas chromatography. Energy Support use for their online device (Fig. 5.45) [110] the same principle as used in the laboratory. This online monitoring equipment is available and is supported by laboratory tests. Online monitors have proven their accuracy and with getting the results just in time, they are very helpful for detecting faults coming into being.

Comparing the test results of different methods, one should be aware of uncertainties of all of them. The most common uncertainty sources are the calibration method (quality and concentration range of the calibration gas), the liquid-sampling, the gas extraction from the liquid and the used partition coefficients. But the biggest uncertainty source is the sampling of the insulating liquid. The more steps must be done to

Fig. 5.44 Online-DGA—Vaisala

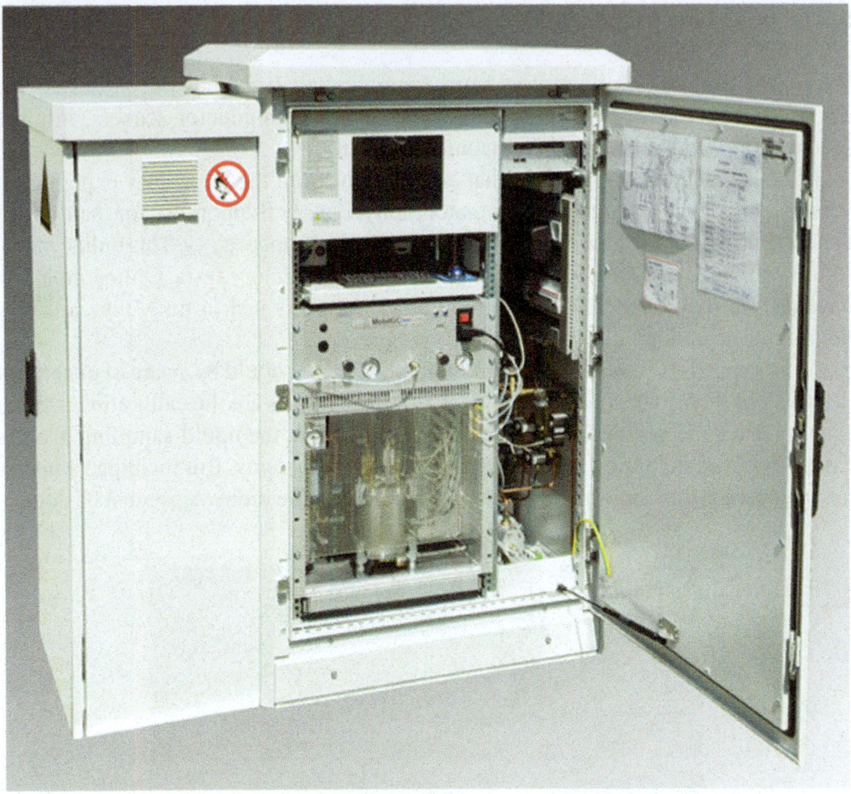

Fig. 5.45 Online-DGA—Energy Support

get the sample to the analyzer, the more errors are possible. In this case, online monitoring has its advantages. The sample should not be stored and transported, which always can affect the quality. During this handling, a significant amount of gases, like hydrogen and carbon monoxide, can escape from the insulating liquid. Otherwise, humidity and ambient air can enter into the sample vessel. All this handling is significantly reduced by using online monitoring. In [101], Martin et al. circulated Envirotemp FR3 liquid from the transformer tank through the monitor for analysis and then return to the tank. The dissolved gases are extracted from the insulating liquid with the aid of a carrier gas, in this case with helium. The online automated system recorded the dissolved gas levels every four hours.

For quick analyzes, for example, during maintenance a mobile gas chromatograph can be helpful Fig. 5.46 [111].

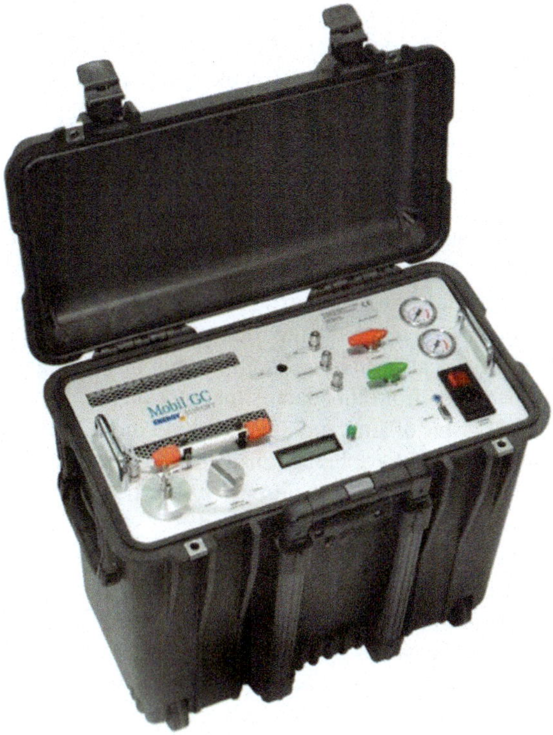

Fig. 5.46 Portable gas chromatograph

5.5.11.4 Data Evaluation

The best information is obtained by viewing trends and, if possible, fleets of transformers should be compared to each other. So, it is useful to take a benchmark sample when a transformer is first energized, or when an oil treatment is performed, and to take further samples at regular intervals so that any variation in quality can be identified to indicate developing faults. Since mineral oil has been used for such a long time, a large database of information is available to enable interpretation of changes to its characteristics and thus predict the possible malfunction of a transformer.

Air or nitrogen saturated insulating liquids can dissolve only small amounts of fault gases compared to degassed insulating liquids. Therefore, the transformer conservator tank sealing principle is of crucial importance for fault gas limit concentrations.

The gases generated in liquid-filled transformers can be used for qualitatively determining fault types since the gases are typical or predominant at various temperatures. After incipient faults are detected, preventive maintenance can be performed, and conditions can be assessed by DGA. There are various conventional and non-conventional DGA methods used by various agencies and utilities to monitor the condition of the cellulose contained in service transformers. Principally, there is not much difference which insulating liquid is used and the experience with biological insulating liquids is not very high. Some small differences have their origin in the base oil and have an impact on the interpretation. When using soybean oil, the higher ethane content, which comes from stray gassing, must be considered. Soybean based biological insulating liquids can produce ethane at moderate temperatures below 250 °C, which may occur in localized hot spot and possible during factory testing. This does not happen in either mineral oil or synthetic ester liquids, and is thought to be related to the unsaturated content in the base vegetable oil.

Table 5.17 shows the result of an experiment done by Hanson et al. [112] in which different seed oils were exposed to an elevated temperature (120 °C) over a period of one day. It is interesting to note that the level of ethane is significantly higher in the biological insulating liquids which have a higher polyunsaturated content. The interference from this data is that low-oleic oil vegetable oils are likely to produce higher levels of stray gassing than high oleic vegetable oils. This does not mean that the oils which have higher unsaturated content are less preferable, it just means that it is useful to know the oil base type when evaluating DGA data, especially when high levels of ethane are indicated.

In mineral oil DGA, the presence of high levels of carbon oxides is usually an indication of the involvement of cellulose paper in a fault. The ratio of CO to CO_2 is also used to try to interpret whether paper is involved when a thermal or electrical fault is indicated by other gasses. According to IEC 60599 [113], the ratio for normal cellulose aging should be $3 < CO_2/CO < 10$. Values below 3 indicate above-average aging of the cellulose, while values above 10 allow conclusion to be drawn about mild overheating of the insulation (> 160 °C). Münster et al. [13] suggest an adaptation of the ratios from IEC 60599 for ester liquids. Higher values are achieved with natural esters and the ratio for synthetic esters tend to behave constant and the values remain above 10. Xiang et al. describe in [102] that when the three-ratio method was applied to analyze the data of simulated electrical faults in mineral oil, the diagnosis results are correct in all cases. However, when it is applied to analyze data in Envirotemp FR3 liquid and camellia oil, it only diagnoses breakdown faults correctly; the diagnosis results are not consistent with simulated fault types under partial discharge.

Table 5.17 DGA results for various seed oils [ppm]

Base oil	High oleic sunflower	Peanut	Soybean	Flaxseed
Polyunsaturated content [%]	0.2	0.2	7	53
H_2	357	282	316	708
CH_4	21	10	10	17
C_2H_6	4	26	563	2371
C_2H_4	8	16	7	16
C_2H_2	0	0	0	0
CO	203	389	408	977
CO_2	876	2232	1330	3212

In biological insulating liquids and synthetic esters, this is somewhat complicated as carbon dioxide will be formed under electrical and thermal faults, even when no paper is present. The other point of note is that with biological insulating liquids, the level of carbon oxide can be higher than the level of carbon dioxide and this is reverse for mineral oil and synthetic ester. This means that the use of the CO_2/CO ratio may be less valid for biological insulating liquids than for mineral oil when determining if a fault involves cellulose. Looking at the chemical structure of esters (Sect. 2.3), it is possible to see how carbon oxides are formed with these liquids, even in absence of cellulose. The ester linkage contains oxygen atoms, not present in the mineral oil structure. In simple terms, high energies can break the bonds between the carbon and oxygen atoms, allowing the recombination into carbon oxides, normally associated with paper degradation in mineral oil. This means that care needs to be taken when looking at carbon oxide numbers in ester DGA to infer whether paper is involved in a fault, and it is unlikely that this will be a reliable indicator for cellulose involvement.

Theuermann et al. found in their investigations [114] with instrument transformers filled with mineral oil that cellulose can adsorb/absorb fission gases like H_2, CO_2 and C_2H_2. Higher temperature and higher level of moisture decreased the adsorption effect. If the ratio of the fission gases at the source of the error does not match that in the insulating liquid, the interpretation of the gas-in-oil test is influenced and can ultimately lead to misinterpretations. With different diffusion rates of the fission gases, after passing the cellulose barrier, a gas pattern can occur in the free insulating liquid that differs from the characteristic gas pattern generated at the fault location. This phenomenon has not yet been researched very well.

The key step in using gas analysis for fault detection is correctly diagnosing the fault that generated the gases. Considerable experience is necessary to make the proper decisions, based on the type of equipment used, its loading, operating and maintenance history, the number of years in service, the nature of the fault involved, and gas formation trends. Judgement based on so many issues depend on the skill and intuition of experienced and expert engineers, who must decide on proper withdraw timing, if equipment is withdrawn from service too early, the fault may be difficult to find by inspection and the resulting costs will be high. Catastrophic failure will be even more expensive if the fault is allowed to develop for too long. Abnormal electrical or thermal stresses cause an insulation system breakdown and to release small quantities of gases. The composition of these gases depends on the insulation system and the fault type. The detection of certain level of gases generated in liquid-filled transformer in service is often the first available indication of a malfunction that can cause a transformer failure if not corrected.

Various DGA methods have been used by organizations and utilities to assess transformer conditions. These DGA interpretation schemes are based on empirical and practical knowledge gathered by experts worldwide. Figures 5.47 and 5.48 present the different graphical representation based on gas ratios with the keys (fault zones):

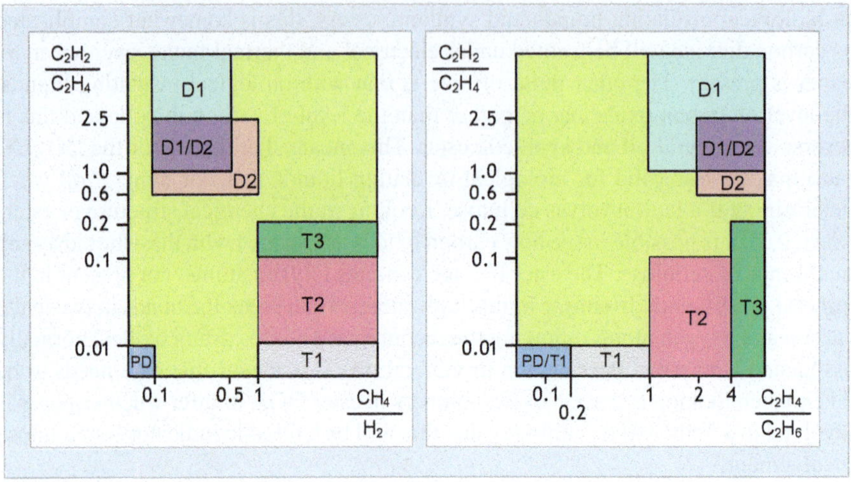

Fig. 5.47 Graphical representation 1 of gas ratios [113]

S Stray gassing region
PD Partial discharges
D1 Discharges of low energies
D2 Discharges of high energies
T1 Thermal fault t < 300 °C
T2 Thermal fault 300 °C < t ≤ 700 °C
T3 Thermal fault t > 700 °C.

Similar deviations, like for the graph representation method of biological insulating liquids, are received using Duval Triangle and Pentagon method.

For new insulating liquids, the data must still be gathered and improved. For soja been based liquids, just the value of ethane must be adapted. Interpretation schemes are generally based on defined principles such as gas concentration, key gases, key gas ratios, and graphical representations. In [115], Duval and Lamarre adapted their Duval pentagons for several ester liquids (Figs. 5.49, 5.50, 5.51 and 5.52). If you compare the individual pentagons, you can see clear differences (for example, the area "S" for stray gassing).

How to Work with Duval Pentagon This Pentagon uses the ratios of five gases (H_2, C_2H_6, CH_4, C_2H_4, C_2H_2) to find the correct fault zone. Several mathematical steps must be done to find the correct position of the spot in the Duval Pentagon [116].

- Calculation of the relative percentage of the five gases (Eq. 5.16).

$$\sum H_2 + C_2H_6 + CH_4 + C_2H_4 + C_2H_2 = 100 \quad (5.16)$$

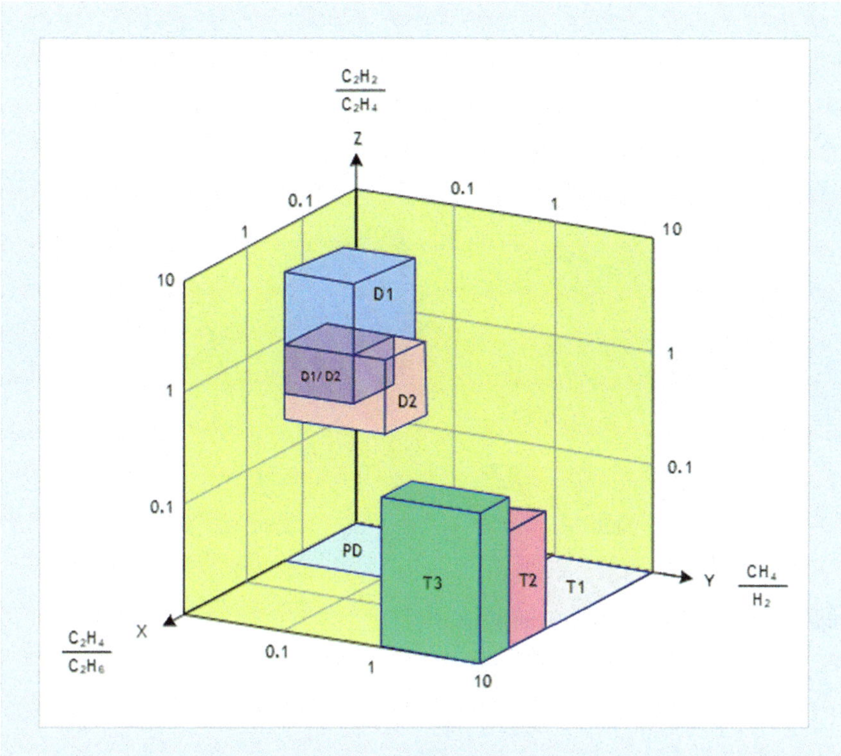

Fig. 5.48 Graphical representation 2 of gas ratios [113]

- Calculation of the coordinates of each of the five points which lie on the lines between the center and the corner points by transforming in the x, y cartesian coordination system (Fig. 5.53).
- Calculation of the centroid point coordinates of the five points. The centroid is also known as the center of gravity or the center of mass (Eqs. 5.17 and 5.18).

$$C_x = \frac{1}{6 * A} \sum_{i=0}^{N-1} (x_i + x_{i+1}) * (x_i * y_{i+1} - x_{i+1} * y_i) \tag{5.17}$$

$$C_x = \frac{1}{6 * A} \sum_{i=0}^{N-1} (y_i + y_{i+1}) * (x_i * y_{i+1} - x_{i+1} * y_i) \tag{5.18}$$

with A (Eq. 5.19).

$$A = \frac{1}{2} \sum_{i=0}^{N-1} (x_i * y_{i+1} - x_{i+1} * y_i) \tag{5.19}$$

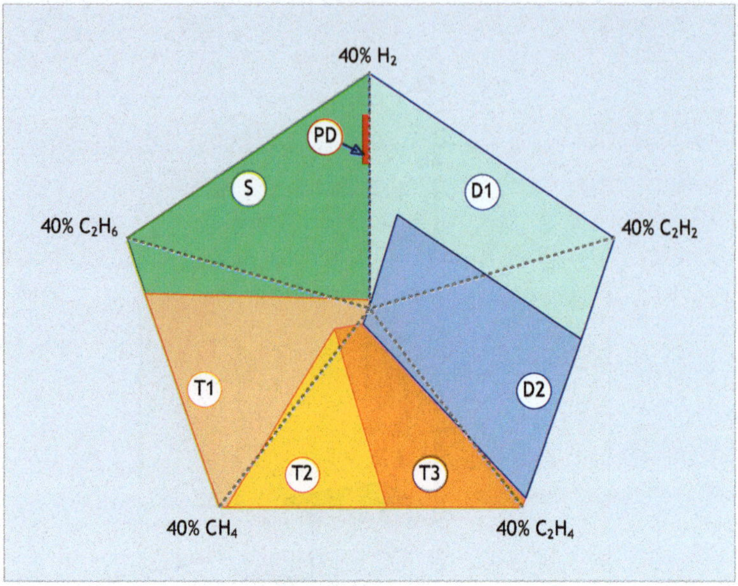

Fig. 5.49 Duval Pentagon for mineral oil liquids

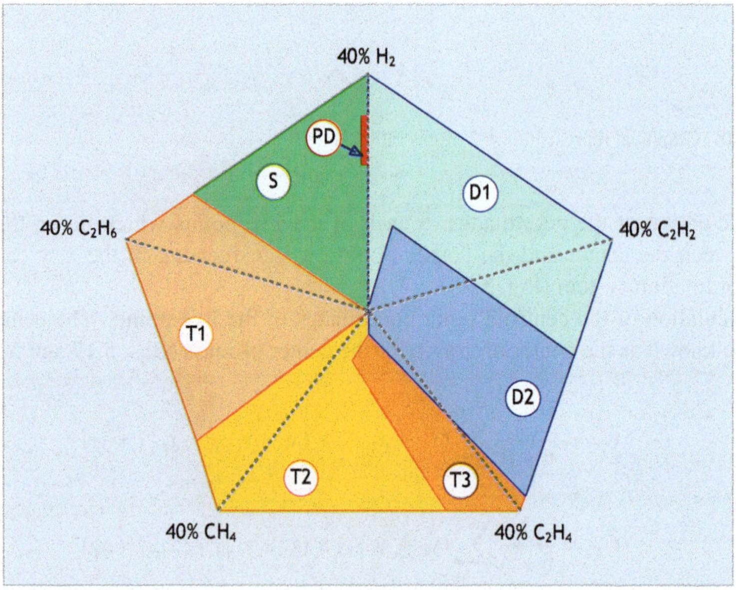

Fig. 5.50 Duval Pentagon for natural ester Envirotemp FR3

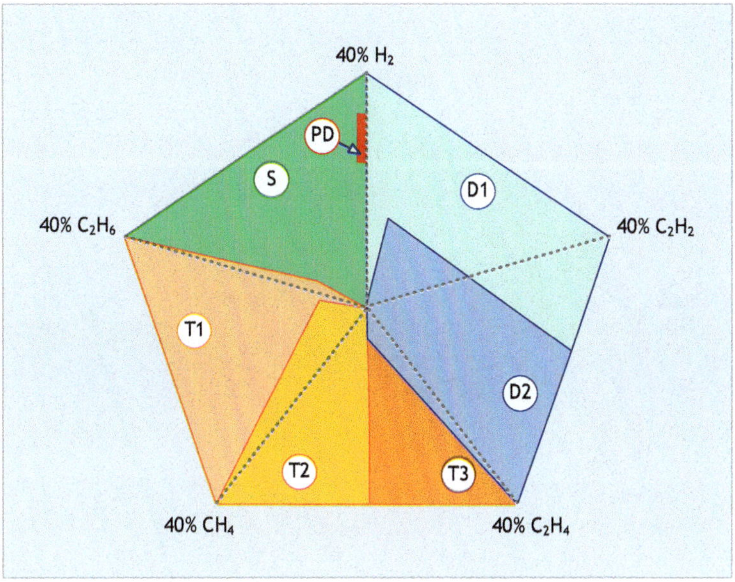

Fig. 5.51 Duval Pentagon for rapeseed oil

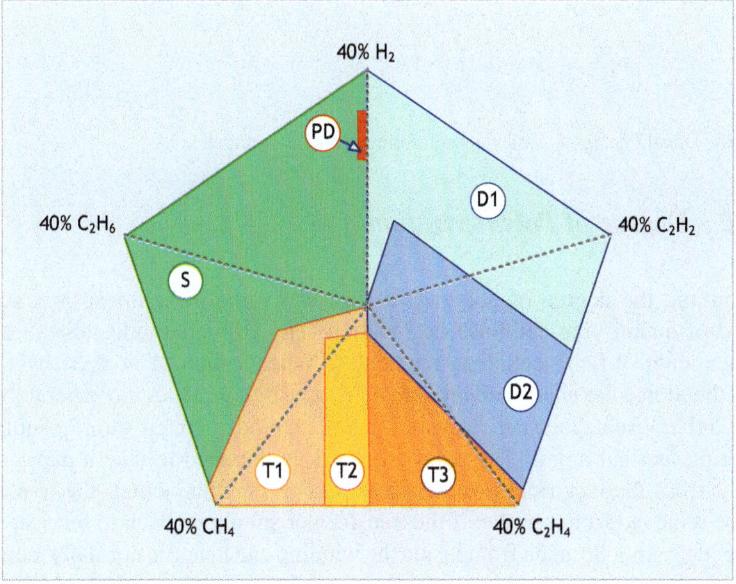

Fig. 5.52 Duval Pentagon for sunflower oil

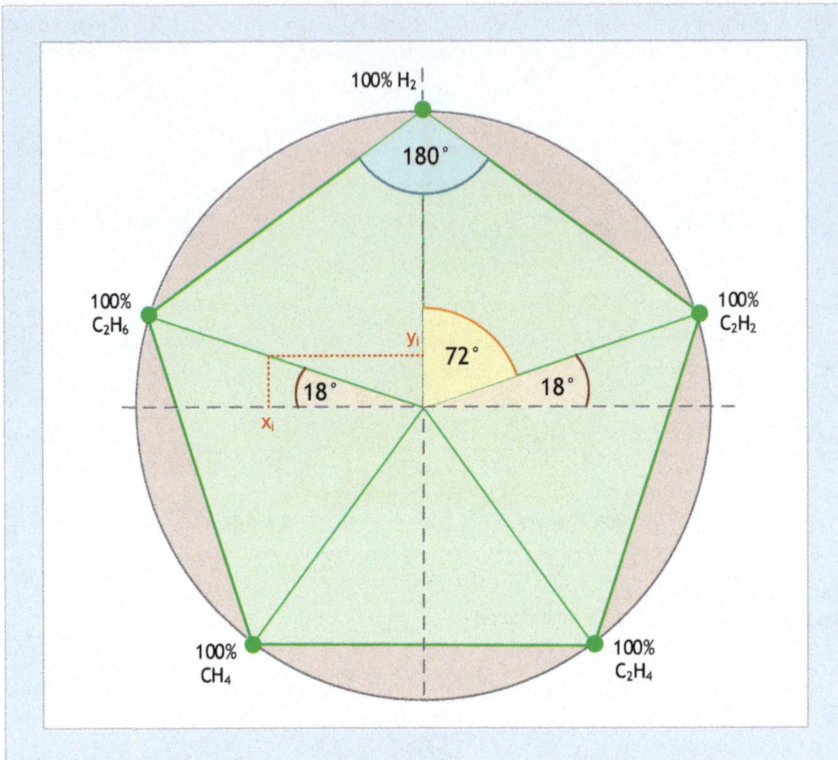

Fig. 5.53 Duval Pentagon—calculation for the x_i and y_i coordinates

5.5.12 Degree of Polymerization

Determining the degree of polymerization (DP) value of cellulose is a standard method of quantifying cellulose degradation. The DP value indicates the average polymer length of the cellulose molecules. When cellulose is aged by thermal stress, the molecular chains are broken. Advanced aging causes the paper to become brittle and to lose its mechanical stability. The method is effective for quantitatively measuring thermal aging. The main drawback is the need to take a paper sample directly from the electrical device (transformer). For this action, the transformer must be removed from service. If the transformer should go back to service, mostly the sample cannot be taken from inside the winding and hence is normally taken from one of the high current leads in the upper part of the transformer, which reduce the informative value. In cellulose materials, aging caused by thermal stress generates furanic compounds as a degradation product. These compounds are a family of chemical substances that differ in stability and production rates. Because these compounds dissolve into insulation liquid, they can be detected and studied by standard analytical methods. The observed relationships between DP and various concentrations of furanic compounds—especially 2-furfural (2-FAL)—enable indirect measurement of aging of the cellulose. Furanic compound analyses are widely used for predicting the lifetime of insulation material.

5.5.12.1 Degree of Polymerization—Test Method

The viscosity of a solution of macromolecules is dependent on the molecular weight at a given (low) concentration. This phenomenon is used for the determination of the DP-value of cellulosic insulating materials. A small amount of de-oiled and fluffed paper or board is dissolved in copper-ethylene-diamine (Cuen). The viscosity of the paper solution is determined using a capillary viscosimeter. From this result, the degree of polymerization is calculated using experimental equations and constant that can be found in the relevant standard IEC 60450 [117].

5.5.12.2 Furanic Compounds—Test Method

Furanic compounds refer to a whole family of compounds, all of which could be described as furanic derivatives. The most abundant is 2-furfural (2-FAL), but 2-acetylfuran (2-ACF), 5-methyl-2-furfural (5-MEF), 5-hydroxymethyl-2-furfural (5-HMF), and 2-furfurylalcohol (2-FOL) have been found in insulating liquid and paper. Whereupon 2-furfural (2-FAL) provides the most relevant information on paper degradation. This test method determines the furanic compounds generated because of the degradation of the cellulose insulation materials, such as paper, pressboard and cotton, used in the solid insulation systems of electrical equipment. Furanic compounds that are soluble to an appreciable degree typically migrate into the insulating liquid. The presence of high concentrations of furanic compounds in samples of the insulating liquid can be an indication of cellulose degradation from aging or incipient fault conditions. After extraction of the furanic compounds from the insulating liquid, high performance liquid chromatography (HPLC) can be used for determining the furanic compounds (ASTM D5837 [118], IEC 61198 [119]).

5.5.13 Vibro-Acoustic Measurement

Low level of the audible sound is a compulsory aspect for transformers today. Therefor it is important to determine the parameters of generated sound in the transformer. Power transformers in operation make noise, produced by an overlapping of different sources. On the one hand, no-load noises are produced by vibrations caused by the magnetostrictive deformation of the core. On the other hand, electromagnetic forces in the windings, tank walls and magnetic shielding predominantly influence the load noise. In addition, cooling systems using fans have a large influence on sound emissions [120]. Generally, the vibration will be attenuated in the process of transmission through the insulating liquid. The temperature of the insulating liquid and its change has a particular impact on vibration transmission. Increasing temperature in insulating liquid will decrease the attenuation degree of vibration transmission in insulating liquid [121]. The determination of sound levels of transformers is regulated in the standard IEC 60076-10 [122].

5.6 Passivators

In the past, there have been several failures of large power transformers and shunt reactors associated with the presence of corrosive sulfur in the transformer insulating liquid. This problem occurs even though the insulating liquids have passed specification test criteria for corrosive sulfur, such as ASTM D1275 [61]. These failures occur because corrosive sulfur in the insulating liquid reacts with copper to form copper sulfide, a conductive compound. The copper sulfide formed at the copper surface can also migrate to the paper insulation. The conductive copper sulfide causes a reduction in dielectric strength of the paper insulation. Failure results when the dielectric breakdown strength of the conductor insulation is exceeded by the voltage stress, which can be influenced by transient voltages. The result is arcing between two or more turns or possibly disks [123]. There is still a population of transformers and reactors that are in service with insulating liquids that have excessive corrosive sulfur or a propensity to form it.

There is not only the cost of the device but also loss of revenue and possible regulatory fines if power is not delivered as specified in contracts. This concern about corrosive sulfur has led several transformer manufactures and oil refiners to recommend the use of passivators by adding them to the insulating liquid. There are different kinds of passivators that can be used (Sect. 2.7.3). For the corrosive sulfur problem, a passivator concentration of 100 ppm is recommended. The passivator decreases the catalytic effect of the copper and the amount of copper dissolved in the insulating liquid. This reduces the oxidation of the insulating liquid and formation of deleterious by-products.

Herath et al. reported in [124] about abnormal increments of hydrogen and carbon monoxide when using Irgamet® 39 (N-amino methylated benzotriazole) as a passivator. This needs to be considered avoiding a misleading interpretation of DGA data in facilitating asset management decisions on transformers.

Once copper sulfide is formed on copper surfaces and deposited in the paper insulation, it is not removed by passivator or other means. The passivator undergoes a variety of reactions in the apparatus and can be degraded or consumed and might need to be added periodically to retain sufficient binding on the copper surface to block the corrosive sulfur reactions. For new insulating liquids, the best approach is to use a liquid that does not have significant amounts of corrosive sulfur compounds. In case where insulating liquids have been used that do not meet these stringent criteria, passivator could be used to retard further corrosion. An alternative approach is to use full or partial retrofill of the insulating liquid with an insulating liquid that does not have significant amounts of corrosive sulfur compounds even, when exposed to very high temperatures. A regeneration of the insulating liquid could be a way to reduce chemical compounds which are responsible for forming copper sulfide.

5.7 Transformer Dielectric Liquid Regeneration

Low values of the dielectric strength of the insulating liquid or high dielectric loses, therefore, do not necessarily mean that the insulation system is compromised, but rather that the insulating liquid is dirty. Besides acids and sludges, many polar oxidation products can be formed during insulating liquid oxidation. Those containing –OH groups (such as phenols and alcohols) are generally more polar than those only containing aldehydes and ketones. The later are oxidation precursors, which often precede the formation of acids and sludge and have a marked influence on the interfacial tension and the dielectric properties of the insulation liquid. Cleaning up the contaminated insulating liquid and drying the solid insulation (when necessary) are often enough to restore both the insulating liquid and the solid insulation to their original quality. Only in the case of thick solid insulation layers and prolonged contact with badly contaminated insulation liquid will cleaning of the solid insulation also be necessary.

During the manufacture of electrical equipment, considerable attention is given to meeting levels of dryness appropriate to the type and voltage class of particular equipment. The residual water in the insulation system redistributes itself between cellulose insulation and insulating liquid as the temperatures changes in operation. The saturation value for water in cellulose is decreased as the temperature increases, while the saturation value in an insulating liquid is increased. Water then moves from paper to insulating liquid as the temperature increases and the system reaches toward a new equilibrium in which the relative saturation of the two phases is the same. When the insulation system temperature falls, water is sucked back into the cellulose. Water from two sources can contaminate equipment during service. Water is the primary product of the degradation of cellulosic paper and press board insulation during aging. Water can also be absorbed from moist air through leaks or in free-breathing equipment. Regardless of its source, water will cycle between cellulose and insulating liquid as the temperature cycles.

The saturation limits of water in dielectric insulating liquids increase with increasing temperatures (see Fig. 4.8). Water is more harmful to dielectric strength in combination with fibers and other particulates, which become more conductive when wet. Water alone affects the dielectric strength as saturation is approached.

The nature of the particles (metal, dirt, fiber, sludge) certainly plays a major role, as well as the nature of the metal (iron, aluminum, copper, etc.) in the particles.

Apart from the products resulting from oxidation and stress decomposition of insulating liquid and cellulose in service, soluble contaminants may result from refining treatments, which may leave traces of sulfates, chlorides, phenols etc. Ferrous metals and fluxes used for brazing, welding, and soldering can dissolve and affect the insulating liquid properties [86].

5.7.1 *Reconditioning, Regeneration-Reclamation Process*

Insulating liquid reclaiming is a major restoration done mainly to increase the life of the insulating liquid in a transformer. It is known that mineral oil degradation is reversible by appropriate treatment (regeneration, reconditioning, and re-refining). Reconditioning is defined as the removal of dissolved water, gasses and solid materials by mechanical means while reclaiming (regeneration) is defined as the removal of acidic and colloidal contaminants and oxidized matter by chemical and absorbent means. Hence, the type of decay particles needs to be known as this allows choosing the best regeneration method and adsorbent for biological insulating liquids. The reclamation aspects of mineral oil used in liquid filled electrical apparatus are reported in the IEEE standard C57.637 [125]. Presently, there is no established standard for the reclamation of ester liquids [126]. The mechanical means that are used for removing water and solids from liquids include several types of filters, centrifuges, and vacuum dehydrators. However, high treatment temperatures influence the oxidation stability of biological insulating liquids and exhibit a notable impact on the weight of the available antioxidants in the bulk of the insulating liquid. In general, water removal filters and vacuum dehydrators should be placed before the final particulate removal filters. If the water is in the form of free water, filter elements utilizing blotter paper have been used effectively. Filter cartridges packed with moisture absorbing media are recommended to help achieve desired dryness. Proper care and storage of the water absorbing filters is essential to help ensure they do not absorb moisture before use. Most types of filters now being used on mineral oil can be used for biological insulating liquids.

Most power equipment today is using transformer insulating liquids for its dielectric properties for cooling, insulating, and protecting the active part. Transformer insulating liquid are highly refined insulating liquid that consists mainly of a mixture of hydrocarbons or vegetable oil-based liquids. Over time, oxidation by-products such as water and acids start to form in the oil. Increases in oxidation by-products result in the increase of acidity (neutralization number), high dielectric losses, and a decrease in the interfacial tension of the insulating liquid. As the chemical composition of natural and synthetic esters is different from that of mineral oil, the aging processes is in some ways different, too. Investigations of Mohan Rao et al. showed in [125] that in case of thermal aging mineral oil produced colloidal particles at temperatures up to 150 °C, whereas in the case of ester liquids they were not evident up to temperatures 175 °C. They found that ester liquids generate less sludge or colloids will have a high scope for the generation of soluble particles. However, it is challenging to comment that ester dissolves sludge or does not generate any sludge with degradation. Thus, one may understand that the treatment of esters may be mostly focused on the removal of soluble particles along with water and dissolved gasses. The author's result is that the soluble decay content mostly governs the degradation in ester liquids and the suitability of Fuller's earth is questionable. One must be aware that it is easier to remove the high molecular weight acids than the low molecular weight acids, but that the normal measurement of acidity does not discriminate between the different groups of acids.

Transformer insulating liquid will discolor as oxidation of the insulating liquid takes place. Once transformer insulating liquid changes from the colorless to yellow color range into the brown and black range, it has degraded to the point where the vital parts of the transformer are being seriously affected. If the insulating liquid has its own color, this must be considered in the assessment. In any case, the color of the liquid itself is not the only defining characteristic of transformer insulating oil. It does, however, work as an indicator of adverse or good changes in the liquid. Figure 5.54 shows the change in the color because of its regeneration process. On the right-hand side, you see the insulating liquid before and on the left-hand side after the regeneration process. It should also be mentioned, and this after just one run.

At this stage, sludge starts to form when mineral oil liquids are used, and the insulating liquid is losing its dielectric properties. Deposited sludge continues to oxidize and harden, blocking vents and insulating cooling fins, causing higher operating temperatures. Insulation shrinkage may take place, and premature failure is possible.

To prevent further deterioration of the insulating liquid and the possibility of damaging the active part of the transformer, insulating liquid needs to be regenerated. Oil regeneration equipment regenerates the insulating liquid in steps:

- Preferably, reclaiming should be done on transformers under load to keep a high temperature and thereby increasing diffusion rate in the cellulose and keeping the concentration of aging by-products in the insulating liquid high.
- At the inlet of the regeneration equipment, insulating liquid is filtered through a coarse filter to prevent any particles from entering the equipment.
- The transformer should be in service at the highest temperature possible, because of the solubility of water. Additional heating to reach the desired temperature to

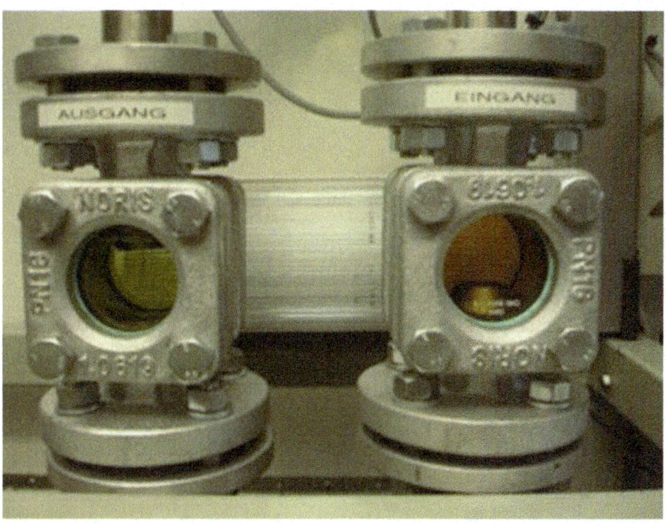

Fig. 5.54 Effectivity of a regeneration process

elevate the regeneration effect should be foreseen. After the insulating liquid has been heated, it enters the regeneration equipment.

- Insulating liquid is pumped through columns with sorbent media (Fuller's Earth). Just as when selecting pumps, care should be taken in selecting a filter for biological insulating liquids. Because biological insulating liquid viscosities are higher than those for mineral oil, larger filters or higher liquid temperature may be required to achieve the same flow rate. Filters of the adsorption type, such as activated Fuller's earth, can be used; however, certain pour point depressant, passivator and antioxidant additives can be removed from the liquid by these filters if the vacuum is too high. It is in this section where the insulating liquid is stripped of impurities and aging byproducts. If the dissolved water content should be lowered, a high vacuum dehydration system may be required. The vacuum dehydrator is an efficient means of reducing the gas and water content of an insulating liquid to a very low value. In addition to removing water, processing natural ester insulating liquid with a vacuum dehydrator also extracts dissolved gases, normally removing any volatile acids. Insulating liquid is then pumped through a vacuum breaking valve into the degassing section where it is dehydrated and degassed. Treated insulating liquid is then pumped back to the transformer by the outlet pump. After a given period, sorbent in the back section of the equipment achieves full saturation and is no longer able to regenerate oil.
- At this stage, the sorbent needs to be reactivated (restored to its original state) to be able to regenerate oil again. Reactivation stage starts by draining the columns in the back section of saturated insulating liquid. After the insulating liquid has been drained, a vacuum is created and maintained throughout the whole reactivation process in the back section.
- Then, by selective use of heating elements on the top parts of individual columns, the reactivation process is initiated. During the process, the impurities are removed from the sorbent, which restores it to its original state.
- This entire process can be repeated many times until the sorbent starts to lose its properties and needs to be replaced.
- After reclaiming the insulating liquid needs to be inhibited—also in the case of uninhibited liquids, as the reclaiming process removes all inhibitors, also the natural one.

Aldehydes and ketones cannot always be removed completely by reclaiming processes, so its content in insulating liquid should be closely monitored. This can be done by infrared spectrometry in the 1700–1780 cm^{-1} wavelength region of carbonyl vibration or, more accurately, by HPLC techniques in the reverse-phase mode.

5.7.1.1 Fuller's Earth

Fuller's Earth can be used to decolorize and neutralize any petroleum and vegetable oil, especially insulating liquids. Fuller's earth treatment is a traditional practice of rejuvenating mineral insulating oil. It excels in neutralizing traces of strong inorganic acid. Due to the relatively large pores in Fuller's earth, it is well adapted to the removal of high molecular weight sulfonates, resins, and asphaltenes.

The Fuller's Earth adsorbent material is a very common kind of clay that has the ability to remove harmful contaminants from vegetable or mineral oils and fats. The Fuller's Earth adsorbent material has been accepted as the highest standard of filtering quality when using technologies designed for the recovery and purification of industrial oils. The active components of Fuller's Earth are widely used in petrochemical industries worldwide. Additionally, Fuller's Earth is quite successfully used in the construction and pharmaceutical industries, as well as thanks to the specific properties that enable the sorbent materials to adsorb and remove a wide range of unwanted fluids and contaminants.

In past centuries, the use of Fuller's Earth was very different from the modern-day usage of this unique and versatile material. The initial application of Fuller's Earth, for the extraction of oil and other fatty impurities from wool fabrics in the finishing process, gave Fuller's Earth its modern name. It also applies to the clay that can absorb harmful particles that serve as an antiseptic or bleach the fabric. Fuller's Earth was used for the first time in the United States for purification of vegetable oils starting in 1878. In modern day industry, Fuller's Earth continues to be used to purify new transformer liquid before being placed into service. The Fuller's Earth adsorbents can remove moisture, gases, and other noxious particles from the insulating liquids. Using a Fuller's Earth regeneration process is a perfect filtering medium for restoring the performance characteristics and properties of used and contaminated transformer insulating liquids.

Apart from Fuller's Earth, molecular sieve filters have been found effective for removing dissolved water from biological insulating liquid. Activated grade 3A or 4A molecular sieves are recommended for water removal from biological insulating liquid and are effective over a broad temperature range [29].

5.8 Economic and Ecological Consideration

Every industrial operation produces remains and discarded yields that may impact the atmosphere, particularly which are categorized as non-renewable and non-eco-friendly. The requirement of clean expertise and residue handling has surfaced as one of the most significant concerns of recent times, demanding explicit activities to inhibit debris production and stimulate recycle and reuse actions to decrease ecological impacts [100].

Vasconcellos et al. write in [127] that the purchase price of a compact green transformer compared with that of a conventional mineral oil transformer is 5–12% higher because natural esters are used. That fits very well with the statements about the distribution and power transformers in Austria installed in the years 2016 to 2019. A higher price range of 5–10% was found here. Their higher price compared to mineral oil has been one barrier in the past to the more widespread adoption of ester liquids.

However, in more recent times, users are realize that there are significant savings to be made in overall installation costs if ester-based liquids are used in place of mineral oil. The fire safety benefits alone mean that fire barriers can be removed, and active fire suppression systems are no longer required, since the safety is effectively built into the transformers. Factoring in the reduction in civil engineering costs and the potential longer life of ester transformers, this solution starts to look very attractive [128]. The benefits that ester can bring in fire protection and increased transformer life can financially outweigh the extra cost of using an ester. But the price of vegetable oil is subject to significant oscillation caused by the food market, since generally, the feedstock is also used in edible oils.

On closer inspection, one sees that the biological insulating liquids are not a direct competitor to the food chain, but rather a supplement. For example, when using soybean oil, there is no direct competition to the food chain.

Soybean cultivation, which has risen sharply worldwide in the last decades, is often the subject of criticism. Very often justified, because the ecological, economic, and social effects are serious. No other cultivated plant has developed so a dynamically in recent years [129]. The main reason for the continuous expansion of global soybean cultivation is the increasing meat consumption because most of the soybean harvest ends up in animal feed (Fig. 5.55). The increase in soybean acreage in Europe will slow down the expansion of acreage in sensitive regions like this in South America. In 2021, soybean cultivation in Germany reached a historic high of around 35,000 hectares. The main areas of cultivation were Bavaria and Baden-Württemberg. Soybeans are already the most important grain legume in cultivation here—ahead of peas, broad beans, and lupins. Market participants expect further strong growth in 2022 and expect a cultivation volume of 100,000 ha in the medium term [130].

Like all legumes, the soy plant has so-called nodule bacteria in its root network. This allows them to supply themselves with nitrogen from the air and is considered as a nitrogen collector. The plant needs this to form protein from it. The nodule bacteria are the only ones that can do this. The soy plant is more efficient than other legumes

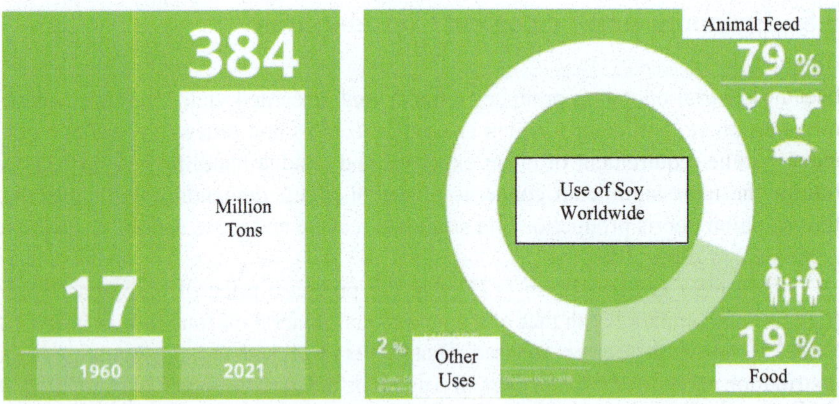

Fig. 5.55 Global soybean production and uses [129]

Fig. 5.56 Renewable energy—closed material cycle

in this regard. Therefore, it requires less fertilizer than other plants. This is directly linked to energy and CO_2 savings. Due to the high protein feed requirement caused by animal feed production, soybean oil is one of the most produced vegetable oils on our planet alongside palm oil. A large part is used to produce biodiesel. This use is also possible after using as an insulating liquid. To do this, the material cycle is opened on the left side of the Fig. 5.56 and as depicted on the right side, the insulating liquid is added as an additional source.

In this context, Jatropha curcas oil points out as a potential source for a new vegetable insulating liquid because of some beneficial characteristics over commercial insulating liquids available. Jatropha curcas oil price does not compete on the food market since the plant is non-edible and is not used for food and animal feeding. The reason for this is that the plant contains a class of toxins known as phorbol esters, which concentration in the seeds depends on the soil and climate conditions [84]. Most part of phorbol esters present in the crude vegetable oil can be extracted with methanol before its processing. In terms of cost saving, even if the biological insulating liquids are more expensive, the removal of ancillary equipment such as fire extinguishers, or reductions in containment, can give big savings and very quickly offset the extra capital expense. Not to forget the argument about the use of renewable raw materials. In addition, there is evidence to suggest that Kraft paper will live much longer if immersed in an ester, when compared to mineral oil, and this additional lifetime can significantly reduce overall cost of an installation considering the whole lifetime.

References

1. Dames, Corné. 2022. Transformer Reliability by Design. *Transformer Technology* (21).
2. IEEE C57.12.00. 2015. IEEE Standard for General Requirements for Liquid-Immersed Distribution, Power, and Regulating Transformers.

3. IEEE C57.154. 2012. IEEE Standard for the Design, Testing and Application of Liquid-Immersed Distribution, Power, and Regulating Transformers Using High-Temperature Insulation Systems and Operating at Elevated Temperatures.
4. IEC 60076-1. 2011. Power Transformers—Part 1: General.
5. IEC 60076-2. 2011. Temperature Rise for Liquid-Immersed Transformers.
6. IEC 62770. 2013. Fluids for Electrotechnical Applications—Unused Natural Esters for Transformers and Similar Electrical Equipment.
7. IEC 61099. 2010. Specification for Unused Synthetic Organic Esters for Electrical Purposes.
8. IEC 60296. 2020. Fluids for Electrotechnical Applications—Mineral Insulating Oils for Electrical Equipment.
9. Transformers Magazine 2016. 3 (3).
10. Yuerekten, S. 2021. *Transformer Components—Properties and Applications*. Birikim Ofset Matbaacilik Ltd..
11. Fellers, C., et al. 1989. Ageing/Degradation of Paper, A literature survey, Report No. 1E, ISSN 0284-5636, Stockholm, September 1989.
12. PTS. 2008. The Paper Technology Specialists; Technologische Darstellung der Alterungsbeständigkeit von Papier und Karton—Vergleich zwischen Baumwollzellstoff und Holzzellstoff; 19 September 2008.
13. Münster, T., et al. 2021. Thermally Accelerated Aging of Insulation Paper for Transformers with Different Insulating Liquids. *Energies* 14: 3036..
14. Lampe, W., and E. Spicar. 1976. *Oxygen-Free Transformer, Reduced Ageing by Continuous Degassing*. Paris: CIGRE, paper 12–05.
15. Ramalhoetal. O. 2008. *Emission Rates of Volatile Organic Compounds from Paper*. Austria: IAQ 2008, April 2008.
16. Kanumuri, D., et al. 2019. Analysis Using Various Approaches for Residual Life Estimation of Power Transformers. *International Journal on Electrical Engineering and Informatics* 11 (2), June 2019. https://doi.org/10.15676/ijeei.2019.11.2.11.
17. Christofidis, S. 2006. *Analytical Methods for Assessing Paper Degradation in Transformers*. Cranfield University.
18. Ortiz, F., et al. 2016. Estimating the Age of Power Transformers Using the Concentration of Furans in Dielectric Oil. In *International Conference on Renewable Energies and Power Quality (ICREPQ'16)*. Madrid. https://doi.org/10.24084/repqj14.557.
19. Li, S., et al. 2019. A New Technique to Estimate the Degree of Polymerization of Insulation Paper Using Multiple Aging Parameters of Transformer Oil. *IEEE Access*. https://doi.org/10.1109/ACCESS.2019.2949580..
20. Prasojo, R.A., et al. 2017. Transformer Paper Expected Life Estimation Using ANFIS Based on Oil Characteristics and Dissolved Gases (Case Study: Indonesian Transformers). *MDPI Energies* 10: 1135. https://doi.org/10.3390/en10081135..
21. Wang, Z., et al. 2012. Transmission Power Transformer Assessment Using Furan Measurement with the Aid of Thermal Model. In *Proceedings of 2012 IEEE International Conference on Condition Monitoring and Diagnosis, CMD 2012*, https://doi.org/10.1109/CMD.2012.6416194.
22. Cigre WG D1.53. 2018. Ageing of Liquid Impregnated Cellulose for Power Transformers. *Brochure* 738.
23. Lundgaard, L., et al. 2004. Aging of Oil-Impregnated Paper in Power Transformers. *IEEE Transactions on Power Delivery*, 19 (1), January 2004.
24. Gasser, H.P., et al. 2011. Aging of Pressboard in Different Insulating Liquids. In *IEEE International Conference on Dielectric Liquids*. Trondheim.
25. Judendorfer, T. 2012. Oil-cellulose Insulation Systems for HVDC Applications, Doctoral Thesis, Graz University of Technology, April 2012.
26. Yuliastuti, E. 2010. Analysis of Dielectric Properties, Comparison Between Mineral Oil and Synthetic Ester Oil, Thesis, Faculty of Electrical Engineering, Mathematics, and Computer Science, Delft University of Technology, June 2010.
27. Sbravati, A. 2014. *A Practical Approach to Designing More Cost-Efficient Transformers, Version 1.0*. Cargill, November 2014.

28. Podesser, J.A. 2015. Vergleich der elektrischen Aufladung alternativer und konventioneller flüssig/fester Isoliersysteme für Leistungstransformatoren, Doctoral Thesis, Graz University of Technology, September 2015.
29. IEEE C57.147. 2018. IEEE Guide for Acceptance and Maintenance of Natural Ester Liquids in Transformers.
30. Pagger, E., et al. 2018. Natural Ester FR3 Insulating Liquid—Very Paper Friendly. In *12th IEEE International Conference on the Properties and Applications of Dielectric Materials*. Xi'an, China.
31. Tenbohlen, S., et al. 2008. Application of Vegetable Oil-based Insulating Fluids to Hermetically Sealed Power Transformers, A2-102, CIGRE.
32. Fink, H. et al. 2011. New and Innovative Smart and Green Transformer Technologies. In *21st International Conference on Electricity Distribution*. Frankfurt, June 2011.
33. Zimmermann, N., and R. Bass. 2014. Consideration of Ester-Based Oils as Replacements for Transformer Mineral Oil. In *Portland State University, Maseeh College of Engineering & Computer Science Department of Electrical & Computer Engineering*.
34. Dielectric Fluids. 2016. Envirotemp™ FR3™ FluidStorage and Handling Guide, S10. Cargill.
35. Ghani, A.S., et al. 2018. Methods for Improving the Workability of Natural Ester Insulating Oils in Power Transformer Applications: A review. *Electrical Power Systems Research*, 163. https://doi.org/10.1016/j.epsr.2017.10.008.
36. Moore, S.P., et al. 2015. Cold Start of a 240-MVA Generator Step-Up Transformer Filled With Natural Ester Fluid. *IEEE Transactions on Power Delivery* 30.
37. Cargill®. 2017. Cold Start Recommendations for Envirotemp™ FR3™ Fluid Filled Transformers, R2120, August 2017.
38. ISO 31000:2018(en). 2018. Risk Management—Guidelines.
39. Hellebuyck, D. 2013. Functional performance criteria for comparison of less flammable transformer oils with respect to fire and explosion risk. In *Report 5424, Department of Fire and Safety Engineering and Systems Safety*. Lund University.
40. Schreiner, Z. 2017. IT-unterstützte Optimierte Instandhaltungsstrategie, Basierend auf Kombination von Zeit, Zustand, Risiko und RCM-Konzept. Oberlausitzer Energie-Symposium, Zittau, Oktober 2017.
41. DelFiacco, G. et al. 2013. Proactive Fire Safety for Transformers, Version 1.0. Cargill, June 2013.
42. IEC 61039. 2008. Classification of Insulating Liquids.
43. IEC 61936-1. 2010. Power Installations Exceeding 1 kV a.c.—Part 1: Common Rules.
44. Da Silva, R.I., and A. Coelho. 2019. Recognition of the Benefits by the Use of Transformers and Reactors Immersed in Insulating Fluid with High fire Point (K-Class): The Case of Fire Department of Sao Paulo state in Brazil. In *CIGRE Canada Conference*.
45. DelFiacco, G. 2013. Retrofilling Transformers: A Financial Perspective, Version 1.0. Cargill, June 2013.
46. Hot Spot Temperature Sensor Installation Guide MAN-00098 R2. 2020. FISO Technologies Inc..
47. IEC 60076-7. 2018. Power Transformers—Part 7: Loading Guide for Oil-Immersed Power Transformers.
48. ASTM D92-18. 2018. Standard Test Method for Flash and Fire Points by Cleveland Open Cup Tester.
49. ISO 2719. 2016. Determination of flash point—Pensky-Martens Closed Cup Method.
50. ISO 2592. 2017. Petroleum and Related Products—Determination of Flash and Fire Points. Cleveland Open Cup Method.
51. ASTM D445-21. 2021. Standard Test Method for Kinematic Viscosity of Transparent and Opaque Liquids (and Calculation of Dynamic Viscosity).
52. ISO 3104. 2020. 3rd Edition, Petroleum Products—Transparent and Opaque Liquids—Determination of Kinematic Viscosity and Calculation of Dynamic Viscosity.
53. ASTM D97-17b. 2017. Standard Test Method for Pour Point of Petroleum Products.

54. ASTM D5949-16. 2016. Standard Test Method for Pour Point of Petroleum Products (Automatic Pressure Pulsing Method).
55. ASTM D5950-14. 2020. Standard Test Method for Pour Point of Petroleum Products (Automatic Tilt Method).
56. ISO 3016:1994. 2017. Petroleum Products—Determination of Pour Point, (DIN ISO 3016).
57. ASTM D1500-12. 2017. Standard Test Method for ASTM Color of Petroleum Products (ASTM Color Scale).
58. ASTM D1298-12b. 2017. Standard Test Method for Density, Relative Density (Specific Gravity), or API Gravity of Crude Petroleum and Liquid Petroleum Products by Hydrometer Method.
59. ISO 3675. 1998. Crude Petroleum and Liquid Petroleum Products—Laboratory Determination of Density—Hydrometer Method.
60. ISO 12185. 1996. Crude Petroleum and Petroleum Products—Determination of Density—Oscillating U-tube Method.
61. ASTM D1275-15. 2015. Standard Test Method for Corrosive Sulfur in Electrical Insulating Liquids.
62. IEC 62535. 2009. Insulating Liquids—Test Method for Detection of Potentially Corrosive Sulphur in used and Unused Insulating Oil.
63. IEC 62697-1. 2012. Test Methods for Quantitative Determination of Corrosive Sulfur Compounds in Unused and Used Insulating Liquids—Part 1: Test Method for Quantitative Determination of Dibenzyl Disulfide (DBDS).
64. IEC 60666. 2010. Detection and Determination of Specified Additives in Mineral Insulating Oils.
65. ASTM D974-14e2. 2014. Standard Test Method for Acid and Base Number by Color-Indicator Titration.
66. IEC 62021-3. 2014. Insulating Liquids—Determination of Acidity—Part 3: Test Methods for Non-mineral Insulating Oils.
67. ASTM D1533-20. 2020. Standard Test Method for Water in Insulating Liquids by Coulometric Karl Fischer Titration.
68. IEC 60814. 1997. Insulating Liquids—Oil-Impregnated Paper and Pressboard—Determination of Water by Automatic Coulometric Karl Fischer Titration.
69. ASTM 1816-12. 2019. Standard Test Method for Dielectric Breakdown Voltage of Insulating Liquids Using VDE Electrodes.
70. IEC 60156. 2018. Insulating Liquids—Determination of the Breakdown Voltage at Power Frequency—Test Method.
71. ASTM D924-15. 2015. Standard Test Method for Dissipation Factor (or Power Factor) and Relative Permittivity (Dielectric Constant) of Electrical Insulating Liquids.
72. IEC 60247. 2004. Insulating Liquids—Measurement of Relative Permittivity, Dielectric Dissipation Factor (tan δ) and d.c. Resistivity.
73. ASTM D3300-20. 2020. Standard Test Method for Dielectric Breakdown Voltage of Insulating Liquids Under Impulse Conditions.
74. OECD 301. 1992. OECD Guideline for Testing of Chemicals, Ready Biodegradability.
75. EPA. 1998. OPPTS 835.3110 Ready Biodegradability, Fate, Transport, and Transformation Test Guidelines.
76. IEC 61125. 2018. Insulating Liquids—Test Methods for Oxidation Stability—Test Method for Evaluating the Oxidation Stability of Insulating Liquids in the Delivered State.
77. ASTM D1524-15. 2015. Standard Test Method for Visual Examination of Used Electrical Insulating Liquids in the Field.
78. ASTM D3612-02. 2017. Standard Test Method for Analysis of Gases Dissolved in Electrical Insulating Oil by Gas Chromatography.
79. IEC 62975. 2021. Natural Esters—Guidelines for Maintenance and Use in Electrical Equipment.
80. ISO 2049. 2001. Petroleum Products—Determination of Color (ASTM scale) (ISO 2049:1996).

81. ASTM D1544-04. 2018. Standard Test Method for Color of Transparent Liquids (Gardner Color Scale), 2018.
82. IEEE C57.152. 2013. IEEE Guide for Diagnostic Field Testing of Fluid-Filled Power Transformers, Regulators, and Reactors.
83. IEC 62961. 2018. Insulating Liquids—Test Methods for the Determination of Interfacial Tension of Insulating Liquids—Determination with the Ring Method.
84. Evangelista, Jr. et al. 2017. Development of a New Bio-Based Insulating Fluid from Jatropha Curcas Oil for Power Transformers. *Advances in Chemical Engineering and Science*, 7, 235–255.
85. Perrier, C. et al. 2006. Improvement of Power Transformers by using Mixtures of Mineral Oil with Synthetic Esters. *IEEE Transactions on Dielectrics and Electrical Insulation.*
86. Bartnikas, R. 1994. Engineering Dielectrics, Electrical Insulating Liquids. III, ASTM, January 1994.
87. Maneerot, S., et al. 2018. Studies of Electrical and Thermal Characteristics of Natural Ester Immersed Transformer Compared with Mineral Oil Immersed and Palm Oil Immersed Transformer. In *International Conference on Condition Monitoring and Diagnosis, CMD*. Perth. https://doi.org/10.1109/CMD.2018.8535807.
88. IEEE C57.93. 2019. IEEE Guide for Installation and Maintenance of Liquid-Immersed Power Transformers.
89. Hajek, J. et al. 2012. Considerations for the Design, Manufacture, and Retro—Filling of Power Transformers with High Fire Point, Biodegradable Ester Fluids. A2_203. CIGRE.
90. Pagger, E. 2013. Alternative Insulating Liquids Compared to the Classic Mineral Oil, Doctoral Thesis, Graz University of Technology, May 2013.
91. Der Houhanession, V. 1998. Measurement and Analysis of Dielectric Response in Oil-Paper Insulation Systems, Doctoral Thesis. ETH Zürich.
92. Lashbrook, M., et al. 2013. Synthetic Ester for Power Transformers at > 100 kV. *ITMA Journal*, July/August 2013.
93. Pagger, E. et al. 2011. Alternative Insulating Fluids in Comparison to Traditional Mineral Oil. VGB Conference, Munich.
94. Pukel, G., et al. 2016. Safe and Environmentally Friendly Large Power Transformers with Ester. In *Electrical Insulation Conference (EIC)*. Montréal, Canada, 19-22 June 2016.
95. Valdo, K. 2014. CPFL Energia Press Relations, CPFL adopts Cargill's Vegetable Fluid in Its Distribution Network. São Paulo, 10 April 2014.
96. Frotscher, R. 2013. Alternative Flüssigkeiten für Stufenschalter, Maschinenfabrik Reinhausen GmbH.
97. Gomna, A., et al. 2016. Review of Vegetable Oils Behavior at High Temperature for Solar Plants: Stability, Properties and Current Applications, ScienceDirect. *Solar Energy Materials and Solar Cells* 200. https://doi.org/10.1016/j.solmat.2019.109956.
98. ISO 6885:2016. 2021. Animal and Vegetable Fats and Oils—Determination of Anisidine Value.
99. Yan, Z.W., et al. 2020. Measuring Low Molecular Weight Acids in Mineral and Ester Transformer Liquids. In *The 8th International Conference on Condition Monitoring and Diagnosis (CMD 2020)*. Thailand, October 2020.
100. Rafiq, M., et al. 2020. Sustainable, Renewable and Environmental-Friendly Insulation Systems for High Voltage Applications. *Review, MDPI Molecules*. https://doi.org/10.3390/molecules 25173901..
101. Martin, D., et al. 2010. Preliminary Results for Dissolved Gas Levels in a Vegetable Oil-Filled Power Transformer. *IEEE Electrical Insulation Magazine*, November 2010. https://doi.org/10.1109/MEI.2010.5585007.
102. Xiang, C., et al. 2016. Comparison of Dissolved Gases in Mineral Oil and Insulating Vegetable Oils under Typical Electrical and Thermal Faults. *MDPI Energies*. https://doi.org/10.3390/en9050312..
103. ASTM D3284-05. 2019. Standard Practice for Combustible Gases in the Gas Space of Electrical Apparatus Using Portable Meters.

104. IEC 60475. 2012. Method of Sampling Insulating Liquids.
105. IEC 60567. 2011. Oil-Filled Electrical Equipment—Sampling of Gases and Analysis of Free and Dissolved Gases—Guidance.
106. TOP TOGA GC. 2016. ENERGY SUPPORT, 02/2016.
107. IEEE C57.104. 2019. IEEE Guide for the Interpretation of Gases Generated in Mineral Oil-Immersed Transformers.
108. IEEE C57.146. 2011. IEEE Guide for Interpretation of Gasses Generated in Silicone-Immersed Transformers.
109. Cigré Brochure 443. 2010. DGA in Non-Mineral Oils and Load Tap Changers and Improved DGA Diagnosis Criteria, December 2010.
110. Online-DGA. 2018. Dissolved Gas Analyzes On Site, Description, ENERGY SUPPORT, 12/2018.
111. MOBILE GC. 2018. Portable Gas Chromatograph, ENERGY SUPPORT, 01/2018.
112. Hanson, D. et al. 2010. Understanding Dissolved Gas Analysis of Ester Fluids—Part 1: "Stray" Gas Production under Normal Operating Conditions, Siemens Transformer Conference.
113. IEC 60599. 2015. Mineral Oil-Filled Electrical Equipment in Service—Guidance on the Interpretation of Dissolved and Free Gases Analysis.
114. Theuermann, et al. 2020. Verbesserung der Zustandsbeurteilung von Messwandlern Mittels DGA, Österreichs E-Wirtschaft.
115. Duval, M., L. Lamarre. 2017. The New Duval Pentagons Available for DGA Diagnosis in Transformers Filled with Mineral and Ester Oils. *IEEE Electrical Insulation Conference, Baltimore*. https://doi.org/10.1109/EIC.2017.8004683.
116. Duval, M., and L. Lamarre. 2014. The Duval Pentagon—A New Complementary Tool for the Interpretation of Dissolved Gas Analysis in Transformers. *IEEE Electrical Insulation Magazine*. https://doi.org/10.1109/MEI.2014.6943428..
117. IEC 60450.2004. Measurement of the Average Viscometric Degree of Polymerization of New and Aged Cellulosic Electrically Insulating Materials.
118. ASTM D5837-15. 2015. Standard Test Method for Furanic Compounds in Electrical Insulating Liquids by High-Performance Liquid Chromatography (HPLC).
119. IEC 61198. 1993. Mineral Insulating Oils—Methods for the Determination of 2-Furfural and Related Compounds.
120. Stirl, T. et al. 2011. New Trends in Noise Reduction of Power Transformers. In: *21st International Conference on Electricity Distribution*. Frankfurt, June 2011.
121. Gao, S., et al. 2020. On-line Monitoring and Data Analysis Based on Vibro-Acoustic Signals of High-Voltage Shunt Reactor. In *The 8th International Conference on Condition Monitoring and Diagnosis (CMD 2020)*. Thailand, October 2020.
122. IEC 60076-10. 2016. Power Transformers—Part 10: Determination of Sound Level.
123. Lewand, L. 2006. *Passivators—What They are and How They Work*. Double Engineering Company, NETA WORLD, Spring 2006.
124. Herath, T., et al. 2020. Observation of Abnormal Gassing Rise After Metal Passivation in Transformers. In *The 8th International Conference on Condition Monitoring and Diagnosis (CMD 2020)*. Thailand, October 2020.
125. IEEE C57.637. 2015. IEEE Guide for the Reclamation of Mineral Insulating Oil and Criteria for Its Use.
126. Mohan Rao, U. et al. 2022. Decay Particles and Regeneration of Ester Dielectric Liquids, A Challenge. *Transformer Technology* 17, January 2022.
127. Vasconcellos, V., et al. 2014. From Prototype to Standardization, 10 Years or Research and Innovation Produces the Compact Green Transformer. *Transmission & Distribution*, November 2014.
128. Lashbrook, M. et al. 2016. *The Development of 400 kV Transformers with Ester-based Dielectric Liquids, ARWtr2016*. La Toja Island, Spain, October 2016.
129. Soja aus Österreich. 2022. https://soja-aus-oesterreich.at/soja-aus-oesterreich/.
130. Deutscher Soja Förderring. 2022. https://www.sojafoerderring.de.

Chapter 6
Advanced Research in the Field of Biological Insulting Liquids

This chapter presents and explains further research in connection with the use of biological insulating liquids. Results based on laboratory tests should be used for later up-scaling and show the limits of biological insulating liquids. The good material compatibility of the biological insulating liquids is demonstrated based on laboratory tests. Possibilities for improving the dielectric properties by using nanoparticles are presented. The use of the good thermal properties of the biological insulating liquids in the course of the volatility of the power grid is shown. Conductivity measurements on impregnated solid insulation should serve the design of HVDC systems.

The improvement of the characteristics of power transformers is an ongoing duty for manufacturers, since this equipment constitutes one of the most expensive and highly strategic components of electric power transmission and distribution systems. Any failure of such components can result in an ecological disaster and/or economic losses. Hence, the manufactures must try to improve the reliability of these devices by reduction of the costs at the same time. Not only the manufacturing costs are to be considered but also the costs in service which incurred during a life cycle. Not to forget the costs that can arise in the event of an incident.

One way of reducing costs is to design a transformer with a smaller size by reducing the insulation gap. Insulating liquids must ensure the integrity of this gap for the existing or even higher voltage levels, and to enable the cooling to be still effective. The most used liquid in power transformers is mineral oil due to its good electrical properties and its low price. However, with the new ecological requirements, mineral oil seems to lose its standing. For that purpose, numerous activities have been initiated to improve the properties of mineral oil or to find other substitute liquids.

The large proportion of research on biological insulating liquids has emphasized their deterioration conduct and compatibility with solid insulation. Even though research investigations endorse the use of these liquids as a substitute for mineral oil

E. P. Pagger et al., *Biological Insulating Liquids*,
https://doi.org/10.1007/978-3-031-22460-7_6

liquids, there are numerous issues and challenges that must be explored. This chapter presents some research work to enhance the understanding of biological insulating liquids.

6.1 Corrosion Caused by the Insulating Liquid (Corrosive Sulfur)

The existence of sulfur or sulfur comprising combinations in the insulating liquid may result in erosion of polished copper coils. This may, however, originate from crude or from the rubber hoses that are used during liquid processing and even it may emerge from gasket materials. Corrosive properties of insulating liquids are a crucial parameter. The recent past, around the turn of the millennium has shown how quickly a chapter that was long believed to be done can come back to life. The corrosive sulfur caused alarming transformer failures in some countries. The main cause was found to be dibenzyl disulfide. However, there are also other sulfur compounds which, through the formation of copper sulfide, reduce the electrical strength of the solid insulation. The sulfur molecules in the insulating liquid can have both negative and, due to their antioxidant properties, positive effects on the insulating liquid. The greater their effect as an inhibitor, the more corrosive they are to copper [1].

Mineral oil-based insulating liquids may contain sulfur compounds that form monovalent copper sulfide with the copper in the windings. Copper-(I)-sulfide is a blue to gray-black powder that conducts electricity very well. The sulfur compounds in the insulating liquids that lead to the copper sulfide deposits are known as "corrosive sulfur". To clarify this fact for the alternative liquids as well, these were subjected to a test according to [2, 3], whereby the samples were pre-treated differently [4]. Biological insulating liquids are extracted from seeds of harvests and normally do not comprise sulfur that leads to the corrosiveness of metallic constituents in the high voltage equipment. Sulfur may be initiated into the transformer through causal ways, for instance, via the application of mismatched hosepipes [5]. Transformers that contain insulating liquids with corrosive sulfur, have unpainted copper conductors, are operated under high load (temperature) and have a closed cooling system (low oxygen content in the oil) are particularly at risk. The copper sulfide precipitated in the paper is electrically conductive and thus reduces the insulating properties of the paper.

Both the copper strip and the paper strip are used to assess whether the insulating liquid contains corrosive sulfur compounds or whether it is potentially sulfur-corrosive.

Investigated Insulating Liquids

- Mineral oil Nynas Nytro® 4000X
- Natural ester Envirotemp® FR3™

- Natural ester Biotemp®
- Natural ester Midel® eN (also known as Midel eN 1204)
- Synthetic ester Midel®7131
- Synthetic ester BecFluid®
- Silicone oil Wacker®TR50.

6.1.1 Sample Treatment with Air and Nitrogen

All samples were treated for 72 h at 150 °C under the following conditions:

1. Determination of the corrosive sulfur with air as a liquid conclusion. In this case, the space between the liquid and the closure is filled with ambient air.
2. Determination of the corrosive sulfur with nitrogen as a liquid conclusion. To simulate a hermetic seal against the atmosphere, the insulating liquids were stripped with nitrogen and sealed under a nitrogen atmosphere.

6.1.1.1 Sample Treatment with Air and Nitrogen—Results

Copper conductors treated with Nynas Nytro® 4000X and with Midel 7131 show a multi-colored surface, which, however, according to [2] cannot be described as potentially corrosive to sulfur, since no changes can be detected on the paper (Tables 6.1 and 6.2).

6.1.2 Sample Treatment with Sulfur Compounds

The insulating liquids were doped with three different sulfur compounds to study the formation of copper sulfide under known conditions. The weight of the sulfur compounds was chosen so that the dosage was 4–5 mg of pure sulfur per 15 ml of insulating liquid. The further test procedure corresponded to the specification [2].

1. Elemental sulfur doping
2. Copper reacts with sulfur to form copper sulfide (Eq. 6.1).

$$2Cu + S \rightarrow Cu_2S \tag{6.1}$$

3. Thioacetamide doping
   Thioacetamide (C_2H_5NS) forms acetamide by absorption of water and releases hydrogen sulfide (Eq. 6.2). This hydrogen sulfide reacts with the copper to form copper sulfide and hydrogen (Eq. 6.3).

Table 6.1 Test—air as liquid closure

Treatment:	**Air as liquid closure**			
Insulation liquid	Mineral oil Nynas Nytro® 4000X	Natural ester Envirotemp® FR3™	Natural ester Biotemp®	Natural ester Midel® eN 1204
Appearance Paper				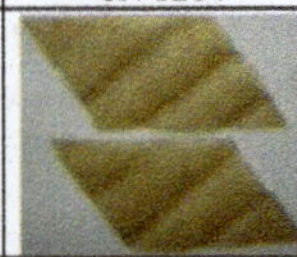
Appearance Copper conductor				
Copper in paper [ppm]	313	0	88	1055
Copper in the insulating liquid [ppm]	61	65	71	59
Insulating liquid	Synthetic ester Midel®7131	Synthetic ester BecFluid®	Silicone oil Wacker® TR50	
Appearance Paper				
Appearance Copper conductor				
Copper in paper [ppm]	0	1355	1718	
Copper in the insulating liquid [ppm]	67	55	19	

Table 6.2 Test—nitrogen as closure

Treatment:	**Stripped and closured with nitrogen**			
Insulating liquid	Mineral oil Nynas Nytro® 4000X	Natural ester Envirotemp® FR3™	Natural ester Biotemp®	Natural ester Midel® eN
Appearance Paper	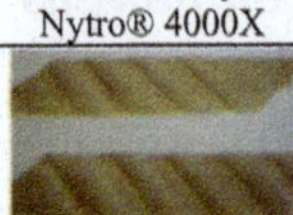	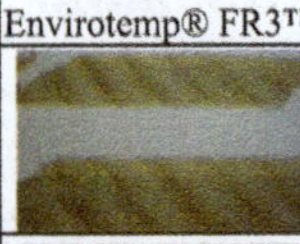	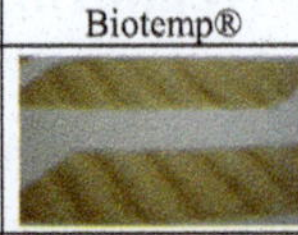	
Appearance Copper conductor		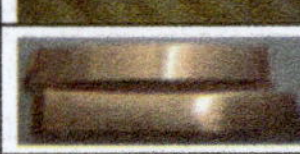		
Copper in paper [ppm]	4601	1861	2222	1192
Copper in the insulating liquid [ppm]	70	50	60	50
Insulating liquid	Synthetic ester Midel®7131	Synthetic ester BecFluid®	Silicone oil Wack-er®TR50	
Appearance Paper	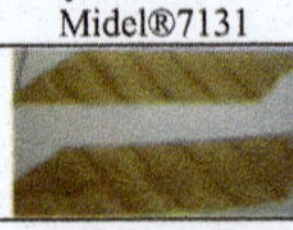			
Appearance Copper conductor				
Copper in paper [ppm]	3784	1503	1754	
Copper in the insulating liquid [ppm]	53	22	113	

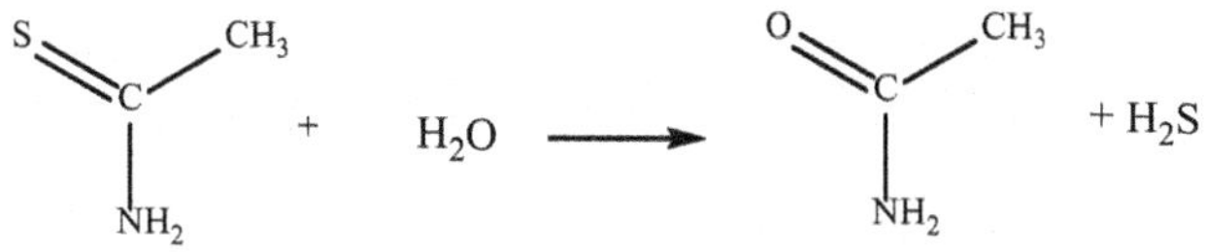

(6.2)

$$H_2S + 2Cu \rightarrow Cu_2S + H_2 \quad (6.3)$$

4. Dibenzyl disulfide doping

Dibenzyl disulfide ($C_{14}H_{14}S_2$) is a substance that is soluble in water, but not in benzene, ether, and hot ethanol. Dibenzyl disulfide is used as an antioxidant and anti-sludge agent for petroleum oils and earth waxes and as a high-pressure additive to cutting oils, as an additive for the formation of boundary layers in lubricants and as an additive to silicone oils. Because of its properties as an antioxidant, dibenzyl disulfide was not necessarily an undesirable substance in the insulating liquids. Only when damage related to dibenzyl disulfide was discovered worldwide [6], this chemical compound was banned from insulating liquids (mineral oils). The corrosive properties towards copper are described in [7]. Dibenzyl disulfide can develop hydrogen sulfide with hydrogen (Eq. 6.4), which in turn reacts with copper to form copper-(I)-sulfide [2, 8].

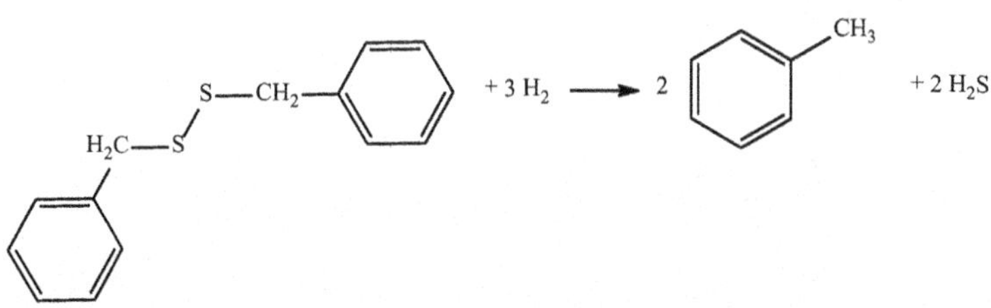

(6.4)

Because of the importance of this chemical compound in connection with insulating liquids, three test series with different concentrations of DBDS (dibenzyl disulfide) were set up. The entire possible concentration spectrum was thus covered.

- Test series with 1050 mg dibenzyl disulfide per kg insulating liquid.
  The sample weight of dibenzyl disulfide in this series of tests was 17–18 mg per 15 ml ($\approx$ 1050 mg/kg) of insulating liquid, which corresponds to a sulfur equivalent of 4–5 mg per 15 ml.
- Test series with 150 mg dibenzyl disulfide per kg insulating liquid.
  In this series of tests, the samples were spiked with 150 mg of dibenzyl disulfide per kg of insulating liquid, which corresponds to a sulfur content of approximately 39 mg per kg of insulating liquid. This concentration can definitely be contained in transformers on service [7].
- Test series with 5 mg dibenzyl disulfide per kg insulating liquid.
  The 5 ppm value is already close to the detection threshold of < 5 ppm [9]. This corresponds to a sulfur content of approximately 1.3 mg/kg insulating liquid.

6.1.2.1 Sample Treatment with Sulfur Compounds—Results

Elemental Sulfur Apart from BecFluid, the copper conductor shows massive chemical attack and significant copper sulfide deposits in all test subjects. The paper samples show different behavior. The sample with the insulating liquid Nynas 4000X shows strong copper sulfide deposits over the entire area (Table 6.3). The samples of the insulating liquids Biotemp, Midel eN and silicone oil show moderately distributed copper sulfide over the entire sample. The samples of the insulating liquid FR3 showed only copper sulfide deposits at those points that were in direct contact with the edges of the copper conductor. The insulating liquid Midel 7131 doped with elemental sulfur does not only result in deposits, but also in a complete destruction of the paper. The samples from BecFluid are hardly influenced. This is surprising because both insulating liquids (synthetic esters) contain the same basic material (pentaerythritol). A correlation between visual assessment and analytically determined copper content can be seen except for silicone oil.

Thioacetamide Massive corrosive attack on the copper conductors regardless of the type of insulating liquid (Table 6.4). The paper samples look very different after the treatment. Treatment of Nynas 4000X resulted in massive copper sulfide deposit.

Table 6.3 Test—elemental sulfur

Treatment:	**Doping with elemental sulfur**			
Insulating liquid	Mineral oil Nynas Nytro® 4000X	Natural ester Envirotemp® FR3™	Natural ester Biotemp®	Natural ester Midel® eN
Appearance Paper	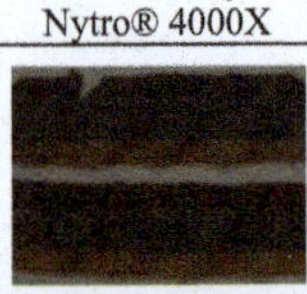		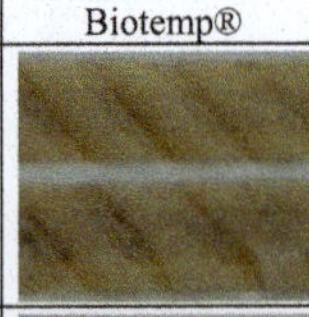	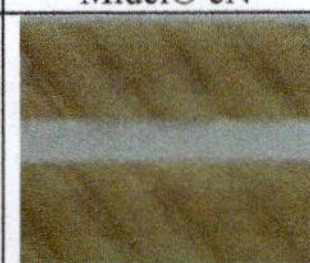
Appearance Copper conductor				
Copper in paper [ppm]	144859	10519	50921	10730
Copper in the insulating liquid [ppm]	46	68	51	53
Insulating liquid	Synthetic ester Midel®7131	Synthetic ester BecFluid®	Silicone oil Wacker®TR50	
Appearance Paper	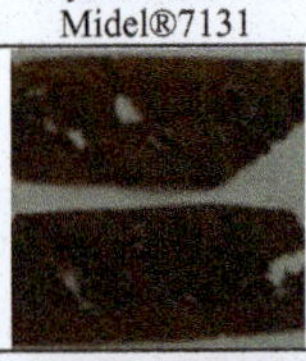			
Appearance Copper conductor		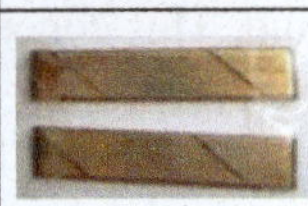		
Copper in paper [ppm]	14153	1230	4065	
Copper in the insulating liquid [ppm]	52	49	52	

The result after treatment with Midel 7131 is particularly dramatic. In addition to a strong deposition of copper sulfide crystals, the structure of the paper was also so damaged that it became brittle and fragile. The paper treated with FR3 is visually moderate and that with Biotemp slightly contaminated with copper sulfide. Neither deposits nor structural changes can be observed on the paper after treatment with Midel eN. Overall, the visual assessment correlates with the copper values found on the paper. For Midel eN, however, the analytically determined value is higher than the visually determined one.

Dibenzyl Disulfide (1050 ppm) The paper samples sometimes behave very differently. The samples treated with Nynas 4000X and Midel 7131 show clear deposits of copper sulfide. The paper sample treated with Biotemp only shows deposits of copper sulfide in the places where the paper touched the edges of the copper conductor. No copper sulfide deposits can be seen on the paper samples treated with FR3, Midel eN, BecFluid and silicone oil (Table 6.5). It is surprising that the paper sample from the silicone oil contains more analytically determined copper than Midel 7131.

Dibenzyl Disulfide (150 ppm) Compared to the test series with 1050 ppm, the samples show significantly lower copper-(I)-sulfide deposits. The copper conductors

Table 6.4 Test—thioacetamide

Treatment:	**Doping with thioacetamide**			
Insulating liquid	Mineral oil Nynas Nytro® 4000X	Natural ester Envirotemp® FR3™	Natural ester Biotemp®	Natural ester Midel® eN
Appearance Paper		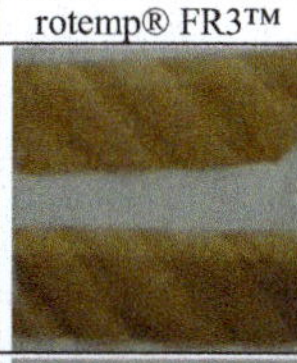		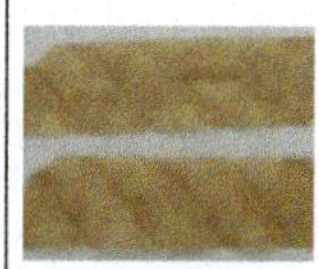
Appearance Copper conductor	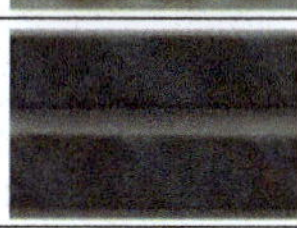		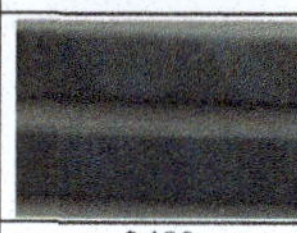	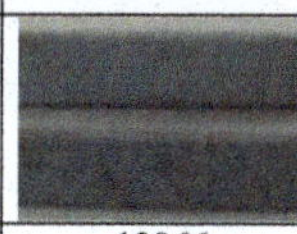
Copper in paper [ppm]	203867	6555	2480	12866
Copper in the insulating liquid [ppm]	85	59	57	65
Insulating liquid	Synthetic ester Midel®7131	Synthetic ester BecFluid®	Silicone oil Wacker®TR50	
Appearance Paper		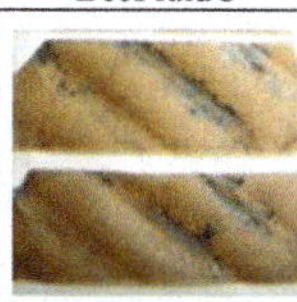	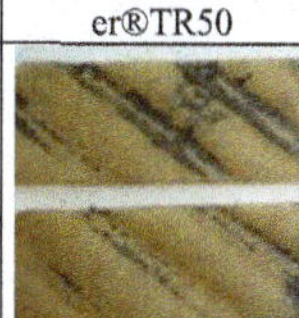	
Appearance Copper conductor		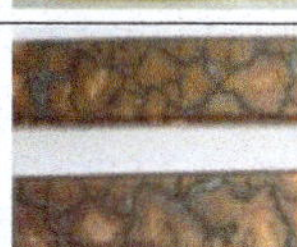		
Copper in paper [ppm]	141735	13858	58833	
Copper in the insulating liquid [ppm]	66	34	35	

Table 6.5 Test—dibenzyl disulfide (1050 ppm)

Treatment:	**Doping dibenzyl disulfide (1050 ppm)**			
Insulating liquid	Mineral oil Nynas Nytro® 4000X	Natural ester Envirotemp® FR3™	Natural ester Biotemp®	Natural ester Midel® eN
Appearance Paper			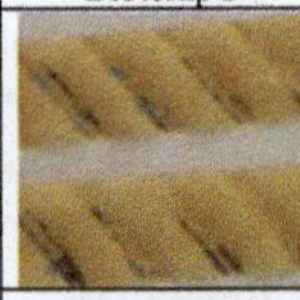	
Appearance Copper conductor				
Copper in paper [ppm]	61060	3892	9533	3286
Copper in the liquid [ppm]	43	41	29	16
Insulating liquid	Synthetic ester Midel®7131	Synthetic ester BecFluid®	Silicone oil Wacker®TR50	
Appearance Paper				
Appearance Copper conductor	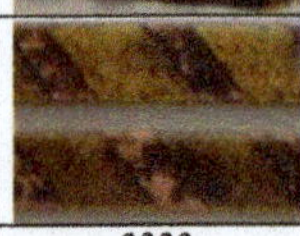			
Copper in paper [ppm]	3333	2905	5484	
Copper in the insulating liquid [ppm]	27	24	48	

of the samples Nynas 4000X, FR3, Biotemp and Midel eN behave very similarly. The copper conductor of the Midel 7131 sample is discolored only, and that of the BecFluid sample showed hardly any change compared to the initial state. The copper conductor from the silicone oil shows no change at all (Table 6.6). From the paper samples, the Nynas 4000X sample shows most clearly copper sulfide crystal precipitates. The sample Midel 7131 shows a stronger brown coloration of the paper, the samples FR3, Biotemp, and Midel eN a weaker or barely perceptible one, respectively. The paper sample of the BecFluid and silicone oil showed no changes. The visual assessment also corresponds widely to the analytically determined amount of copper.

Dibenzyl Disulfide (5 ppm) Visible copper sulfide deposits on the paper of the mineral oil and the ester Midel 7131. Surprisingly, the analytical copper determination for the Midel 7131 sample shows a higher copper value than that of the mineral oil, although this cannot be seen optically (Table 6.7). In any case, the limit value of 5 ppm must be handled with care, even if these test series were carried out at a much higher temperature (150 °C), not occurring in the equipment under normal conditions. Except for BecFluid, the other copper conductors are attacked, too. A

Table 6.6 Test—dibenzyl disulfide (150 ppm)

Treatment:	**Doping dibenzyl disulfide (150 ppm)**			
Insulating liquid	Mineral oil Nynas 4000X	Natural ester Envirotemp® FR3™	Natural ester Biotemp®	Natural ester Midel® eN
Appearance Paper			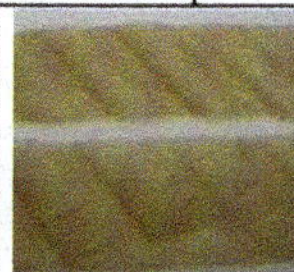	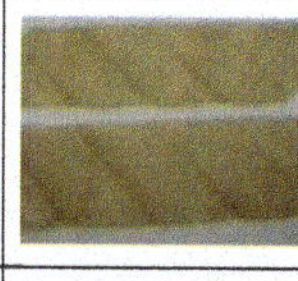
Appearance Copper conductor				
Copper in paper [ppm]	12313	1590	2950	3307
Copper in the insulating liquid [ppm]	63	31	42	39
Insulating liquid	Synthetic ester Midel®7131	Synthetic ester BecFluid®	Silicone oil Wacker®TR50	
Appearance Paper		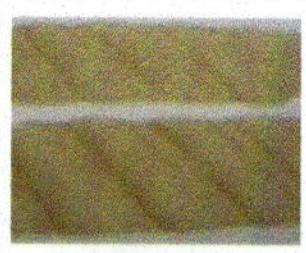	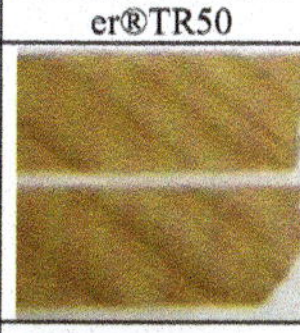	
Appearance Copper conductor		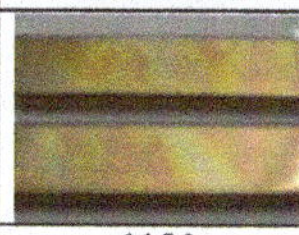		
Copper in paper [ppm]	4167	1156	2250	
Copper in the insulating liquid [ppm]	38	29	25	

determination of the copper content in the insulating liquid was not carried out in this test series.

In almost all samples, the copper content in the paper is below that of the mineral oil. A clear indication that mineral oil as an insulating liquid behaves significantly worse in terms of corrosive behavior if the insulating liquid contains sulfur-containing substances or these are introduced into the insulating liquid from outside. In general, it can be said that in the presence of corrosive sulfur in connection with paper-wrapped copper conductors, ester liquids—especially, the natural esters—have an advantage.

6.1.2.2 Corrosive Sulfur—Test Method

The tests were carried out according to standard IEC 62535 [2]. A piece of copper conductor wrapped in Kraft paper is immersed in the insulating liquid and heated to 150 °C for 72 h in a sealed headspace vial. The copper conductor is visually examined for signs of color change and the paper for any copper sulfide deposits. A separate method was developed to determine the copper content in the insulating liquid.

Table 6.7 Test—dibenzyl disulfide (5 ppm)

Treatment:	**Doping dibenzyl disulfide (5 ppm)**			
Insulating liquid	Mineral oil Nynas Nytro® 4000X	Natural ester Envirotemp® FR3™	Natural ester Biotemp®	Natural ester Midel® eN
Appearance Paper				
Appearance Copper conductor	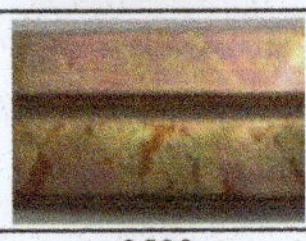			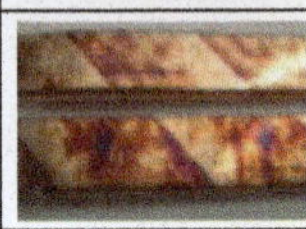
Copper in paper [ppm]	2500	1765	1656	1420
Insolating liquid	Synthetic ester Midel®7131	Synthetic ester BecFluid®	Silicone oil Wack-er®TR50	
Appearance Paper				
Appearance Copper conductor		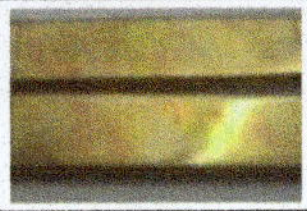		
Copper in paper [ppm]	8077	1683	1630	

6.1.2.3 Determination of Copper in the Insulating Liquid—Test Method

Dissolved copper compounds are active oxidation catalysts and can strongly influence insulating liquid properties, such as conductivity, interfacial tension, and dielectric losses. Determining the presence of copper in insulating liquid, therefore, may be useful not only in the case of corrosive sulfur tests but also to explain changes in insulating liquid properties and general behavior.

As it was expected that, in addition to the dissolved copper, there would also be suspended copper and copper sulfide, as well as copper and copper sulfide adhering to fibers in the insulating liquid. A test method was developed that takes this into account. Furthermore, due to the expected inhomogeneous distribution of the copper and copper sulfide, a larger sample amount had to be considered so that the scatter of the results is not too great.

The insulating liquid is incinerated in a calorimeter bomb under defined conditions. For this purpose, the samples are weighed in a quartz crucible, distilled water is placed in the bomb, and the bomb is filled with oxygen. The sample is ignited from the outside via an ignition wire (pure iron wire) inserted into the sample. The procedure is like that described in ASTM D129 [10]. After the combustion, the bomb is rinsed with distilled water and the water is transferred to a volumetric flask. The quartz crucible, which contains practically the entire residue from the combustion,

Fig. 6.1 Silicone oil—self-locking combustion

is additionally treated with dilute nitric acid in the heat and this liquid is also transferred to the volumetric flask. It was not possible to undertake this handling with silicone oil. The formation of a crust over the liquid phase (Fig. 6.1) prevents the supply of oxygen required for combustion, the combustion comes to a standstill and an undefined portion of silicone oil remains unburned in the crucible. The same phenomena describe Hellebuyck in [11] as in a square shallow cup, all samples of insulating liquid burnt out except for the silicone oil where the formation of a white crust on the surface impeded the combustion process. For silicone oil, the samples were incinerated in the muffle furnace (Fig. 6.2). The residue was dissolved in dilute nitric acid and collected in a volumetric flask as well.

The copper content in the liquid was determined photometrically. The entire process of copper determination for all insulating liquids is monitored and checked with an internal copper in oil standard.

6.1.3 Degradation of Dibenzyl Disulfide (DBDS) Through Thermal Treatment

By determining the degradation of dibenzyl disulfide in the insulating liquid, it should be determined whether the decrease in dibenzyl disulfide behaves differently in the various insulating liquids and whether there is a correlation between

Fig. 6.2 Combustion of silicone oil in the muffle furnace

dibenzyl disulfide degradation and copper sulfide deposition. For this reason, from the samples—dibenzyl disulfide (1050 ppm) and dibenzyl disulfide (5 ppm)—of the test runs (6.1.2), the content of dibenzyl disulfide has been analyzed. The determination was carried out with a device consisting of a gas chromatograph and an electron capture detector (ECD) for detection [9]. Figure 6.3 shows a corresponding chromatogram. The chromatogram shows a peak (dibenzyl mercaptan) at approximately 5 min after injection, which can be identified as a decomposition product. The main peak appears at a retention time of approximately 33 min. The evaluation shows that there is no correlation between the copper sulfide deposition and the decrease in DBDS and is independent of the insulating liquid. At the high starting value (1050 ppm), the degradation rate was approximately 50% and at the low starting value (5 ppm) between 90 and 99%.

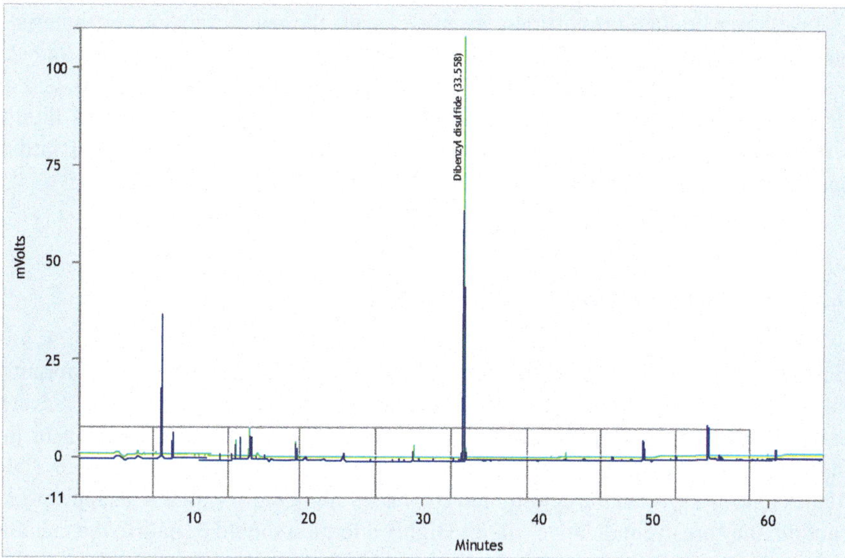

Fig. 6.3 Determination of DBDS with GC-ECD

6.2 Interaction with the Solid Insulation (Paper) During an Aging Test

The insulation system in a liquid filled transformer is composed of insulating liquid and cellulose (paper, pressboard). Since the transformer liquid is replaceable, the service life of transformers is mainly affected by the service life of the solid insulation. Solid insulation or impregnated pressboards in transformers supports the mechanical forces and are commonly used between transformer windings as insulation liquid barriers for breaking up large liquid gaps. Liquid impregnated pressboard that consists of liquid insulation and solid insulation (pressboard, Kraft paper) forms a kind of composite insulation patterns. It is the major insulating material in the electrical equipment due to its high insulation performance. Kraft paper contains cellulose ($\approx 90\%$), hemicellulose ($\approx 7\%$), and lignin and inorganic salts ($\approx 3\%$) [12].

The investigation was carried out with mineral oil and with esters, which are increasingly used in transformers, but knowledge of their aging in the transformers is not yet as profound as for mineral oil liquids. The literature describes great advantages of esters over mineral oil in connection with paper aging. In [13], it is reported that the paper insulation lasts five to eight times longer when using natural ester Envirotemp FR3. In [14], a distinction is made between "degradation"—a temporary deterioration in the condition and "deterioration"—a permanent change—when the condition of the equipment decreases due to aging mechanisms. The investigations listed in this section deal with the permanent change in the state of the insulation system.

Insulating liquids, proportionately mixed with the usual transformer materials, have been stored at 140 °C for a period of 14 days. The furan content and the neutralization number of the aged insulating liquids have been determined, and an IR spectrum has been recorded. Since only a limited number of insulating liquids was aged, not all insulating liquids listed at the beginning were available for further analytical evaluations in this case.

6.2.1 Furans Production Due to Aging

Besides methanol, furans are currently the only markers that can be used to monitor the state of the solid paper insulation when the transformer is energized. Several empirical formulas offer the possibility of deduce the furan concentration in the insulating liquid (currently only for mineral oil) on the degree of polarization (DP) of the cellulose. Even if the determined DP values are being assessed with appropriate caution, they are just indicative values. Highly attention should be paid to the fact that only an average DP value can be obtained with this method and that the minimum DP value for a safe operation of the transformer is unknown. Figure 6.4 shows that under this laboratory conditions the content of 2-FAL which stems from the decomposition of the paper is at least more than three times higher by using mineral oil compared to the ester liquids. If the natural ester is very much degraded, the detection by HPCL can be influenced, showing a to high value of furanic compounds [15].

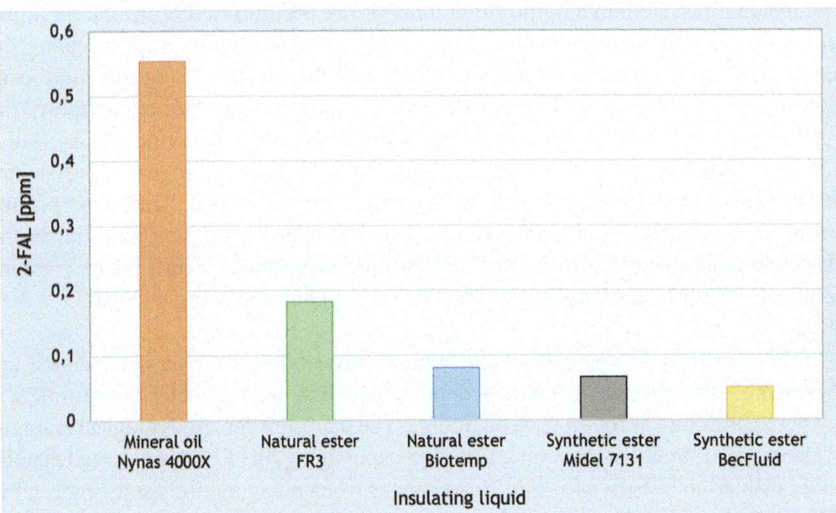

Fig. 6.4 Furan production because of aging

In [16], Augusta and Martins report about aging studies with the biological insulating liquid of Biotemp and an inhibited mineral oil in Kraft paper under temperature condition up to 190 °C. Both mineral oil and vegetable oils were analyzed by HPLC for five furanic compounds: 2-furfuraldehyde (2-FAL), furfuryl alcohol, 2-acetylfuran, 5-methyl-furfuraldehyde, and 5-hydroxymethyl-2-furfuraldehyde (5HMF). However, they found only 2-FAL and 5-HMF above quantification levels, even in the insulating liquids aged at 190 °C. The study showed that the relative concentration of 2-FAL to 5-HMF is approximately 10:1 and that both concentrations increase with the aging temperature. However, 2-FAL and 5-HMF concentrations were approximately three to four times higher in mineral oil than in Biotemp, which correlates very well with the investigation in [4]. Augusta and Martins detected 2-FAL and 5-HMF only in the insulating liquid samples containing paper, when the samples were aged at 130 °C or higher. The concentration of 2-FAL in mineral oil was nearly an order of magnitude higher (about 2 ppm) than the corresponding concentration in biological insulating oil when both liquids were aged at 180 °C. For 5-HMF the difference was about factor 3.

Meixner found in his aging tests of paper and pressboard [17] significant differences in 2-FAL formation between mineral oil and ester liquids (Midel 7131, Midel 1204). In the case of natural Kraft paper, the 2-FAL concentration is up to seven times higher in mineral oil compared to natural ester and does not correlate with the reduction of DP-value at different aging temperatures. Thermal upgraded paper and pressboard have little effect in 2-FAL production.

6.2.2 Furans Transport and Distribution Due to Aging

Lin et al. studied the correlation of moisture and furfural distribution ratio under aging condition [18]. Mineral oil and Kraft paper were used in their experiment. In this experiment, the thermal aging test was carried out at 130 °C for 20 days, thus, similar to that under Sect. 6.2.1. For investigation of the effect of moisture on 2-FAL distribution in the oil-paper system, the aged samples were stored under different moisture concentration in the paper. The results showed that the 2-FAL distribution ratio in mineral oil is positively correlated with the moisture concentration. This phenomenon implies that the increase in moisture concentrations changes the 2-FAL distribution ratio in oil-paper system, causing additional 2-FAL to be diffused from paper to mineral oil.

Aged insulating liquids can contain not insignificant amounts of aging products with different polarities, which may influence the transport and diffusion of the aging markers (furans). Pagger and Scala studied in [19] the influence of long chain acid, short chain acid, and water on the diffusion and transport phenomenon of furans. Furan-containing cellulose was stored for a period of 7 days at a temperature of 60 °C in.

(a) Nynas Nitro 4000X + stearic acid

(b) Nynas Nitro 4000X + stearic acid + water
(c) Nynas Nitro 4000X + acetic acid
(d) Nynas Nitro 4000X + acetic acid + water
(e) Natural ester Biotemp + stearic acid
(f) Natural ester Biotemp + toluene
(g) Natural ester Biotemp + acetic acid
(h) Synthetic ester Midel 7131 + stearic acid
(i) Synthetic ester Midel 7131 + acetic acid

Since the ester liquids already contain enough water (80–200 ppm), no further water doping was done. The addition of toluene as a solvent mediator was necessary for the natural ester, since otherwise the stearic acid could not be dissolved. In the insulating liquids, furans could only be detected in mineral oil. The furan content in the mineral oil was lowest when the long-chain acid was added and increased significantly when the short-chain acid was used. Furthermore, the presence of water in mineral oil increases the mobility of the furans.

The effect of moisture on furans distribution behavior can be explained from two aspects. First, the with rising of moisture in paper, the hydroxyl groups on cellulose chains are more likely to form hydrogen bonds with moisture molecules instead of furans molecules, so that furans absorption capacity of cellulose is weakened. Second, the polarity of the insulating liquid increases with the rising of moisture and dissociates acids in the insulating liquid. As a result, the furans absorption capacity of mineral oil is enhanced. For ester liquids, this phenomenon could not be confirmed. For mineral oil, Lin et al. come to the same result. Hence, the increase of moisture in the insulating liquid-paper system promotes the 2-FAL diffusion from paper to mineral oil (Fig. 6.5).

6.2.3 Change of Total Acid Number Due to Aging

Another important aging product for assessing the condition of the insulation is the acid number of the insulating liquid. Aging of the liquid-paper insulation results in the formation of various acids. IEC 62021-3 [20] is the standard method for non-mineral oil-based liquids to determine the acid content. This is a common method for determining the acid number. However, the disadvantage is that the total amount of acid in the liquid is determined, but it is not possible to distinguish the types of acid. During aging, high and low molecular weight acids are formed by hydrolysis and oxidation processes. Due to thousands of different molecules in mineral insulating liquids, the theoretical number of different acids that can be formed is even larger. Acid is formed mainly by hydrolysis processes and, to a lesser extent, by oxidation processes. The high molecular acids in mineral oil liquids are, for example, stearic and naphthenic acids. The low molecular acids are formic, acetic and levulinic acids. The types of acid have different influences on the aging process of the insulation. For example, it is known that low molecular acids are mainly absorbed by the cellulose

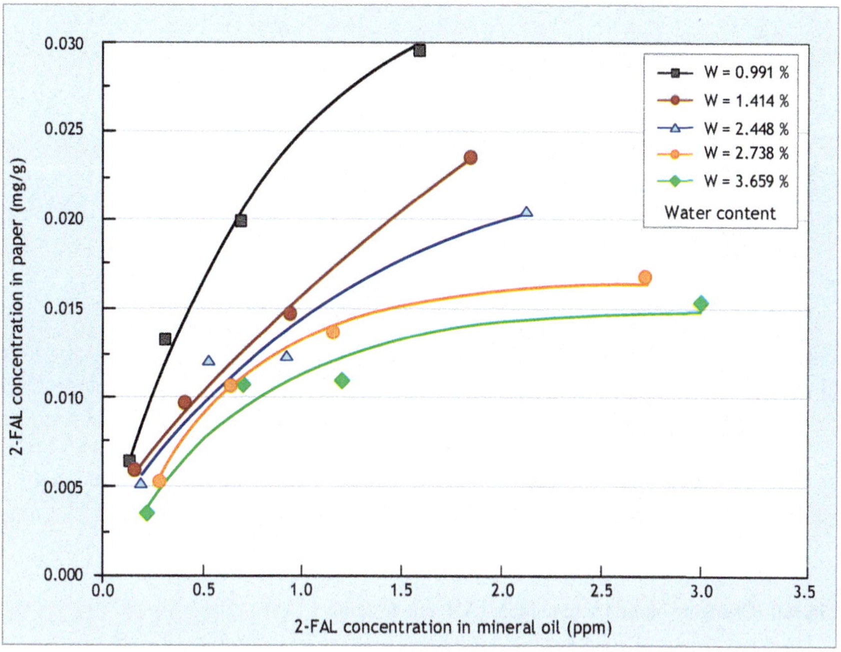

Fig. 6.5 2-FAL partitioning diagrams in the mineral oil-paper system with different moisture concentrations [18]

insulation. This is due to their solubility in water. In the case of high molecular acids, the majority remains in the insulating liquid and thus has no negative effect on the paper aging. In biological insulating liquids, numerous acids can be formed by hydrolysis and oxidation. Most of these acids are stearic, palmitic, oleic, linoleic, and linolenic acids.

Due to the thermal aging, it was expected that the total acid number (AN) of those samples containing the proportionate transformer materials would increase compared to the blanks (same test conditions, but without transformer materials). But the opposite has happened (Fig. 6.6). This figure shows the relation of AN with and without transformer materials. There is nearly no change in the acid number of the mineral oil with and without transformer materials. The value of the total acid number of both samples stayed nearly constant. And the total acid number of the ester liquids with transformer materials compared with the blank sample decreased by approximately 12–50%—an indication of transesterification, respectively esterification [21, 22]. In this case, the hydroxyl groups of the cellulose are esterified by fatty acids and thus acids are consumed. By hydrolytic cleavage of the esters, originating from the insulating liquid, fatty acids can be replenished. Thus, on the one hand water, which behaves damaging to cellulose will be consumed, and on the other hand the new generated ester builds a layer over the cellulose and protects the

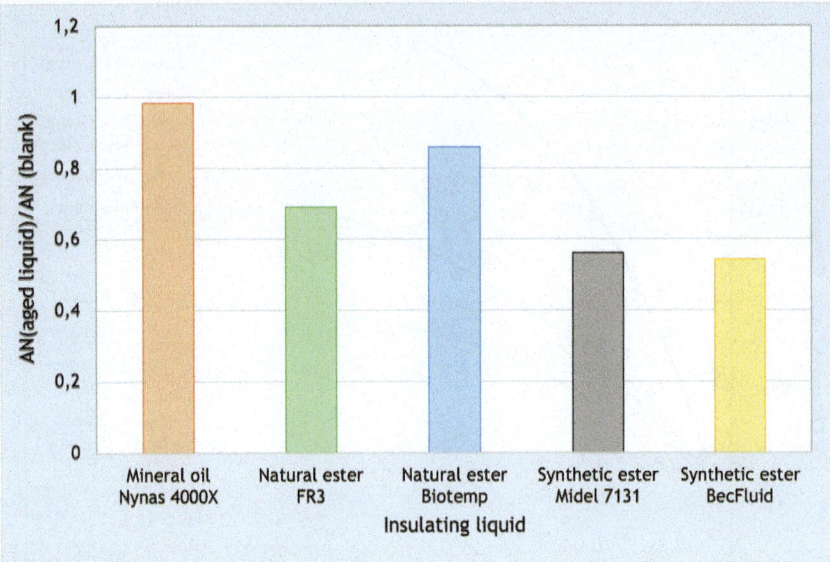

Fig. 6.6 Changing of total acid number (AN) due to aging

cellulose from further chemical attack. Hence, acid number is not a suitable marker for evaluating the aging rate in a closed system.

The splitting of the ester bond in the molecule results in long-chain fatty acids, which is caused by hydrolysis in both natural and synthetic esters but have a minor effect on paper aging [23].

6.2.4 Modifications in IR Spectrum Due to Aging Test

Infrared spectroscopy can be used for both qualitative and quantitative analysis. The particularly useful bands for the structure elucidation of the insulating liquids are in the range of 4000–700 waves per cm. In Figs. 6.7, 6.8, 6.9, 6.10 and 6.11, the FTIR spectrums of the insulting liquids with and without transformer materials are depicted. The mineral oil showed after ageing without transformer materials a massive peak at wavenumber 1740 cm^{-1} (Fig. 6.7, red line), which cannot be seen in the spectrum before aging test (Chap. 2, Fig. 2.1). That means during aging, carbon acids must be formed. All other ester liquids don't show this effect. As the peak of the mineral oil aged with transformer materials is much smaller (Fig. 6.7, purple line) there must be some interactions between the transformer material—above all the paper with mineral oil. The new peak at wavenumber 2800–2900 cm^{-1} could be an indication of this. That a massive degradation of the paper took place shows the high concentration of 2-FAL (Fig. 6.4) and even the CO_2 peak at wave number

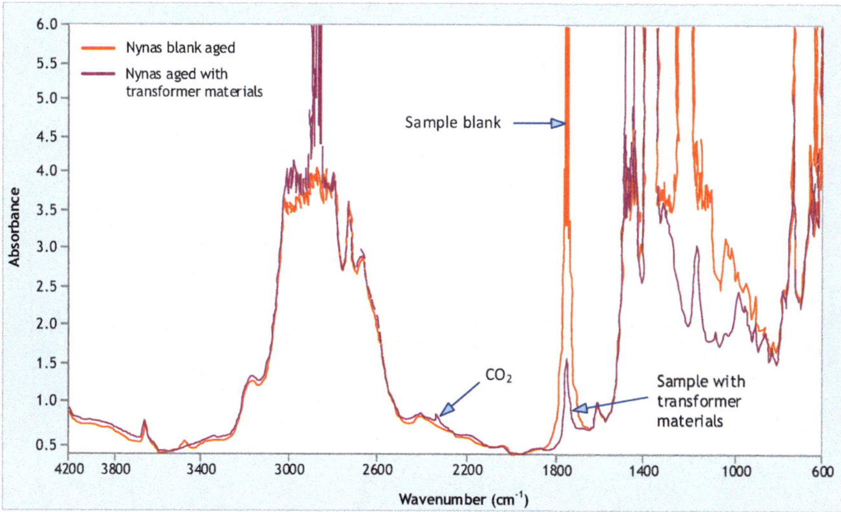

Fig. 6.7 FTIR spectrum—Nynas 4000X—aged with and without transformer material

2337 cm^{-1} which in Fig. 6.7 could be found. As the total number of acids with and without transformer materials stayed constant, new acids must be formed. For the ester liquids, Figs. 6.8, 6.9, 6.10 and 6.11, the spectrograms show a decreased peak around wavenumber 1700–1800 cm^{-1} (carboxylic acid) after aging with transformer materials. The reduction of this peak is in common with the reduction of the total acid number. That means in any case an esterification of paper by consumption of acid take place and if new esters are formed because of hydrolyzing the existing one by water stemming from the cellulose, we can speak from transesterification. This phenomenon is also described in [24].

6.2.5 *Change of Degree of Polymerization Due to Aging Test*

Aging tests performed by Meixner [17] at temperatures in the range of 110–150 °C showed a lower decrease of DP-value of cellulose when using ester liquids (synthetic ester Midel 7131, natural ester Midel 1204) compared to mineral oil.

6.2.6 *Change of Breakdown Voltage Due to Aging Test*

Hence, the transformer insulating liquid is replaceable but not the solid insulation. It is very important to know the breakdown voltage behavior of the insulating liquid/solid system. In the work of Münster et al. [23], they aged liquid-paper insulation at a

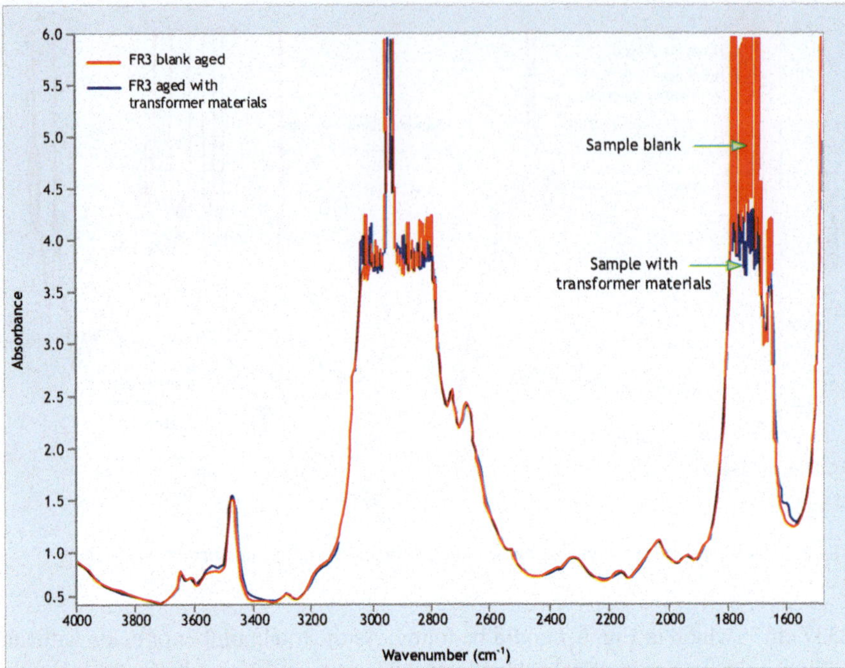

Fig. 6.8 FTIR spectrum—FR3—aged with and without transformer material

temperature of 130 °C. The values show volatile behavior, regardless of the system (open or closed) and insulating liquid. At the end of aging, the breakdown voltage of the ester liquids is slightly higher than that of the mineral oil liquids.

6.2.7 Change in Interfacial Tension Due to the Aging Test

Münster et al. [23] give information about the change in interfacial tension due to thermal treatment. The initial values of the ester liquids are only half of the ones for the mineral oil liquids. For the esters, over the aging time, the interfacial tension remains practically constant, which can be observed for both the synthetic ester and the natural ester. In the esters, the acid number is ten times higher in some cases, but this does not affect the interfacial tension. The formation of long-chain fatty acids in both natural and synthetic esters seems to not influence the interfacial tension as those acids are different from those in the mineral oil where the correlation is higher.

Maneerot and Pattanadech [25] performed an experiment with single-phase transformers rated 2200/230 V and a load of 30 kV. The one filled with mineral oil, the other one with palm oil, respectively. The transformers were operated at 80% load. Samples taken in the first 6 months show an increase of interfacial tension values

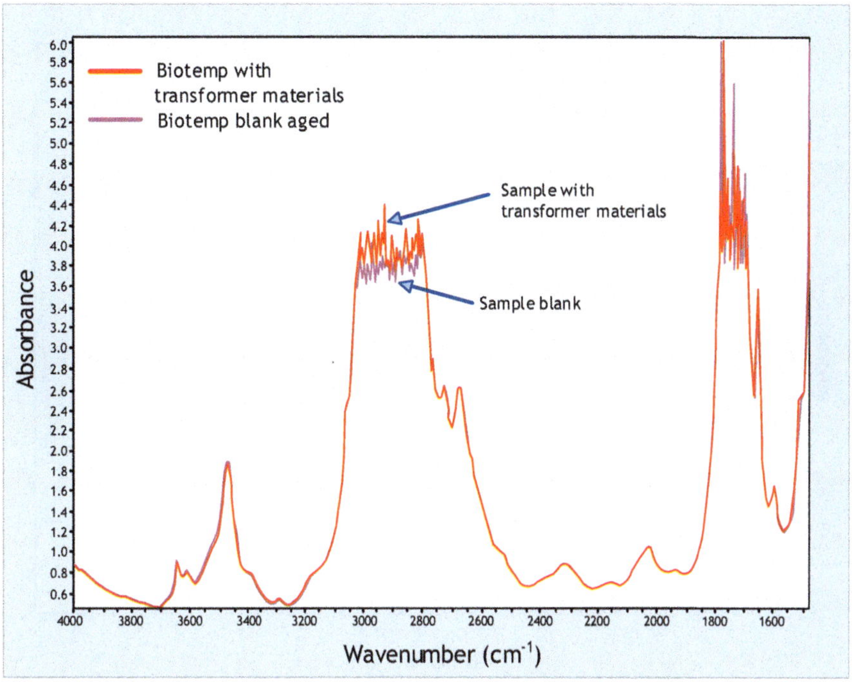

Fig. 6.9 FTIR spectrum—Biotemp—aged with and without transformer material

for mineral oil about 35% and for palm oil 29%, respectively. This positive trend is probably because of transesterification, which is combined with the consumption of acids. Yuerekten reports in [26] that pressboard absorbs substances which have negative effects on criterions like interfacial tension and dielectric dissipation factor.

6.2.8 Change in Viscosity Due to the Aging Test

As any changes in viscosity can better be seen at low temperature, Münster et al. [23] chose a temperature of 10 °C to determine the viscosity in their study. The results for mineral oil show that the most significant changes are observed in the open system, where there is almost a linear increase in viscosity over the aging time.

The significantly higher viscosity of the esters compared to the mineral oil is also noticeable. The relative change in the viscosity for the esters is smaller compared to the mineral oil in the closed system, with the viscosity remaining practically constant. This was also shown by Maneerot and Pattanadech in [25] that's due the low influence of oxidation processes on account of the closed system. Oxidation processes in natural

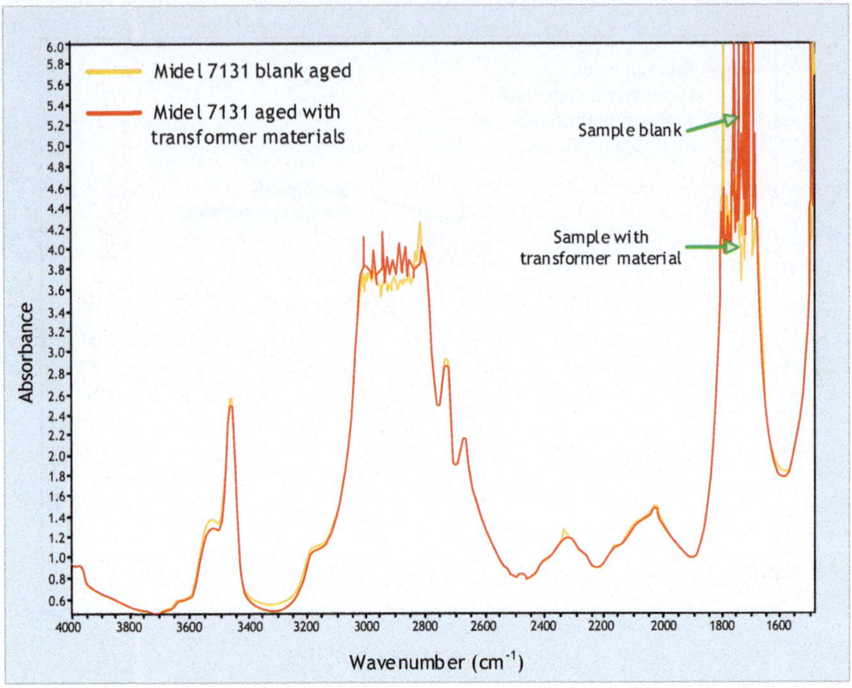

Fig. 6.10 FTIR spectrum—Midel 7131—aged with and without transformer material

esters attack predominantly unsaturated parts of the carbon chain in the molecular structure, leading to an increase in viscosity. For this reason, biological insulating liquids are only suitable for use in hermetically sealed transformers. However, this process of increasing viscosity needs not to be accompanied by acid formation. Since viscosity of ester liquids does not change under normal aging condition, it's not a suitable marker for evaluation the quality of the insulating liquid.

6.3 Moisture Transport Between Insulating Liquid and Solid Insulation

Moisture is the natural enemy of each paper insulation system and influences the end of life most. Moisture can come from outside—free breathing system—or it stems from the paper ageing. In any case, if the aging process is started once it runs iterative and will never stop. The presence of moisture accelerates paper aging. Both, electrical and mechanical strength will be reduced. Water can be present in the insulation system of paper/insulating liquid

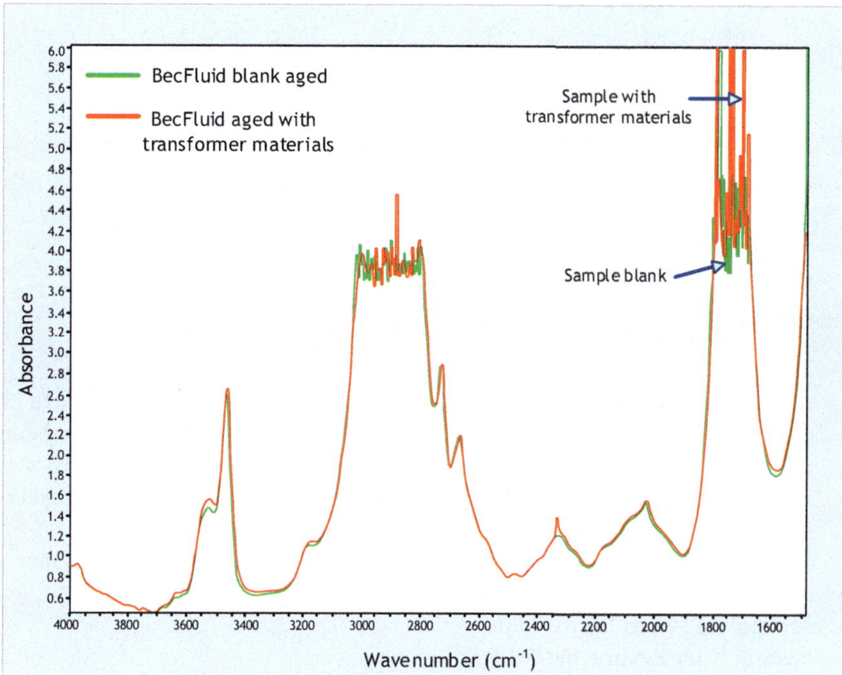

Fig. 6.11 FTIR spectrum—BecFluid—aged with and without transformer material

- in dissolved form.
- bounded to foreign atoms and -molecules, which come from the aging.
- as free water.

Basically, cellulose and the insulating liquids behave diametrically in terms of water absorption. Cellulose is hydrophilic and when temperature increases, the water absorption decreases. The insulating fluids are mostly more hydrophobic. That means if the temperature increases, the water solubility increases, too. According to [27] the thermal decomposition of the paper is proportional to the water content. An increased humidity in the transformer reduces the life expectancy massively [28]. To compare the behavior of the different insulating liquid in the system of paper/insulating liquid/water, tests were done by Pagger [4]

- without adding water.
- by donating 1% water on the weight of the solid insulation.
- by donating 1,5% water on the weight of the solid insulation.

The used transformer materials were:

a. Pressboard 2 mm
b. Nomex® Pressboard 994 4 mm. Nomex® is a synthetic aromatic polyamide polymer [29] (Fig. 6.12) that provides high levels of electrical, chemical, and

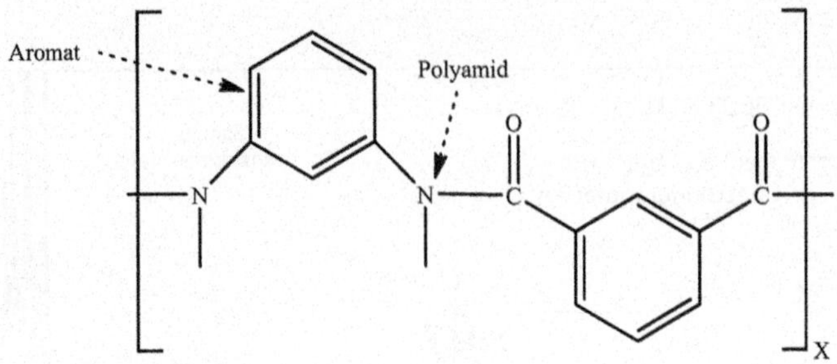

Fig. 6.12 Structural formula of polyamide

mechanical integrity over a wide range of temperatures and does not give off any by-products up to a temperature of 300 °C. Nomex® shows less moisture absorption, which is an advantage when impregnating transformers filled with insulating liquid during the drying process. Nomex® Pressboard 994 is said to be very stable against large loads [30]

c. With four layers of paper-wrapped copper conductors (Kraft paper, 0.08 mm × 11.2 mm) copper conductor (0.18 mm × 6 mm)
d. Mineral oil Nynas Nytro 4000x
e. Natural Ester Envirotemp™ FR3™
f. Natural Ester Biotemp®
g. Natural Ester Midel® eN
h. Synthetic Ester Midel® 7131
i. Synthetic Ester BecFluid® 9902
j. Silicone oil Powersil® Fluid TR 50.

Before testing, pressboard and the paper-wrapped copper conductors were dried under vacuum (1300 Pa) at 105 °C for 24 h. The used materials had the following initial moisture before drying:

- Pressboard: $W = 3{,}61\%$, $\delta = \pm 0{,}45\%$
- Nomex: $W = 2{,}24\%$, $\delta = \pm 0{,}20\%$
- Paper (Kraft paper): $W = 3{,}95\%$, $\delta = \pm 0{,}75\%$.

6.3.1 Impregnation of the Solid Insulation

As the next step the solid samples were impregnated with insulating liquids mentioned above. Due to the higher viscosity of the ester liquids over the mineral oil, a slower intake of the liquid in the solid insulating material has been expected. The basis for the absorption of the insulating liquid in the capillaries of the solid insulation is the Hagen-Poiseuillesche law [31], which is well applicable for Reynolds numbers up to 2000 (Eq. 6.5).

$$dV = \frac{\pi * r^4}{8 * \eta * l} * (p_1 - p_2) * dt \tag{6.5}$$

with Eq. (6.6)

$$V = r^2 * \pi * l \tag{6.6}$$

you get the differential equation (Eq. 6.7)

$$r^2 * \pi * dl = \frac{\pi * r^4}{8 * \eta * l} * (p_1 - p_2) * dt \tag{6.7}$$

and by substituting the initial conditions t = 0, l = 0 the equation (Eq. 6.8)

$$t = \frac{l^2 * 8 * \eta}{r^2 * (p_1 - p_2)} \tag{6.8}$$

Because of the adhesive intermolecular forces, the surface tension at the capillary edges must be considered, which causes an additional pressure, resulting in the following corrected pressure term (Eq. 6.9)

$$p = (p_1 + p_C - p_2) \tag{6.9}$$

Reference [32] describes the correction value p_C as follows (Eq. 6.10)

$$p_c = \frac{2 * \sigma_T * \cos\theta}{r} \tag{6.10}$$

If the surface tension determined by the tensiometer and the capillary height (h) are known, the contact angle (θ) can be calculated by formula (Eq. 6.11)

$$\theta = \cos^{-1}\left(\frac{h * r * g * \rho}{2 * \sigma_T}\right) \tag{6.11}$$

p_1 Pressure capillary (pore) start value (Pa)
p_2 Pressure capillary (pore) end value (Pa)
p_C Pressure capillary (Pa)
r Capillary radius (m)
η Dynamic viscosity (Ns/m^2)
l Capillary height (mm)
g Gravitation acceleration (m/s^2)
σ_T Surface tension determined by tensiometer (mN/m)
ρ Density of the insulating liquid (kg/m^3).

Table 6.8 shows the result of the different insulating liquids.

Table 6.8 Calculation of the contact angle

Insulating liquid	Capillary height (mm) (80 °C)	Density (kg/m^3) (80 °C)	Surface tension (mN/m) (80 °C) Capillary method	Surface tension (mN/m) (80 °C) Tensiometer	Contact angle (θ) (°) (80 °C)
Nynas Nytro 4000X	18.0	821	21.0	23.0	24
Envirotemp® FR3™ Fluid	18.0	880	22.5	29.1	39
Biotemp®	17.0	874	21.1	27.1	39
Midel® eN	18.5	879	23.1	26.6	30
Midel® 7131	16.5	927	21.8	23.8	24
BecFluid® 9902	16.0	931	21.2	23.7	26
Powersil® Fluid TR 50	12.5	904	16.1	17.6	24

The following assumptions were made to determine the theoretical impregnation time:

- Pore radius (r): Because of his own microscopic measurement on the pressboard, the radius was assumed with 2.7×10^{-6} m.
- The pore radius for the pressboard was also used for paper and Nomex.
- As the paper insulation consists of four layers, the thickness of one layer was simply multiplied by four. The phase transitions between the individual layers have not been considered.
- For (l) in Eq. (6.8) half of the thickness of paper, pressboard, Nomex was used.
- The insulating liquids were placed in the vacuum drying oven together with the dried samples. As the liquid didn't come from outside, the pressure difference between the final vacuum and ambient pressure was not included in the calculation. The pressure term consists exclusively of the capillary pressure.

Figure 6.13 shows how many times longer the theoretical impregnation time compared to mineral oil would last. Because of several assumptions regarding the pore size and the pore distribution, the evaluation must be regarded as an approximate estimate. The impregnation time is above all driven by the viscosity (see Sect. 4.2.2).

Test results done by Maneerot et al. [33] showed that the polarization and depolarization current of natural ester impregnated pressboard directly depends on the impregnation period and the electric field stress duration. The polarization and depolarization current decrease with increasing impregnation time and increase with duration of electric field stress.

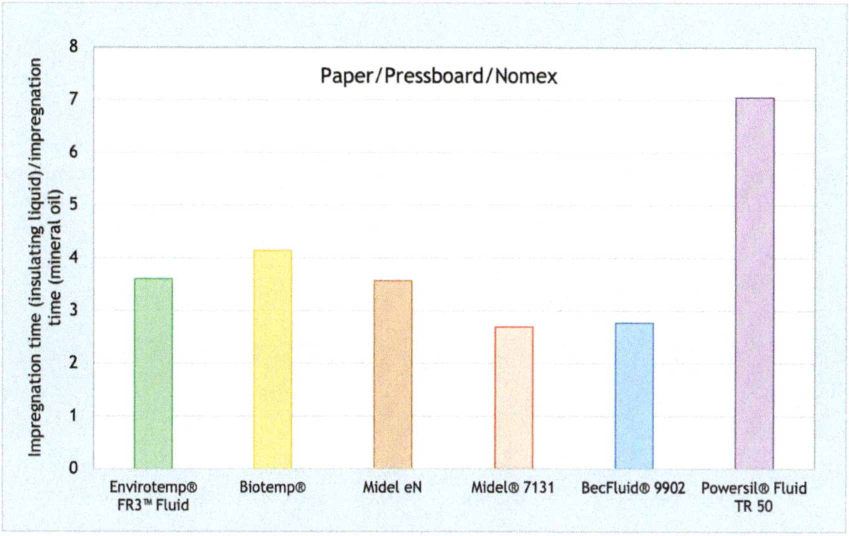

Fig. 6.13 Theoretical impregnation time compared to mineral oil

6.3.2 Preparing and Treatment of Samples After Impregnation for Moisture Transport

The impregnated samples were placed in headspace vials, then the appropriate amount of water was added by microliter syringe. The headspace vials were filled up with the respective insulating liquid and closed. For the distribution and homogenization of the doped water, these headspace vials were placed in an ultrasonic bath (Struers Metason 120, 70 W) and sonicated for 15 min.

The samples thus prepared were stored in a drying oven for a period of 168 h to equilibrate at 80 °C. To prevent any possible exchange of moisture through the septum, the headspace vials were stored in a horizontal position, so that the septum was completely surrounded by liquid. The headspace bottles were rolled from time to time.

After storage in the oven, at first the moisture was determined in the insulating liquid. For this purpose, the insulating liquid was taken from the headspace vial by syringe. The water content determination was carried out by Karl Fischer titration [34–36]. This is also in practice the method of choice to refer from the moisture of the insulating liquid to the moisture of the solid insulation, as it is practically the only method that can be performed on the device in operation.

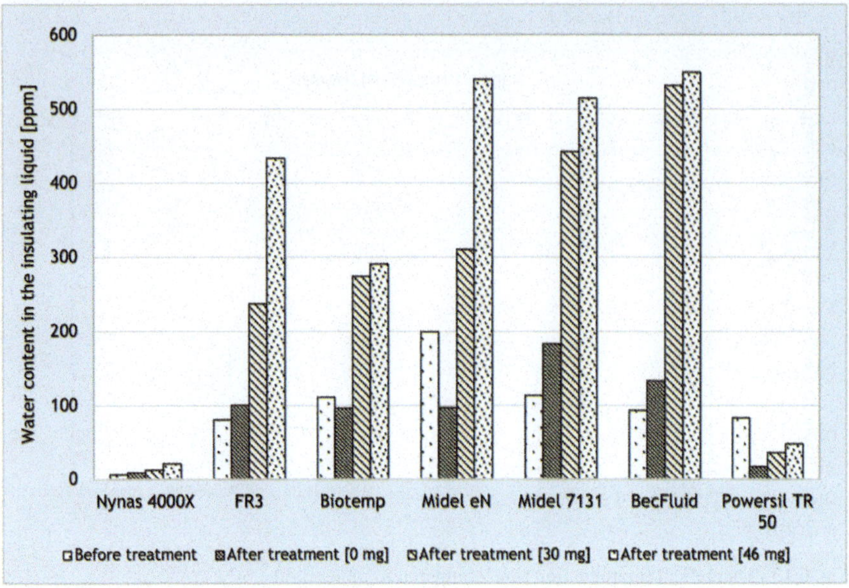

Fig. 6.14 Water content of insulating liquids

6.3.3 Results of Moisture Tests

With the samples obtained from Sect. 6.3.2 several analyzes were done.

6.3.3.1 Water Content in the Insulating Liquid

The results for which no water was added deviate significantly from one another before and after the treatment in the case of the insulating liquids with a higher water content. Tests with the addition of 30 mg of water show, except for silicone oil, a more or less realistic increase in the water content. Also, the tests with the addition of 46 mg of water, apart from the silicone oil, show a further increase in the water content in the insulating liquids. The results with silicone oil show that the affinity of water to the transformer materials is significantly more pronounced than to the insulating liquid, even without treatment (Fig. 6.14).

6.3.3.2 Water Content in the Solid Insulation

To be able to examine the layers of the wrapping paper separately, they were unwound individually. These four layers were designated from the outside inwards towards the copper conductor with layer 1 to layer 4.

The evaluation of the average water content for each test series, regardless of the insulating liquid, shows that the water content in the individual paper layers does not differ significantly (Fig. 6.15). The greatest deviation is around the starting value, i.e., when no water is spiked. With an increase in the water content, the values come closer and closer. Based on the investigations and evaluations, it can be assumed that there is a linear relationship between water doping and water content in the individual paper layers and that, for such an investigation, it is not necessary to split the insulating paper into the individual layers. The pressboard and Nomex samples also show a clear linear relationship, but with a much lower starting value and a lower slope. The starting value is significantly influenced by the hygroscopic properties of the materials and shows that paper is much more hygroscopic than pressboard and Nomex.

Figures 6.16, 6.17 and 6.18 show the water content in the individual solid samples based on the insulating liquids and the water doping. Because of the knowledge already given above, the arithmetic mean was formed from the four values of the individual paper layers. It is noticeable that the water content of the BecFluid with a doping of 46 μl of water, both in Nomex and pressboard, is significantly below the value with the doping of 30 μl (Figs. 6.16 and 6.17). A comparison of the solid insulating materials shows that the highest water absorption occurs in the paper and that the increase in the water content in the paper correlates with the water doping (Fig. 6.18).

But what we really want to know is where we will find all the water stemming from the

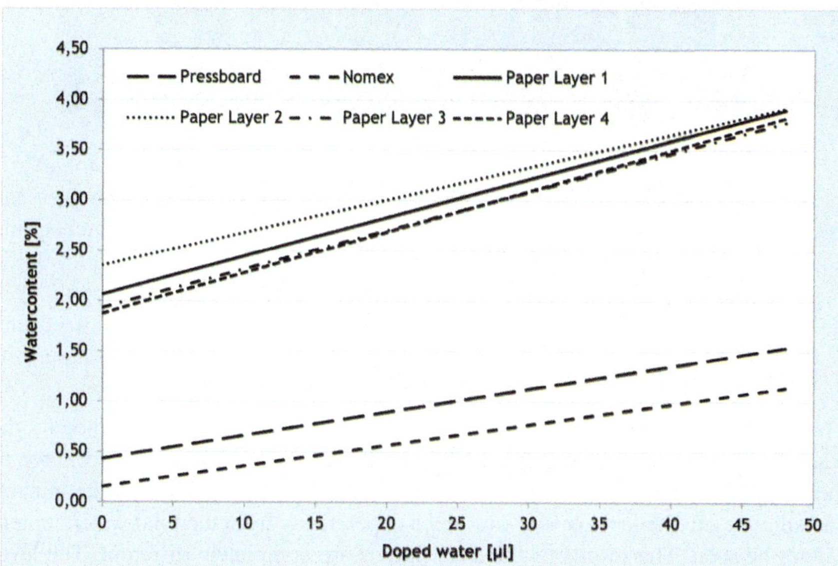

Fig. 6.15 Average water content in the solid insolation independent of the insulating liquid

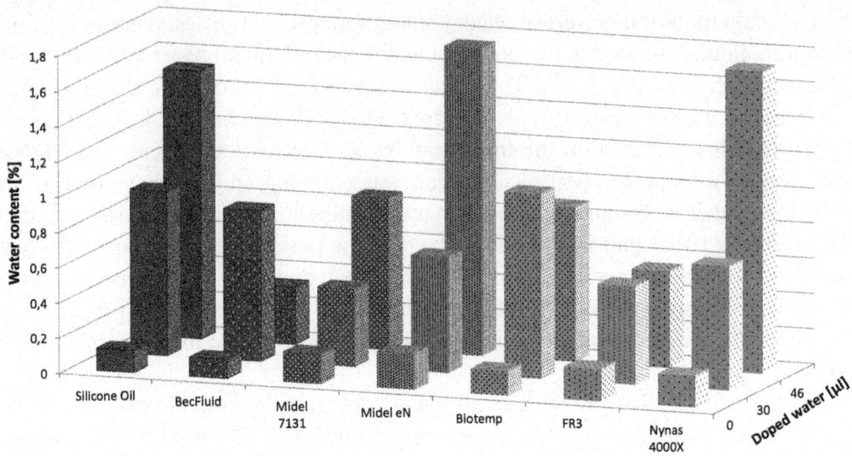

Fig. 6.16 Water content—nomex

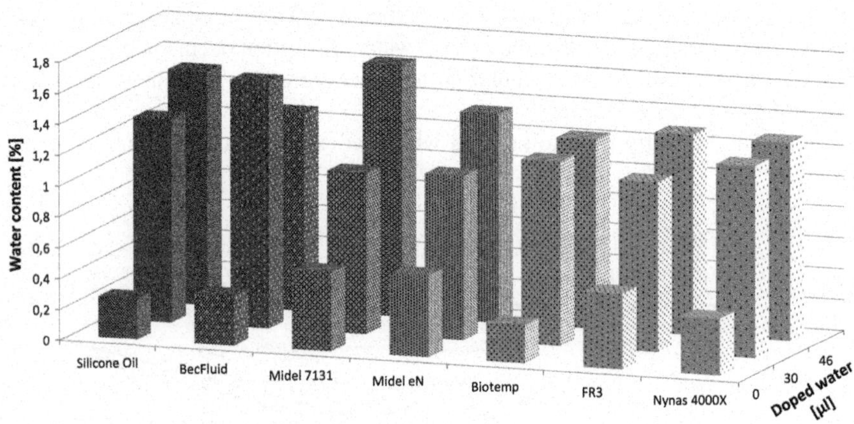

Fig. 6.17 Water content—pressboard

- in situ water content of the materials?
- doped water?
- and from degradation process under this test condition?

As the main reason was to find out how much moisture can be absorbed by the insulating liquids a mass balance was done. Figure 6.19 shows the percentage of the total water which is contained in the different insulating liquid. For the mineral and silicone oil, the level is very low and a dependency from the total water content cannot be seen. The results of the ester liquids are completely different. The level of absorbed water is much higher and the dependency of the amount of doped water can mostly be seen very well.

6.3.3.3 Modelling of Water Transport in Pressboard

The results of the moisture test were also used for modelling the transport of the water in the solid samples. Fick's second law (Eq. 6.12) was used as diffusion equation [37]. With p (Pa) as the vapor pressure in pressboard at the place z (mm) and time t (s).

$$\frac{\partial p}{\partial t} - D * \frac{\partial^2 p}{\partial z^2} = 0 \tag{6.12}$$

The diffusion equation is a differential equation second order with respect to location and first order with respect to time. For solving the equation, you need three boundary conditions.

For getting the parameter D, the first Fick's law (Eq. 6.13) law is useful.

$$\dot{n} = -D * \frac{dc}{dx} \tag{6.13}$$

After converting the equation, D can be calculated by (Eq. 6.14).

$$D = \frac{\dot{n} * d}{2 * c_1} \tag{6.14}$$

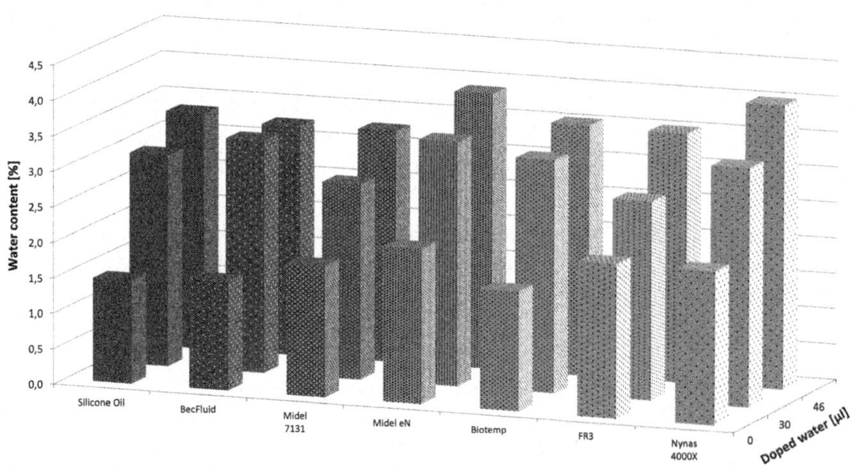

Fig. 6.18 Water content—paper

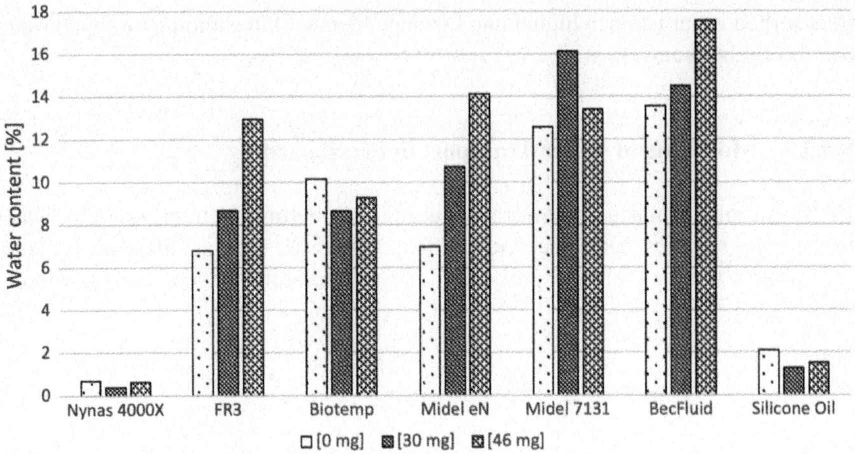

Fig. 6.19 Water content of the insulating liquids after treatment

$\dot{n}$ Particle flux density ($kmol/m^2*s$).
c Concentration ($kmol/m^3$).
x Diffusion length (m).
D Diffusion coefficient (m^2/s).

Musai describes in [38] that the main reason for the deviations between the model and reality lies in the inaccuracy of the diffusion coefficient. The value of the diffusion coefficient depends, for example, on the water content and on the temperature Eq. (6.15).

$$ln\left(\frac{D_T}{D_0}\right) = 0.5 * [\%]_{H_2O,Pressboard} + E_a * \left(\frac{1}{T_0} - \frac{1}{T}\right) \tag{6.15}$$

With an E_a value of 8074 K, the following data were calculated for pressboard with different water doping and a treatment temperature of 80 °C (Table 6.9).

In spite of the statement in [39] that in case of transformers the equation form of Fick's law may not fully describe the diffusion process and that the diffusion coefficient is dependent on the water content, the experimentally determined data and fits very well with [27], where a diffusion coefficient for oil-impregnated pressboard at 70 °C is indicated with 4.7×10^{-12}.

Table 6.9 Calculated diffusion coefficients at 80 °C

Water content (%)	Diffusion coefficient (m^2/s)
0.5	1.168×10^{-11}
1.0	1.499×10^{-11}
1.5	1.925×10^{-11}

After solving the differential equations, you get results for pressboard immersed in the different insulating liquids. The envelopes are shown in Figs. 6.20, 6.21, 6.22, 6.23, 6.24, 6.25 and 6.26. Comparing the figures, the results show that under the completely same condition the water transport in the pressboard is very different and is more or less inverse to the water saturation level of the insulating liquids.

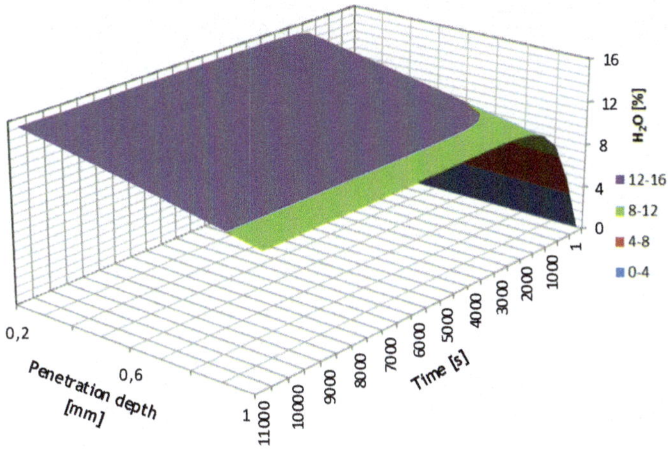

Fig. 6.20 Nynas Nytro® 4000X

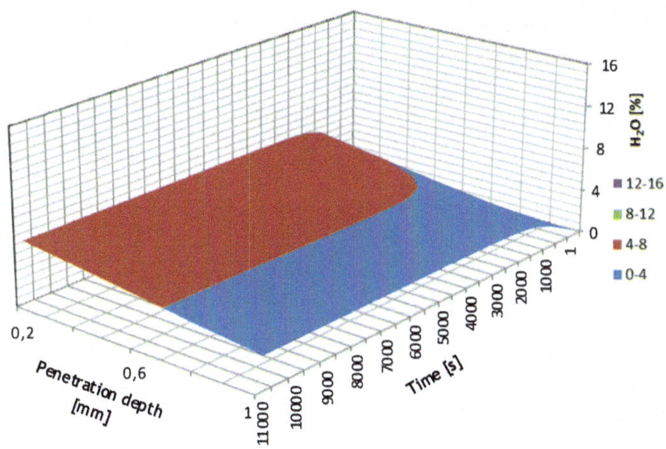

Fig. 6.21 FR3™

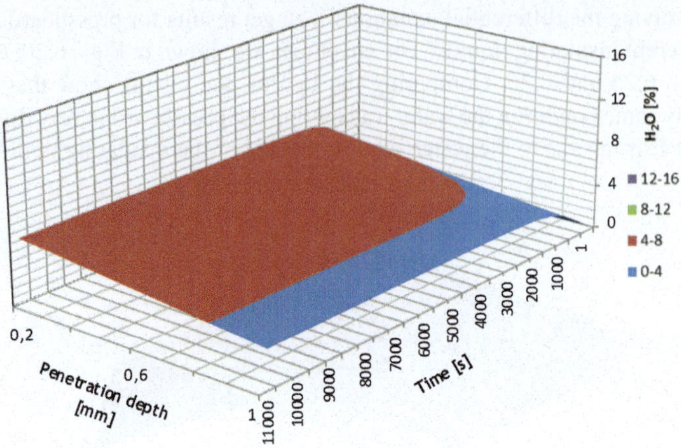

Fig. 6.22 Biotemp®

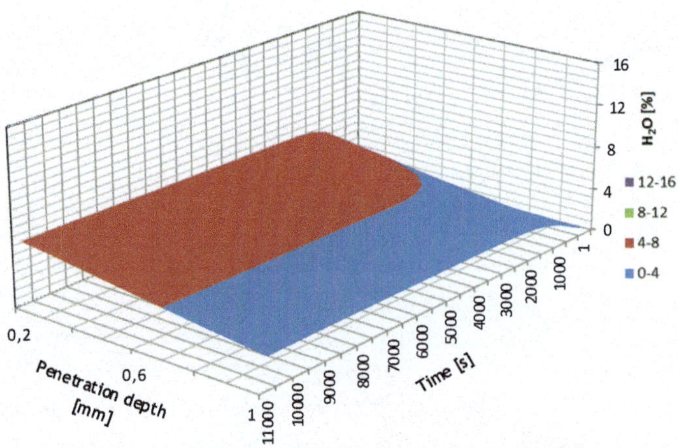

Fig. 6.23 Midel® eN

6.4 Vibration and Noise

With the continuous expansion of power networks and urban development, it is becoming more and more common for transformers to be installed in residential and commercial areas. These transformers generate noise pollution, affecting the lives of residents nearby. Therefore, vibration and noise reduction have become a special concern in the transformer industry nowadays. In the last decade transformer core sound levels are considerably lower than those built 30–40 years ago [26]. According to research studies, transformer noise is mainly caused by the vibration of the transformer and its cooling system, and its intensity is associated with the

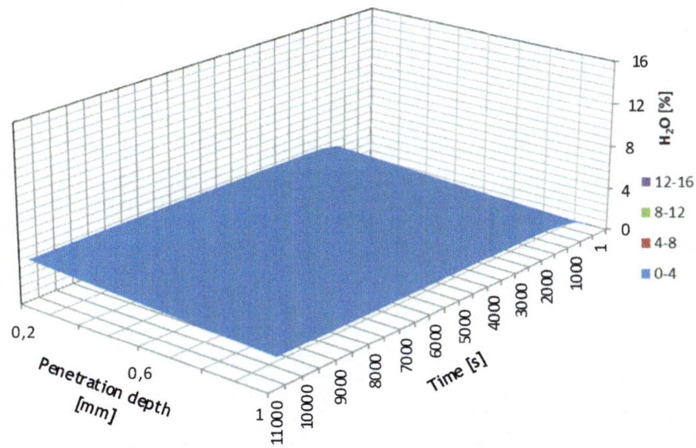

Fig. 6.24 Midel® 7131

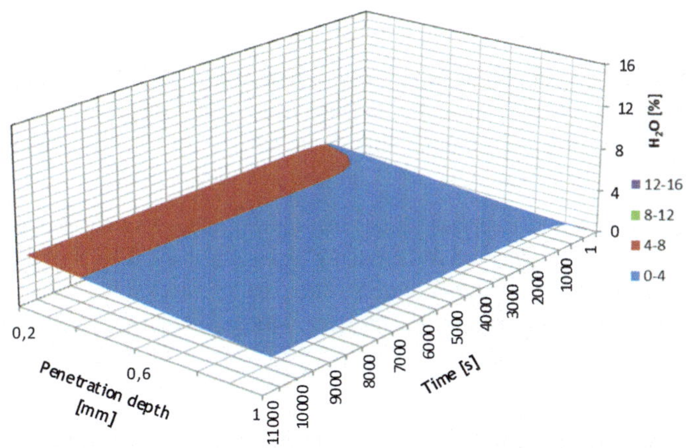

Fig. 6.25 BecFluid® 9902

transformer capacity, magnetic flux density, silicon sheet materials, core structure, and vibration performance of the cooling unit. But above all, magnetostriction is the root cause of core noise. Magnetostriction is a significant source of no-load sound. In recent times, power loss and magnetostriction of grain oriented electrical steel have been considerably reduced. Yürekten reports in [26] that by applying high quality of silicon steels or laser treated material, it is possible to reduce the sound level of core for 4–5 dB. Simply way to reduce the sound is the reduction in core flux density, but this method has a negative effect on the cost and size of the transformer. Reduced core flux density increases load losses.

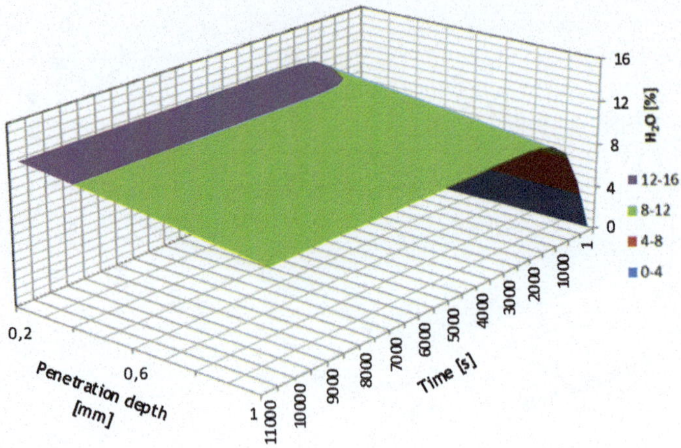

Fig. 6.26 Powersil® Fluid TR 50

Causes of noise development in a transformer can be:

- The core vibration due to the magnetostriction of the silicon steel sheets.
- The longitudinal force caused by the distortion of the magnetic line of force at the joint of the iron core and the side thrust caused by an uneven magnetic flux distribution in the iron core between silicon steel and sheets.

As all insulating liquids are more or less incompressible, much difference in sound propagation cannot be expected. When using biological insulating liquids, the transformer may be operated at higher temperatures, which means that the external fans are not switched on or only switched on later, which affects noise emissions.

6.5 The Effect of Nanoparticles in Biological Insulating Liquids

The current trend of using nanoparticles to improve the dielectric properties of transformer insulating liquid started around 2006. Transformer oil-based nanofluids are a colloidal suspension of nanoparticles in transformer oil. The average size of these particles is in the range of few nanometers. For transformer oil-based nanofluids under external electric field, both transformer liquid molecules and nanoparticles will be polarized. Positive and negative surface polarization charges will gather at two sides of a nanoparticle, making the particle charged. The charged nanoparticle under electric field has the similar motion tendency with that of a polar molecule, namely, random thermal motion and orientation along the electric field, which forms orientational polarization. Considering that pure insulating liquid is a kind of weak polar liquid, and the polarization in pure insulating liquid under external electric field is mainly electronic polarization. Therefore, it can be assumed that there is no relaxation polarization in pure transformer liquid, namely, instantaneous displacement polarization is the only polarization for transformer liquid molecules. So, it can be believed that nanoparticle orientational polarization is the only kind of relaxation polarization.

It has been proven that the addition of nanoparticles improves the lightning impulse strength and thermal conductivity of natural ester insulating liquids [40]. According to Zhong et al. [41], there is an enhancement in the electrical properties of the biological insulating liquids due to the increase of shallow trap density in the nanofluid. Du et al. show in [42] that with TiO_2 semiconductive nanoparticle modified transformer liquid AC and positive impulse breakdown voltage increased by 19% and 24%, respectively, compared to pure insulating liquid. In this context, nanoparticles represent metal particles (mostly metal oxides like TiO_2, Al_2O_3, ZnO, MgO, Fe_3O_4, $BaTiO_3$) or nonmetal oxides like SiO_2. The range of the size lies between 1 and 100 nm. Nanoparticles can vary in size, conductivity (conductive, semi-, or non-conductive) and in their concentration in oil. Their positive effect on dielectric performance (enhancement of breakdown voltage, resistivity, and dissipation factor) was proved by several authors. What is yet unknown is long-term behavior in a transformer in contact with paper insulation. Another factor to consider is the agglomeration and sedimentation of nanoparticle dispersed in oil. Totzauer and Trnka showed in [43] that MgO in the range of 1–1.5 g/l lowered the tan δ value of rapeseed oil significantly.

For the nanoparticles to be effective, they must be kept in suspension. In Fig. 6.27, it is shown which forces act on such a particle. For this purpose, the particle is idealized as a sphere. The necessary velocity of the liquid can be calculated by Eqs. (6.16)–(6.18). Hence, the viscosity is strongly dependent on temperature. This happens for the velocity of the liquid, too.

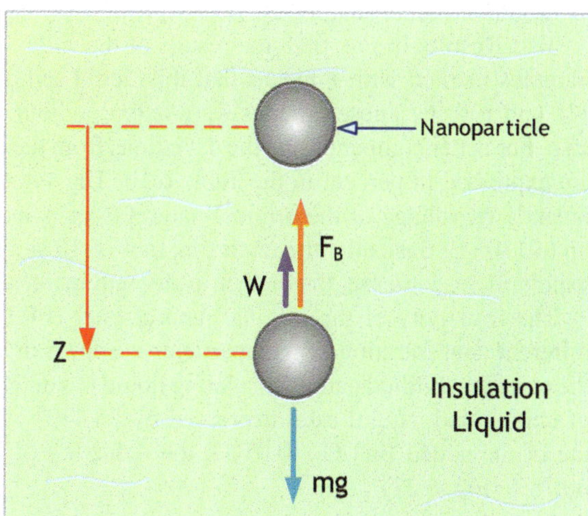

Fig. 6.27 Balance of power on a floating nanoparticle

$$m * g = w + F_P \tag{6.16}$$

$$\rho_P * g * \frac{4}{3} * r^3 * \pi = 6 * \eta * \pi * r * v + \rho_L * g * \frac{4}{3} * r^3 * \pi \tag{6.17}$$

$$v = \frac{2}{9} * \frac{g * r^2}{\eta} * (\rho_P - \rho_F) \tag{6.18}$$

ρ_P Density of nanoparticle (kg/m^3)
g Acceleration of gravity (m/s^2)
r Radius of particle (m)
η Dynamic viscosity of the insulating liquid (kg/m*s)
v Velocity (m/s)
ρ_L Density of insulating liquid (kg/m^3).

6.5.1 *Effects on AC Breakdown and Resistivity*

Maneerat et al. studied in [44] the effect of barium titanate ($BaTiO_3$) and titanium dioxide (TiO_2). For their research, they used TiO_2 and $BaTiO_3$ nanoparticles with a diameter size less than 100 nm. The nanoparticle concentration in the biological insulating liquid (FR3) was 0.01, 0.03 and 0.05%. To decrease the dissolved water content and thereby to improve the constancy of the specimens, the prepared nanofluid samples were heated at a temperature of 110 and 130 °C in an oven for 3 days in a vacuum state. The electrical characteristics of unmodified biological insulating liquid compared with the liquid mixed with nanoparticles were studied by the dielectric breakdown test according to ASTM 1816 [45] and the electrical resistivity measurement according to IEC 60247 [46].

Test Results In Fig. 6.28 the results of the AC breakdown voltage are depicted. Samples marked with * means that this liquid sample was not stressed by heat at 110 and 130 °C. These samples were only dried out at 80 °C in a vacuum oven for 15 h. For better comparability, the deviations from the unmodified natural ester (FR3) are expressed in percent in the Table 6.10. The values are quite scattered. The AC breakdown voltage of the original natural ester is increased by 12.28% by heating up to 130 °C. Basically, the matrix in Table 6.10 shows that as the temperature and nanoparticles increase, the dielectric strength increases.

The resistivity of the original natural ester (FR3) and the nanofluids with the different concentration of nanoparticles is shown in Table 6.10. It can be seen that the pure, unmodified natural insulating liquid has deteriorated by heat. The resistivity of unmodified natural ester decreased by 76.44% when heated to 130 °C. In case the of the added $BaTiO_3$ (0.05%), the reduction of the resistivity by 130 °C was 60.78% (Fig. 6.29).

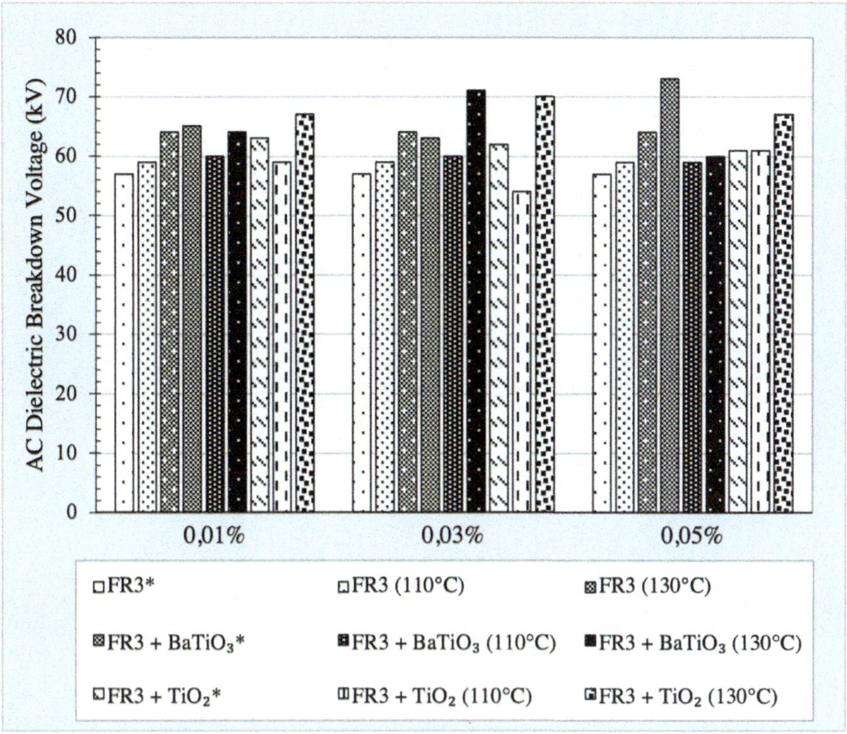

Fig. 6.28 AC dielectric breakdown voltage of unmodified natural ester and nanofluids

6.5.2 Effects on Partial Discharge Characteristics

Maneerat et al. [47] present the partial discharge inception voltage (PDIV) and partial discharge extinction voltage (PDEV) of palm oil based on nanofluids with different concentration of ZnO and $BaTiO_3$. Palm oil as a biological liquid shows a high potential to be used as a transformer insulation liquid. For example, in the south of Thailand, there are many sources of palm oil, and the price of palm oil is quite cheap. Hence, one way would be to use this liquid as transformer insulating liquid. To improve palm oil insulation characteristics, nanoparticle material is an interesting matter which has high potential to enhance palm oil dielectric characteristics, including other physical and chemical properties of palm oil. Partial discharge is one of the most important characteristics of the insulation. PDIV and PDEV are important parameters and must be investigated for the new insulation materials. Maneerat et al. studied the PDIV and PDEV behavior of unmodified palm oil compared with palm oil based on ZnO and $BaTiO_3$ nanoparticles. First, nanofluids were prepared using physical methods through magnetic stirring followed by ultra-sonication to disperse the nanoparticles in the liquid.

Table 6.10 Deviation from the unmodified natural ester in percent

	FR3* [0.01%]	FR3 (110 °C) [0.01%]	FR3 (130 °C) [0.01%]	FR3* [0.03%]	FR3 (110 °C) [0.03%]	FR3 (130 °C) [0.03%]	FR3* [0.05%]	FR3 (110 °C) [0.05%]	FR3 (130 °C) [0.05%]
FR3*	–	−3.39	−10.94	–	−3.39	−10.94	–	−3.39	−10.94
FR3 (110 °C)	3.39	–	7.81	3.39	–	7.81	3.39	–	7.81
FR3 (130 °C)	12.28	8.47	–	12.28	8.47	–	12.28	8.47	–
$BaTiO_3$*	**14.04**	10.17	−1.56	**10.53**	6.78	−1.56	**28.07**	23.72	14.06
$BaTiO_3$ (110 C)	5.26	**1.69**	−6.25	5.26	1.69	−6.25	3.51	**0.00**	−7.81
$BaTiO_3$ (130 C)	12.28	8.47	**0.00**	24.56	20.33	**19.94**	5.26	1.69	**−6.25**
TiO_2*	**10.53**	6.78	−1.56	**8.77**	5.08	−3.13	**7.02**	3.39	−4.69
TiO2 (110 C)	3.51	**0.00**	−7.81	−5.26	**−8.47**	−15.63	7.02	**3.39**	−4.69
TiO2 (130 C)	17.55	13.56	**4.69**	22.81	18.64	**9.38**	17.54	13.56	**4.69**

Fig. 6.29 Partial discharge test vessel

The PDIV investigation was performed in the test vessel with the needle—plane electrode, as shown in Fig. 6.28. The tungsten needle electrode with the tip radius of 10 μm was used to be the high voltage electrode and the brass plane electrode with 75 mm diameter was used as the grounded electrode. The gap distance of the electrode system was set up at 40 mm. Partial discharge, corona in the insulating liquid was produced because of the high eclectic field stress around the tip of the needle. Figure 6.30 shows the test circuit arrangement with the components:

(1) Testing Transformer 75 kV, 40 kVA
(2) Coupling Capacitor
(3) Test Object
(4) PD Measuring Device.

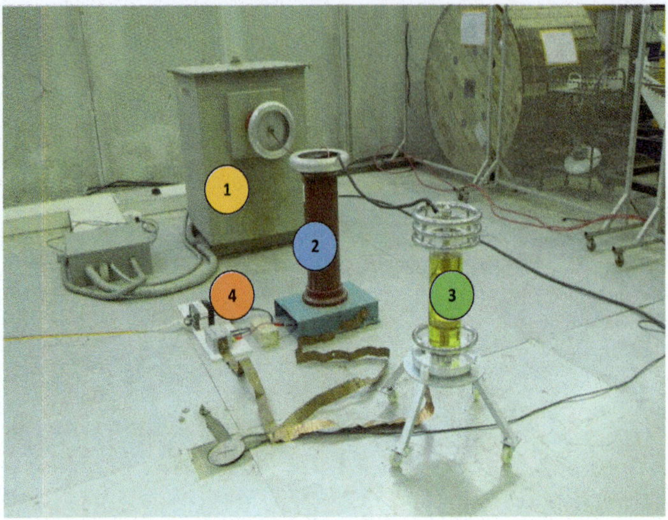

Fig. 6.30 PD test circuit arrangement

Figure 6.31 presents the results of partial discharge inception voltage and Fig. 6.32 the partial discharge extinction voltage, respectively. The test results show that palm oil nanofluid with 0.01% $BaTiO_3$ provides the highest PDIV and PDEV compared to that of other specimens. All other nanoparticles and concentrations with except of 0.03% $BaTiO_3$ have a negative effect of PDIV and PDEV in case of palm oil.

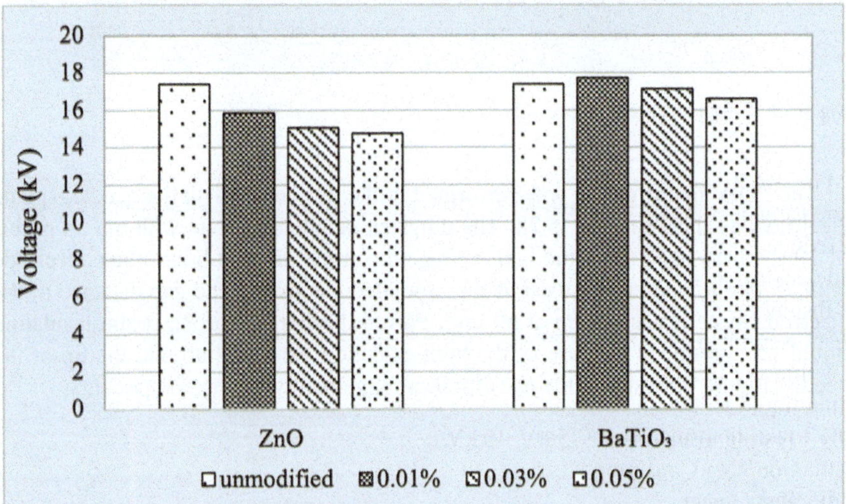

Fig. 6.31 Partial discharge inception voltage

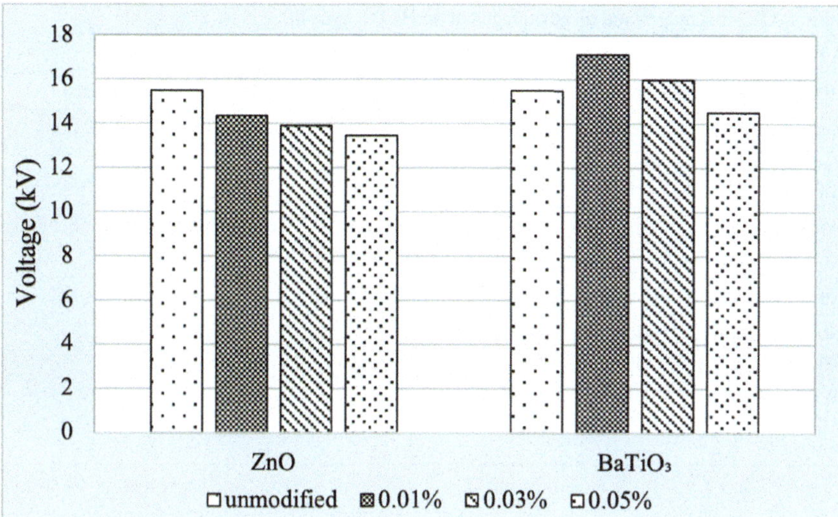

Fig. 6.32 Partial discharge extinction voltage

6.6 Dielectric Behavior of the Liquid Board Insulation Under Direct Voltage Stress

The combination of insulating liquid and cellulose achieves very good electrical insulation strength and forms an inseparable insulation system for every power transformer. In the case of the insulating liquid, it's not only the excellent insulating property. The cooling properties of the insulating liquid also ensure a good dissipation of the internal heat to the outside. This insulation system was developed decades ago for transformers in AC operation, but with direct or mixed field stress, as occurs with HVDC converter transformers, there is greater electrical, mechanical, and thermal stress.

For HVDC systems, the efficiency of electrical energy transmission increases with increasing operational voltage at the same nominal power, as the electrical current can be reduced simultaneously (and so are the losses). Table 6.11 opposes the advantages and disadvantages of an HVDC system to an HVAC system [48].

Insulation design of an HVDC converter transformer is strongly different from AC power transformers: In AC power transformers, most of the electrical field stress is within the insulating liquid gaps. As the dielectric withstand behaviors of insulating liquid is lower than of cellulosic material and the permittivity is also around 50% lower than the one of paper and pressboard, insulation design is focusing on insulating liquid and insulating liquid gaps, respectively. Solid insulation is used additionally for field grading (angle rings, shielding rings, etc.). As the permittivity of cellulose is around twice as high as the one of insulating liquid, the introduction of more solid material is counter effective, as the insulating liquid gaps are stressed even

Table 6.11 Advantages and disadvantages of an HCDC system

Advantages	Disadvantages
Higher power transportation	Expensive converter stations necessary
Reduced losses compared to AC systems	Presently, only point-to-point connections feasible
Reduced number of necessary conductors compared to AC	
Working voltage is equal to rated voltage	Insulation system design is more challenging than for AC systems
No issues with reactive power	
No skin effect	Less operational experience than for AC systems
No (theoretical) limitations of transmission line length	
System-inherent redundancy at bipolar systems (Operation with only one pole and halved power)	

more when overall dimensions are kept the same. Therefore, it aspires to use only as much solid material as required. At DC stress the situation is exactly the opposite: As the cellulose has a much higher specific resistance than insulating liquid ($\sigma_{Liquid} < 10^{-13}$ S/m; $\sigma_{Board} < 10^{-15}$ S/m @20 °C), the electrical field is mainly concentrated within the pressboard and the paper. So, to avoid breakdown or high dielectric stress, more cellulosic material is introduced for field grading purposes. In liquids, electrical conductivity is caused by ions, electrons, and the movement of charged macroscopic particles or molecule assemblies. Which charge carriers predominantly occur depends on the composition of the insulating liquid. Those ions that are involved in the transport of charge carriers are mainly created by the dissociation of aging products (acids) and impurities. The water content also affects the conductivity. Liquid insulating materials have a specific conductivity of $\geq 10^{-10}$ S/m. Conductivity values of 10^{-18} S/m can be achieved through thorough cleaning [49].

When insulation systems are stressed with DC fields or mixed electrical stress, the electrical conductivities of the used insulation materials play a vital role in the distribution of electrical field stress within the insulation system. At pure AC stress, only the (relative) permittivity ε_r is responsible for electrical field distribution. This important material parameter is well known for the materials which are relevant in the sense of (high voltage) electrical engineering. Furthermore, relative permittivity is quite constant over time (=ageing), temperature and within the frequency range of power systems. For decades, AC apparatus design engineers had to focus on the insulating liquid gaps in insulating liquid-board insulation systems, as the insulating liquid gaps have been the (electrically) weaker parts [48].

For the design of HVDC equipment, the electrical conductivity is the dominating parameter. Typically, the ratio of σ_{Liquid} to $\sigma_{Pressboard}$ is higher than 50 at 20 °C. In these cases, the solid insulation (pressboard) is stressed strongly by the electric field, whereas the stress in the insulating liquid is comparable low (at steady state). However, relative permittivity is still important at DC, but only during transient stages. This leads to completely different electrical field distribution within an insulation system.

The viscosity influences the ion mobility in the liquid. There is a direct relationship between viscosity and electrical conductivity [48]. In general, the relationship $\eta * \mu$ can be assumed as constant. According to [48], the mobility of positive and negative ions can be linked to viscosity as (Eqs. 6.19 and 6.20).

$$\mu_+ = \frac{1.5 * 10^{-11}}{\eta} \tag{6.19}$$

$$\mu_- = \frac{3 * 10^{-11}}{\eta} \tag{6.20}$$

μ Ionic mobility ($m^2/V*s$)
η Dynamic viscosity of liquid ($N*s/m^2$).

6.6.1 *Electrical Conductivity in Liquid Immersed Cellulose Insulation System*

When a step DC-Voltage is applied to a liquid immersed-cellulose insulation system, the typical response can be classified into three stages:

- In the first few milliseconds after voltage application, a capacitive current flows, as the (uncharged) dielectric is charged up.
- After this initial behavior, polarization processes govern the current in an intermediate stage. This stage can last for several hours or even days until a steady state is reached and all polarization processes have been decayed.
- The time dependent current decreases asymptotically to the steady-state current, which is the actual conduction current.

6.6.1.1 Influence of Temperature on Electrical Conductivity

Temperature has a vital influence on electrical conductivity. For insulating liquids and cellulose, an increase of temperature generally results in an increased electrical conductivity. This increase can be explained by an elevated charge carrier mobility and further by the increase of available charge carriers, e.g., due to thermal activation. The influence of temperature on the (conduction) current can be seen in Fig. 6.33. Temperature variations of around 30 K can influence the conductivity of insulating immersed paper already by one order of magnitude. For practical applications, this effect can be overlapped by field effects.

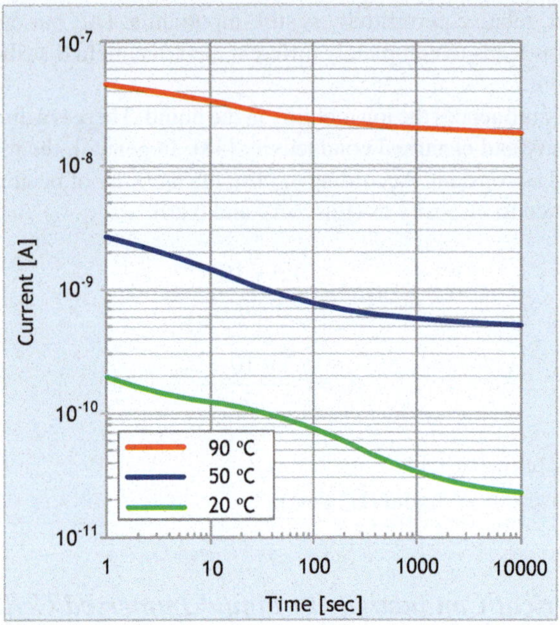

Fig. 6.33 Influence of temperature on electrical conductivity [48]

6.6.1.2 Influence of Electrical Field Strength on Electrical Conductivity

For certain insulation materials and within certain ranges, the electrical conductivity depends significantly on the applied electrical field strength. A general dependence of electrical conductivity on electrical field strength is shown in Fig. 6.34. Here, the determination of electrical conductivity is assumed to be made at constant temperature and within a constant interval after voltage application. Then, four characteristic areas (A to D) can be classified:

Area A Ohmic behavior—is the range of low electrical field strengths.
Area B Electronic conduction processes are overlapping ionic conduction and conductivity increases.

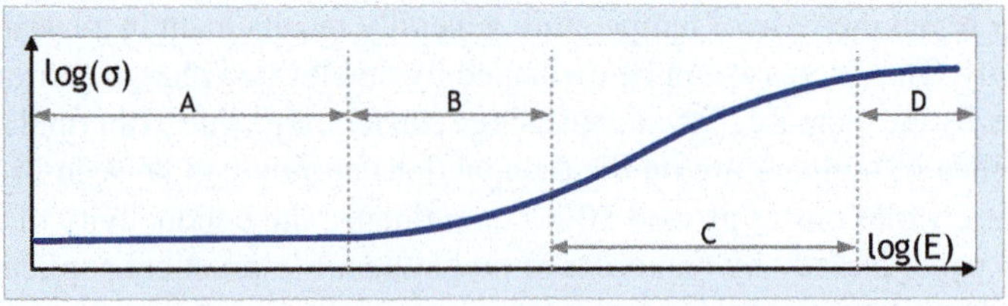

Fig. 6.34 Influence of electrical field stress on electrical conductivity [48]

Area C With increasing field strength, electronic conduction is dominating the conduction mechanism, which leads to a further increase of electrical field strength.
Area D If no discharge processes occur, the conductivity experiences a saturation effect when the field strength is raised further.

6.6.1.3 Influence of Moisture on Electrical Conductivity

The presence of water inside insulating materials is disadvantageous, as it increases losses and electrical conductivity and can significantly reduce electrical strength. Increased moisture content can be a sign of ageing and it can furthermore increase ageing rates.

Generally, it can be expected that with an increased moisture content in insulating liquid, the electrical conductivity will rise. But this is also a temperature related effect as it depends how much water is dissolved (in solution) and which quantity is in suspension (free water). Water which is present in solution does not ionize and therefore does not contribute (significantly) to electrical conductivity. The minimum conductivity is reached when all water is in solution, which depends on

- temperature,
- moisture,
- and the kind of insulating liquid.

The electrical conductivity of cellulose rises with increasing moisture content, too. In insulating liquid immersed cellulose, several layers exist which all contain small insulating liquid films. If the conductivity of these oil layers is increased, either by increasing moisture content or/and using an insulating liquid with higher conductivity, the conductivity of the paper increases significantly as well. Pressboard does not show such a strong increase up to a certain moisture level. However, if the moisture content is raised to higher levels (e.g., > 3%), the electrical conductivity of both insulating liquid immersed paper and pressboard increases steadily [48]. Judendorfer studied the electrical conductivity of pressboard in the range between 0.2 and 5% (Fig. 6.35). It can be clearly seen that the electrical conductivity rises for each percentage point of moisture in one order of magnitude. It is the ultimate objective to keep HVDC equipment as dry as possible under any circumstances during their whole equipment life. Not only that moisture can speed up ageing processes, it can also increase the conductivities of insulating liquid and cellulosic components, of which the former will be stronger affected. The drier an insulation system can be kept, the nearer design values in terms of electrical field strengths and field distribution can be obeyed.

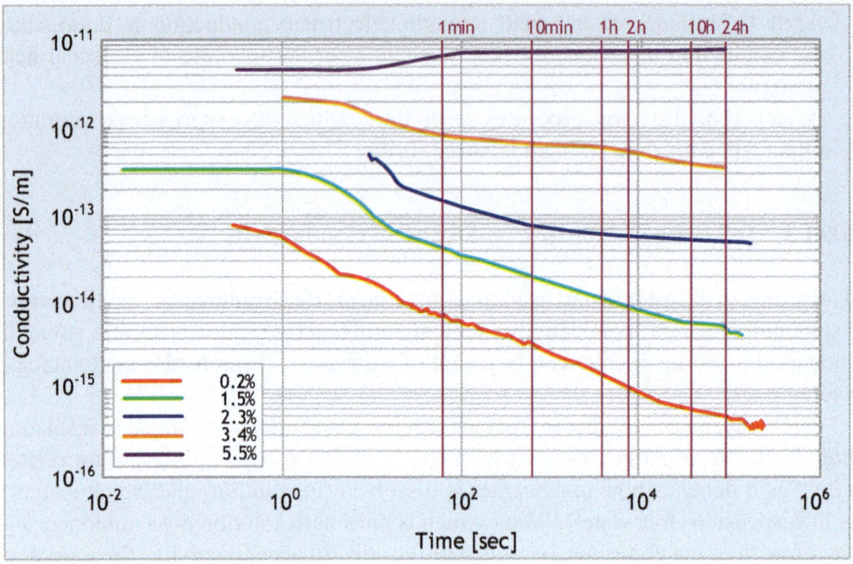

Fig. 6.35 Influence of moisture content on electrical conductivity of pressboard (immersed with Nynas Nitro 4000X) [48]

6.6.2 Measurement of Electrical Conductivity

The determination of electrical conductivity can be made with several different methods and measurement set-ups. When measuring the conductivity (volume conductivity) of solid insulation (including solids impregnated samples), usually there is guidance in some international standards e.g., IEC (IEC 62631 [50] and IEC 60247 [46]) and ASTM (ASTM D257 [51] and ASTM D1169 [52]).

All these methods are based on the measurement of the electrical current, which flows through the investigated material. Ohm's Law (Eq. 6.21) can be used to determine electrical resistance by measuring voltage and current.

$$U = R * I \tag{6.21}$$

U Potential (V)
R Resistance (Ω)
I Current (A).

However, care must be taken, as Ohm's Law is only valid under certain preconditions. Commonly, it can be applied for DC, if temperature and pressure (on the investigated material) are constant during measurement time. Furthermore, a linear relationship between voltage and current must be given, otherwise errors might occur (field strength depended on resistivities). In materials with a high resistivity, measurement time also plays an important role.

6.7 Operation of Equipment with Biological Insulating Liquids

One of the top three causes of power transformer failures is cellulose insulation failure, followed by On-Load Top Changer and bushing failures. Cellulose insulation may be the weakest link of the transformer because the degradation of cellulose insulation is irreversible.

6.7.1 Sustainable Peak Load Transformer

The electric transmission effectiveness depends on the capability to decrease losses by decreasing transmitted current, which has been probable by the manufacturing of high volt apparatus. The grid of the future must consist of contemporary equipment with innovative design. Modern insulating substances and novel expertise to assure safer, consistent, stable, and green electricity [5].

The high fire point of biological insulating liquids means they can be run at higher temperatures, increasing the amount of power distribution without the need to expand the size and weight of the transformer unit. This also permits OEMs to develop smaller ester-based transformers that produce the same power output as larger mineral oil alternatives, further saving space [53].

6.7.1.1 Effects of the Energy Transition on the Energy Transformation

Usually, the transformers in the electrical network are designed according to the largest possible load flow. This means that the installation of large amounts of regenerative energy, which is often only available for a short time, determines the size of the transformer. However, this load is usually not continuously available. The electricity production patterns of photovoltaic and wind power are characterized by large fluctuations. Figures 6.36 and 6.37 gives information about the daily and yearly feed-in fluctuation of electrical power generated by photovoltaic systems in alpine regions in Austria. Due to the lack of solar radiation and varying levels of cloudiness, maximum utilization of the systems is not possible for many hours. With wind power, the situation is similar. But in this case, the daytime and time of the year play a subordinate role (Figs. 6.38 and 6.39).

However, the power consumption is also subject to large fluctuations. First off all because of the seasonal seasons and the climate condition of the countries. But during the daytime also because of charging the vehicles, which are getting more and more [54] (Fig. 6.40).

Basically, the examples given show that if the distribution transformers are designed for the peak load, they are oversized for most of the time. In general, transformers have very different load profiles. According to an EU study, the average

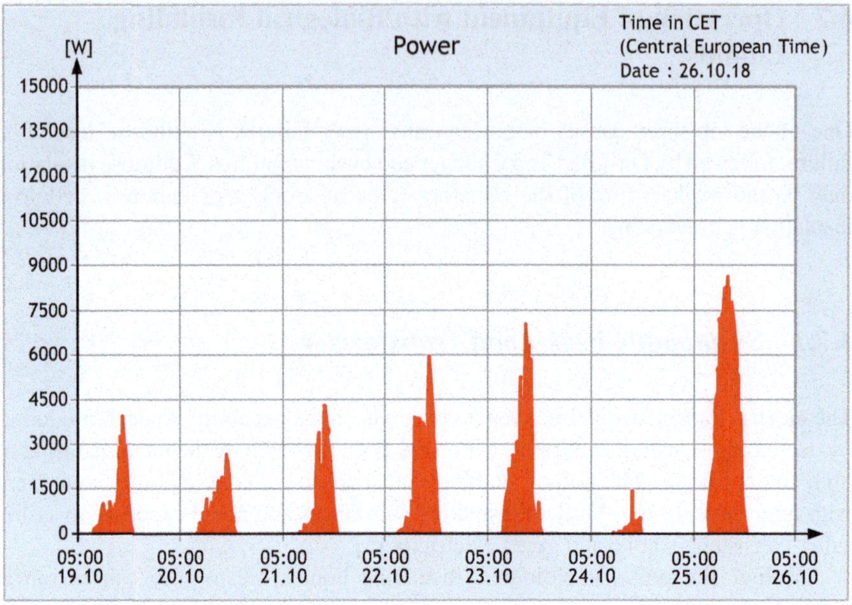

Fig. 6.36 Daily fluctuations in electricity production from photovoltaic systems

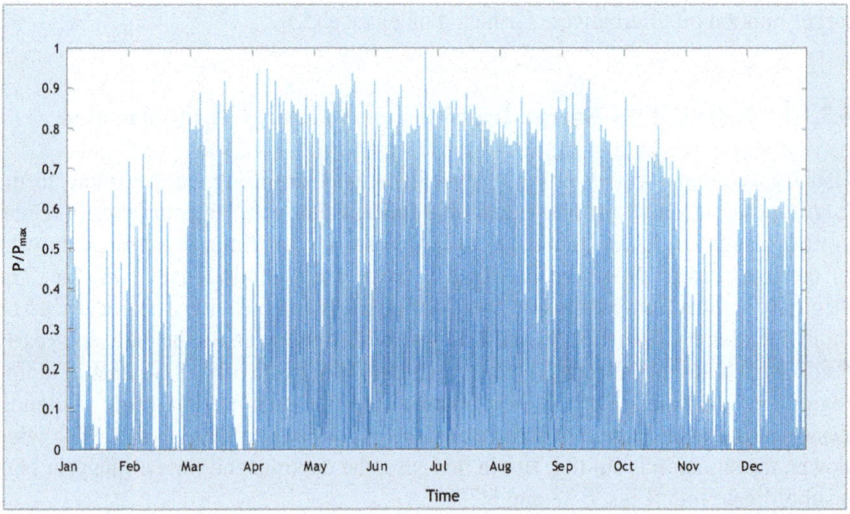

Fig. 6.37 Volatility of PV-feed in of electrical power during a year in a region in Austria

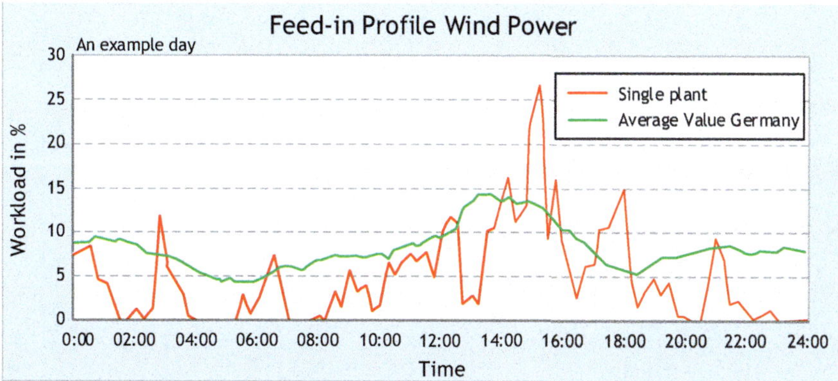

Fig. 6.38 Feed-in profile from wind power distributed over one day in Germany

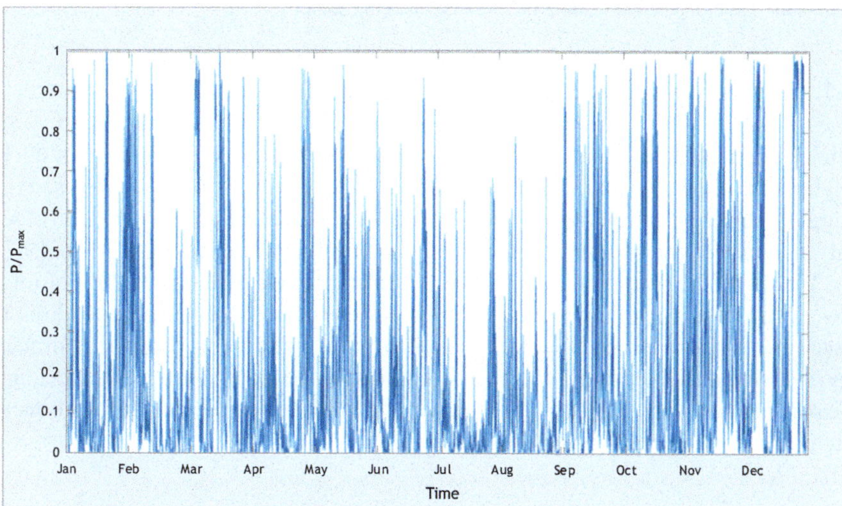

Fig. 6.39 Volatility of wind power feed-in during a year in a region in Austria

load transmitted by the distribution transformers is 18.9% of the design value [55]. WESTRAFO writes in [56] that based on the experience of major utilities around the globe, the yearly average loading of distribution transformers in the electric grid is between 10 and 25%. At such low loads, the no-load losses dominate, while the load losses are less important. As a result, they cause unnecessary losses both in production (higher use of materials) and in operation.

The electricity network can partly mitigate this problem by using sustainable peak load transformers, as they introduce flexibility into the system. These transformers support the transformation of the power supply to renewable sources of energy, accompanied by a substantial increase in energy efficiency.

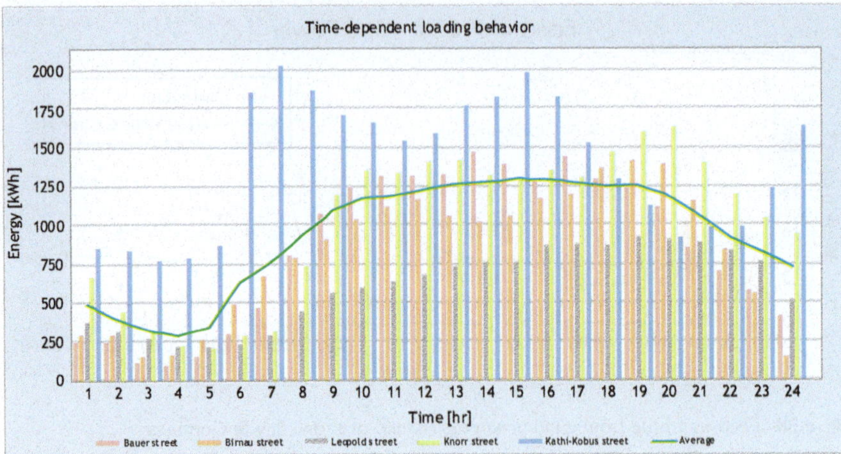

Fig. 6.40 Time-dependent loading behavior on charging stations

6.7.1.2 The Concept of Sustainable Peak Load Transformer

Biological insulating liquids allow transformers to operate at temperatures 20 K higher than allowable for mineral oil (see Sect. 5.2.1). This property will be used for the sustainable peak load transformer. Hence, these insulating liquids have also other advantages that can be used together with this transformer concept [15, 57–62].

Thanks to the increased load ability, a utility can save as much as 20% of total ownership cost, due to the reduction of the no-load losses, reducing the total dissipated energy for one year. Users will be able to rationalize the cost of the transformers by choosing a smaller transformer to be installed in the grid. Whenever the loading peak is reached during the day, the transformer will be able to sustain it for a typical short duration of peak loading time [56].

6.7.1.3 The Implementation of the Sustain Peak Load (SPL) Concept

- Initially, an analysis of the existing or expected load currents should be carried out as a function of time. Figure 6.41 shows, for example, the load of a transformer of a large Brazilian grid operator, set up in a tourist region, as a percentage of the design specification.
- Calculation of no-load and short-circuit losses. While the no-load losses depend more or less linearly on the power (Fig. 6.42), the short-circuit losses increase with the square of the current (Fig. 6.43). This is naturally higher for smaller transformers with the same work. The position of the intersection point of the losses is the decisive variable for transformer selection. For example, Fig. 6.44 shows that the losses of a 160 kVA transformer are higher compared to a 100 kVA transformer if the load is less than 44 kVA.

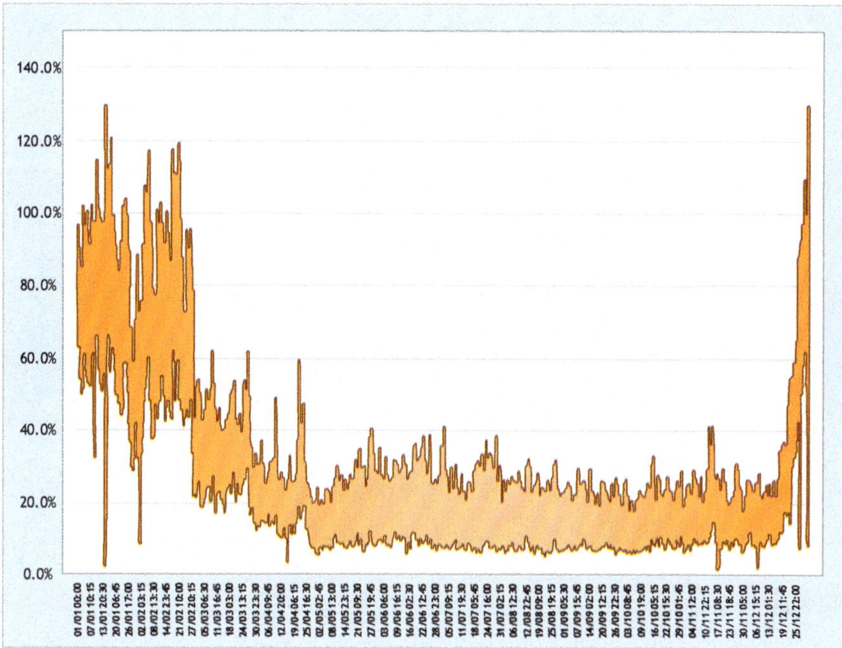

Fig. 6.41 Quarter-hour recording of the transformer power over the period of one year (Santa Catarina, Brazil)

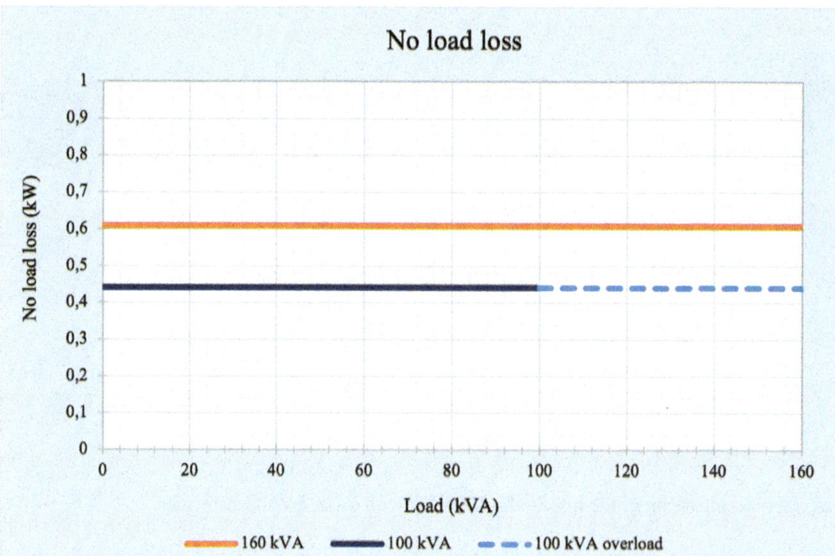

Fig. 6.42 No load losses

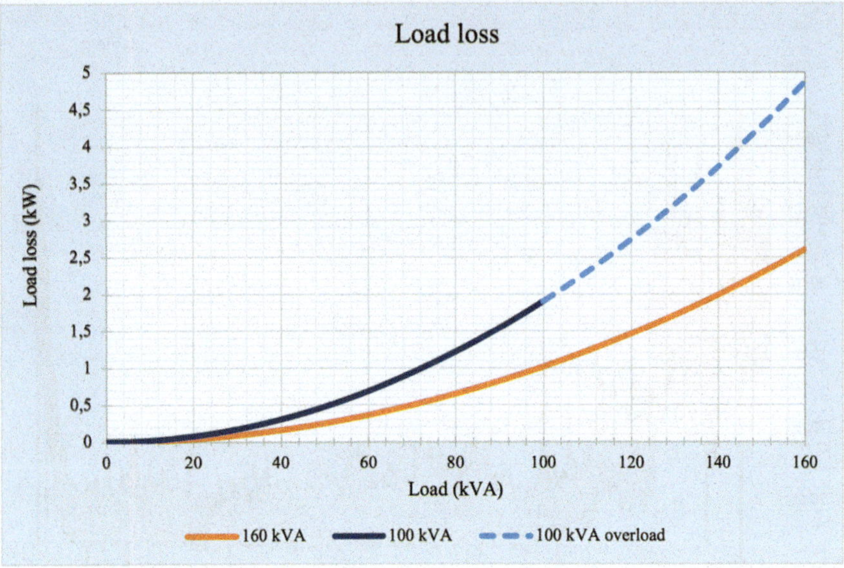

Fig. 6.43 Load losses

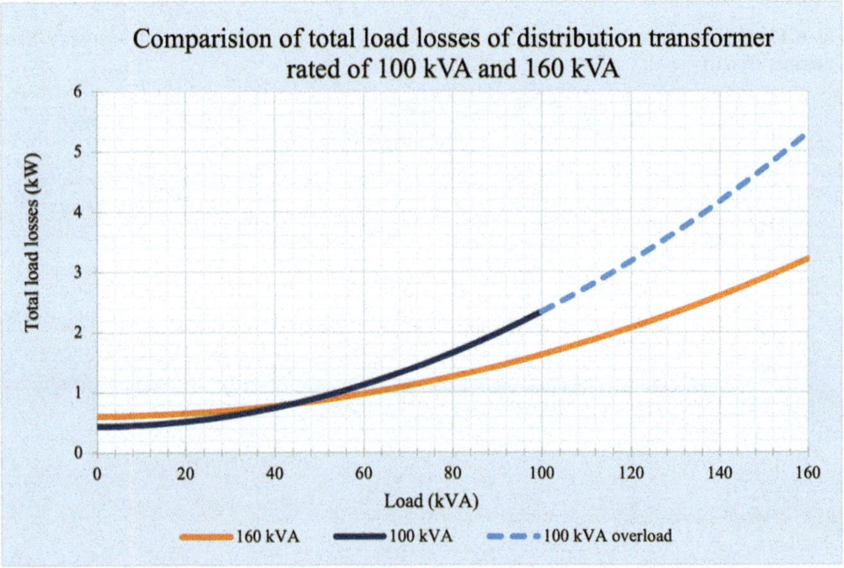

Fig. 6.44 Comparison of the losses of a 100 kVA with a 160 kVA transformer

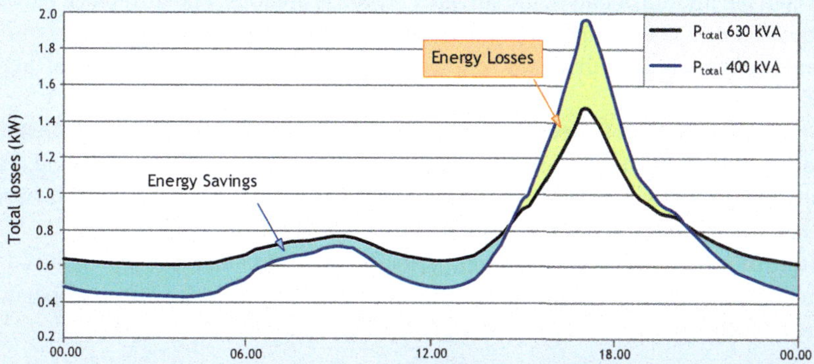

	630 kVA MO	400 kVA FR3 Liquid	Savings	
No-load losses (NLL)	600 W	430 W	170 W	28.3 %
Average load Losses (LL) for loading profile	174.6 W	304.5 W	-129.9 W	-74.4 %
Average total losses for loading profile	774.6 W	734.5 W	40.1 W	5.2 %
Energy dissipated per day	18.59 kWh	17.63 kWh	0.963 kWh	5.18 %

Fig. 6.45 Comparison of the energy gains and losses of a 630 kVA mineral oil filled transformer with a 400 kVA biological insulating liquid filled transformer

- Now, the decisive factor is over what period what load is required. Figure 6.45 from [63] shows the summation of the energy losses and gains of a 630 kVA mineral oil filled transformer compared to a 400 kVA filled transformer.

6.8 Examination of Electrically Stressed Insulating Liquids

Many works in the literature deal with the effects on the different insulating liquids at various electrical loads. For streamer propagation, investigations concentrated on studying streamer conduct and characteristics comprising streamer swiftness, streamer extent, streamer profile, charge and current.

6.8.1 Streamer Propagation in Case of Lightning Impulse

Sitorus et al. tested the streamer shape of jatropha curcas methyl ester oil (JMEO), presented in [64]. The two types of insulating liquid which they have tested were JMEO, obtained by alkali base catalyzed esterification process and mineral oil.

Streamer Shape The negative streamers in mineral oil are more filamentary and luminous than those in JMEO. The branching position is also different. In JMEO, streamer branching position is at the beginning of the streamers, while in mineral oil it is in the middle or end of the streamer. There is obviously no difference between the positive streamer shapes in both liquids, but it seems that the light spots in the streamer channels in JMEO are brighter than in mineral oil, especially at higher voltage levels.

Stopping Length of Streamers It appears that the stopping length of streamers increases with the voltage, and it is slightly longer in JMEO than in mineral oil. For a given voltage, stopping length of streamer decreases when increasing the electrodes gap in both liquids.

Streamers Velocity The average velocity of a streamer can be deduced by knowing the stopping length of the streamer and the corresponding propagation time deduced from the current oscillograms (Eq. 6.22).

$$Velocity\, of\, streamer = \frac{Length\, of\, streamer[km]}{Propagation\, time[s]} \tag{6.22}$$

The streamers velocities in both mineral oil and JMEO are close and vary between 0.2 and 1.2 km/s for the negative streamers and in the range of 0.9 and 1.8 km/s for the positive streamers, respectively.

6.8.2 DGA from Switching and Lightning Impulse Test

In the course of switching impulse tests with the insulating liquid natural ester FR3, the liquid before and after the tests (70 breakdowns), were sampled and analyzed for fission gases [65]. Before the tests, the insulating liquid was stored under a nitrogen atmosphere [4]. Furthermore, a lightning impulse stressed synthetic esters (Midel 7131) was investigated. Both the Toepler- and the headspace method were used for analyzes.

Test Results All results are related to the insulating liquid volume and are expressed in ppm. With the analysis method—headspace—syringe—no results are available for nitrogen, propene, and propane. The reason is that nitrogen is used as the transfer gas and that the determination limit for propene and propane is not reached with this method.

Results of the under nitrogen atmosphere stored FR3 liquid are presented in Table 6.12. Storage under a nitrogen atmosphere shifts the gas ratio in the insulating liquid in the direction of nitrogen. If the ratio of solubilities in air between nitrogen and oxygen for natural esters is around 2 [53], the value increases to around 8 when stored under a pure nitrogen atmosphere.

Table 6.13 shows the change in gas composition due to switching impulse. The results show a slight increase in the fission gases of hydrogen, carbon monoxide, carbon dioxide, methane, ethylene ethyne and propane.

Table 6.14 shows the results after a lightning impulse exposure of the synthetic ester Midel 7131.

Table 6.12 Stored FR3 with nitrogen pad

Gas content: 4.3%	Toepler pump	Headspace	Headspace/toepler pump
O_2	4050	–	–
N_2	34,480	–	–
H_2	40.2	27	0.672
CO	33.6	39	1.161
CO_2	344	370	1.076
CH_4	10.1	6	0.594
C_2H_2	37.3	25	0.670
C_2H_4	10.4	8	0.769
C_2H_6	1.4	1	0.714
C_3H_6	3.3	–	–
C_3H_8	0.7	–	–

Table 6.13 Test results—switching impulse

	Before test	After test
Gas content (%)	1.8	3.8
O_2	1590	5650
N_2	14,610	28,540
H_2	19.6	67.2
CO	19.3	57.7
CO_2	189.5	340.8
CH_4	4.3	9
C_2H_2	12.5	15.3
C_2H_4	4.4	0.9
C_2H_6	0.5	2.7
C_3H_6	2	0.4
C_3H_8	0.3	3.1

Table 6.14 Test results—lightning impulse

	After test
Gas content (%)	3.6
O_2	17,860
N_2	48,620
H_2	88.3
CO	74.4
CO_2	507
CH_4	10.5
C_2H_2	50.9
C_2H_4	15.2
C_2H_6	1.0

Hamid et al. analyzed in [66] the generated gases produced by lightning impulse breakdown tests of palm oil and rice brown oil (see next section). Their experiments showed that for both vegetable oils the highest produced key gases are H_2 and C_2H_2. However, palm oil generates more amount of C_2H_2 as compared to rice brown oil. For DGA interpretation, they used ratio methods of Dornenburg, Roger and IEC, as well as the Duval triangle method. There is much difference in the interpretation of the results and not all methods gave the correct result (arcing):

- DGA analysis of palm oil is considered valid under the Dornenburg key gas ratio, but not for rice brown oil.
- For the Rogers ratio method, most of the ratios are either out of range or don't fit Roger's table. This interpretation method is not suitable to be used for palm oil and rice brown oil in diagnosis the electrical fault.
- The IEC ratio method predicts correctly for the impulse breakdown of palm oil. However, the results are valid for both cases (discharge of low energy and high energy). For rice brown oil, the IEC ratio method doesn't give a correct diagnosis.
- Duval triangle method correctly diagnoses electrical stress faults in palm oil and rice brown oil in the simulated situation.

6.8.3 AC Breakdown Tests with Palm Oil

Suksagoolpanya et al. compared in [67] the behavior of palm oil and mineral oil in case of AC breakdown, positive and negative lightning impulse test.

The AC breakdown voltage test was performed by testing devices with the hemisphere electrodes for applying the voltage with 2 kV/s incremental rate voltage according to IEC 60156 [68]. In the experiment, three different gap spacings with 1.0, 1.5, and 2.0 mm were used.

For the positive and negative lightning impulse test, a needle sphere electrode with different gap spacing of 10, 15, and 20 mm was installed. The lightning impulse tests were applied according to IEC 60897 [69]. Figure 6.46 shows the lightning impulse testing diagram and Fig. 6.47 the corresponding test setup.

The breakdown voltage values of the mineral oil are greater in all gap spaces than that of the palm oil under the electrical phenomena as shown in Fig. 6.48.

Experiments done by Hamid et al. [66] showed very similar results for palm oil and rice brown oil. They carried out the lightning impulse breakdown voltage test per IEC 60897 [69] using a gap distance of 6 mm. The breakdown voltage was investigated using negative polarity 1.2/50 μs lightening impulses. The mean breakdown voltage values of the palm oil and rice brown oil after 50 measurements are −125 kV and −106 kV, respectively.

Similar studies were carried out by Azis et al. in [70]. The aim in this work has been to evaluate the performance of lightning breakdown voltages of RBDPO (refined bleached and deodorized palm oil) and coconut oil as compared to mineral oil under non-uniform field with consideration on different voltage polarities and testing methods at small gap distances. They found only a small influence of voltage

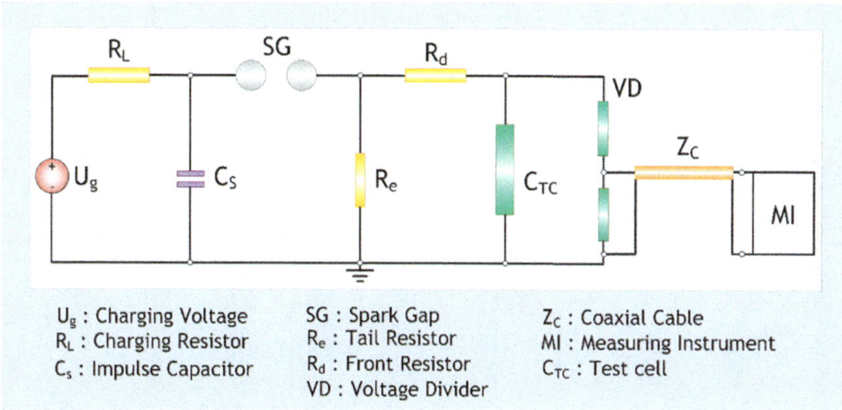

Fig. 6.46 Lightening impulse breakdown test—testing diagram

Fig. 6.47 Lightning impulse breakdown test—test setup

polarities on the breakdown voltages of all samples at all gap distances (2.0, 3.8, 6.0 mm) where the highest percentage of differences is only 11%. Under positive lightning impulse the breakdown voltages of RBDPO and coconut oil were comparable with mineral oil where the highest percentage of difference was less than 11%. Under negative lightning impulse, breakdown voltages of RBDPO and coconut oil was lower than for mineral oil, where the highest percentage of difference can be up to 22.5%. Mineral oil had the highest breakdown voltages followed by RBDPO and coconut oil. At gap distance of 2.0 mm, the breakdown voltages of RBDPO and coconut oil are slightly lower than mineral oil. However, as the gap distances

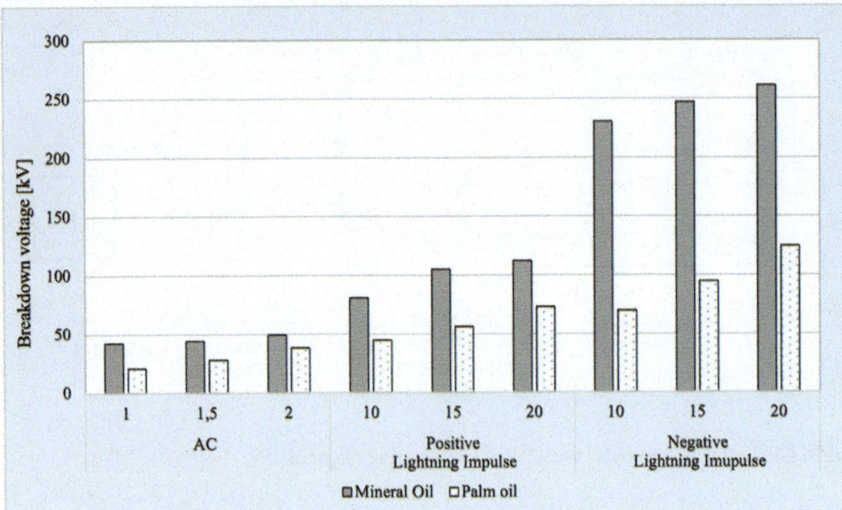

Fig. 6.48 The average values of breakdown voltage tests under different gap spacing

increase to 3.8 and 6.0 mm, the differences in breakdown voltages between RBDPO and coconut oil versus mineral oil increase, too. In all investigations, coconut oil had the lowest breakdown voltages at all gap distances under both polarities, with values between 43 and 69 kV.

6.8.4 Electrical Tests of Natural Ester Impregnated Pressboards

Siriworachanyadee and Pattanadech tested the AC and lightning impulse (positive and negative impulse) breakdown voltage characteristics of natural ester (FR3) impregnated pressboards with different impregnation period [71].

AC breakdown voltage test is one of the destructive tests used to examine the properties of insulation. According to IEC 60243-1 [72], this method is determined the short time electric strength of impregnated pressboard or solid insulation material at frequency between 48 and 62 Hz. The impulse breakdown voltage test, according to IEC 60243-3 [73], determines the electric strength of impregnated pressboard or solid insulation under 1.2/50 μS impulse voltage strength.

Sample Preparation and Test Procedure The samples were impregnated for different lengths of time (no-impregnation, 8, 16, and 24 h). These samples were subjected to electric field stress for different periods (one month, three months). Before the AC and lightning impulse breakdown voltage test were performed, impregnated pressboards were prepared as follows:

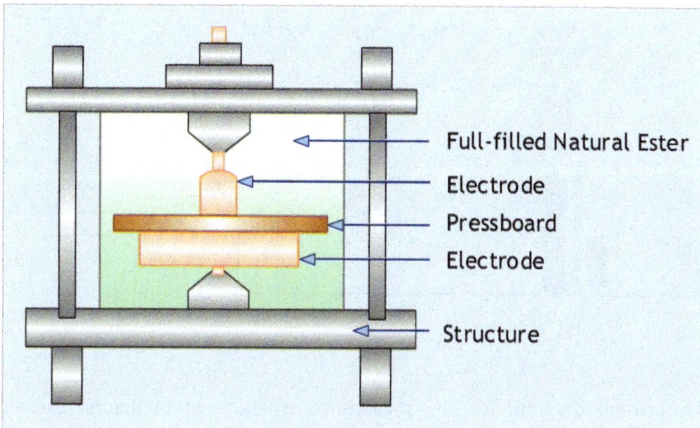

Fig. 6.49 Test cell for breakdown testing of natural ester impregnated pressboard

- Heating the pressboards with temperature at 80 °C for 12 h under 200 mbar pressure for removing the moisture within the pressboard.
- Fulfilling the pressboard with the prepared natural ester and heating with temperature at 60 °C for 0, 8, 16 and 24 h respectively under 200 mbar pressure.
- Applying the electric field stress with a supply voltage of AC 24 kV having the pressboard samples between the electrodes. The test cell is full filled with natural ester (Fig. 6.49) to simulate the same condition of pressboard as the inside of the transformer and prevent flash over along with the pressboard surface.
- At the end of the first month, half of the pressboard samples were taken from the test cell for further testing.
- Finally, at the end of the third month, another half of the pressboard samples were taken from the test cell to perform AC and lightning impulse breakdown voltage test as well.

The AC breakdown voltage tests were carried out according to IEC 60243-1 [72] and the natural ester impregnated pressboard was tested as follows:

- The circuit is prepared as illustrated in Fig. 6.50.
- The voltage is applied to the pressboard from 0 V until breakdown.
- The breakdown voltage values are recorded and analyzed.

In the case of impulse breakdown voltage tests, both the positive and negative impulse breakdown voltage characteristics of the natural ester impregnated pressboard were performed. The circuit diagram for positive and negative impulse breakdown voltage testing is illustrated in Fig. 6.51.

Test Results of Breakdown Voltage Measurement The results of breakdown measurement are shown in Fig. 6.52. The AC breakdown voltage strength of the impregnated pressboard increase with the duration of impregnation. On the other hand, the AC breakdown voltage of the pressboards experienced with longer electric

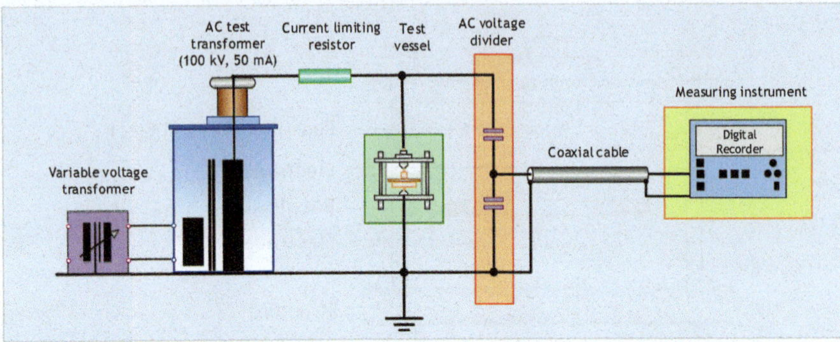

Fig. 6.50 Test circuit diagram for AC breakdown voltage test of natural ester impregnated pressboard

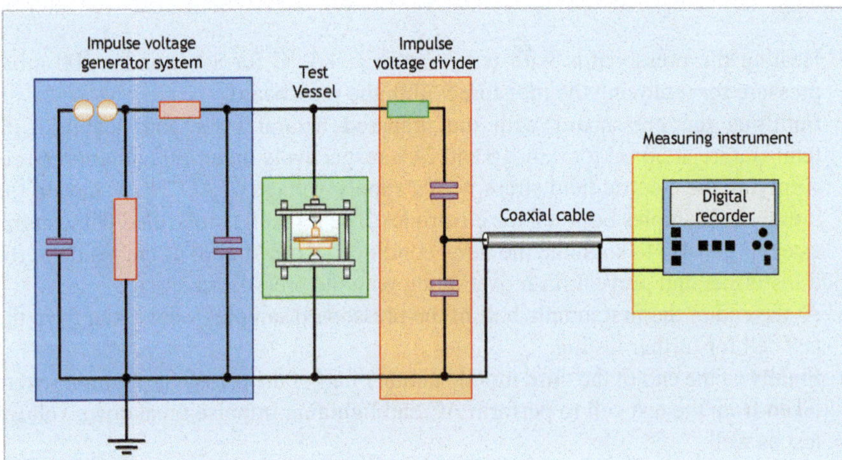

Fig. 6.51 Impulse breakdown testing diagram

field stress period (3 months) was lower than these of the pressboard experienced with shorter field stress period (1 month).

It can be explained that the positive and negative impulse breakdown voltage of the pressboard for a long impregnation period is higher than the positive, negative impulse breakdown strength of the pressboard for short impregnation period. The longer electric field stress period reduces the positive and negative impulse voltage of the pressboard (Fig. 6.52). From the result, it can conclude that

- the period of impregnation of pressboards obviously effects the AC and lightning impulse breakdown voltage of pressboard.
- the electrical aging period effects the AC and lightning impulse breakdown voltage of pressboard negative.

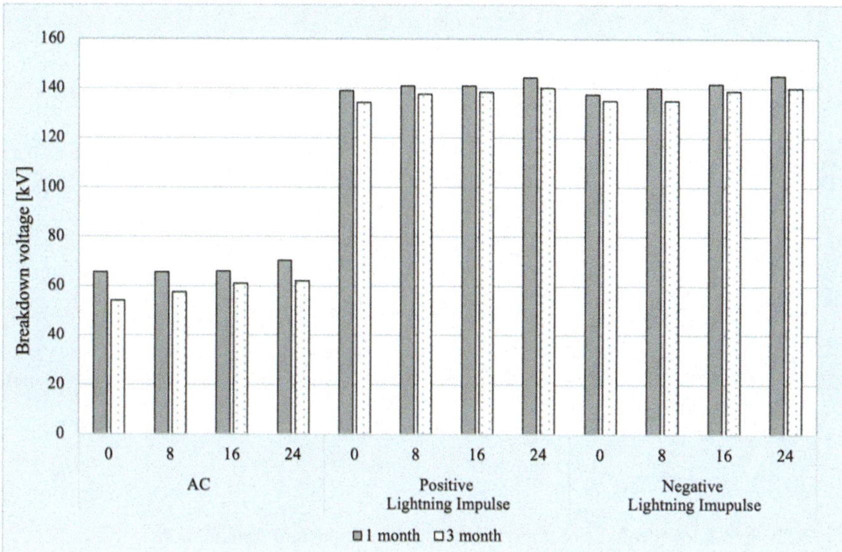

Fig. 6.52 The results of AC, positive lightning impulse, and negative lightning impulse breakdown voltage

6.8.5 Breakdown Voltage Tests Under Cold Condition

In [74], Eberhardt tested the behavior of the dielectric strength of impregnated transformer boards at low temperatures. The use of cellulose as a barrier significantly increases the electrical strength of the insulation system compared to the pure insulating liquid section. From the point of view of breakdown mechanics, the cellulose itself does not contribute to the electrical strength through a higher breakdown strength but through the many microscopic barriers resulting from the fiber structure filled with insulating liquid. He compared transformer boards impregnated with the insulating liquids Nynas 4000X, Midel 7131 and Envirotemp FR3. Except for the natural ester FR3 he used temperatures until −25 °C. None of the liquids shows a significant decrease in electrical strength in conjunction with cellulose and cooling to −10 °C and −25 °C, respectively, compared to the values at room temperature. Figure 6.53 shows the mean values and the standard deviation of the transformer boards. For the transformer board, heights of 10 and 20 mm, which also specify the electrode spacing, were used.

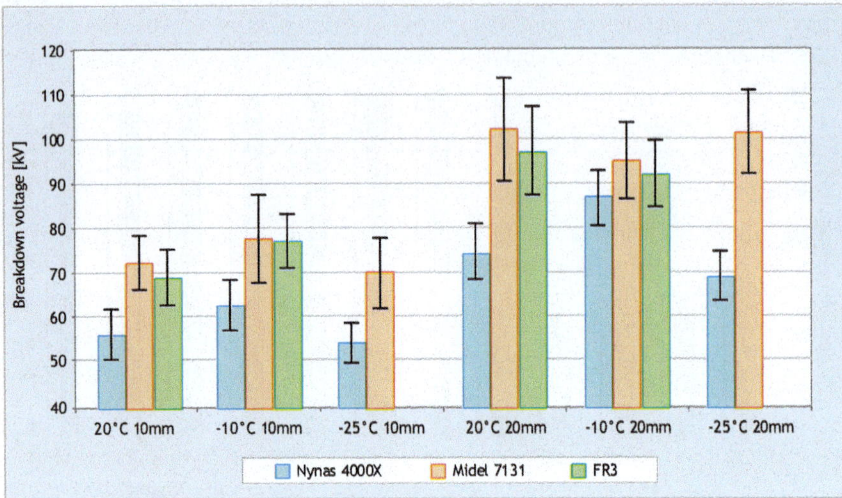

Fig. 6.53 Breakdown voltage of transformer board with standard deviation

6.8.6 Dissolved Gas Analyzes of Electrical Fault Simulation of Natural Ester Insulating Liquids

Chen et al. [75] performed discharge experiments with natural ester insulating liquids (soybean oil and camellia seed oil) and mineral oil to analyze and compare the type and percentage of gas generated under the typical discharge faults of different insulating liquids and paper insulation.

In this study, biological insulating liquids (soy-based natural ester insulating liquid and camellia-seed insulating liquid) and mineral oil were used for the electrical fault simulation tests to analyze and compare the types and percentages composition of the generated gases. Based on the transformer specific electrical fault types, namely partial discharge, spark discharge, and breakdown discharge were tested.

Sample Pretreatment The insulating liquid samples and pressboard samples were placed in a constant temperature and vacuum environment under 50 Pa, and 70 °C to remove moisture and other impurities. The pressboard samples were then liquid immersed in a vacuum environment. The prepared insulating pressboard were stored and sealed in dry tanks for the electrical fault tests. Then the insulating liquid samples were collected and measured by gas chromatography.

Test Results There will be defects in insulation system during the preparation process or after a long run, which cause it prone to breakdown and other failures under high electric field strength. Partial discharge may occur during the operation of the insulation system. These weak discharges will produce a cumulative effect and gradually deteriorate the dielectric properties of the insulation system to enlarge the local defects and finally lead to the breakdown of the entire insulation system. The dissolved gas contents in pure mineral oil, in soy-based natural ester and camellia oil show after needle-plate breakdown discharge that:

- Both mineral oil and biological insulating liquids produced H_2, CH_4, C_2H_4, C_2H_6, C_2H_2, CO, and CO_2 after the breakdown discharge fault. The CH_4, C_2H_4 and C_2H_2 gas productions of mineral oil were higher than that of biological insulating liquids, of which, C_2H_2 gas production was very much higher than that of biological insulating liquids, while CO gas generation of biological insulating liquids were much higher than of mineral oil. Under the condition of breakdown discharge fault, the main gas production of biological insulating liquids was CO_2, CO and C_2H_2 and the main gas production of mineral oil was CO_2, C_2H_4 and C_2H_2.
- Both liquid immersed paper—mineral oil and biological liquid—produce more CO and CO_2. The main reason is that, under the action of high electric field energy, C–O bonds in cellulose rapidly break down producing O_2. At high temperature O_2 reacts with C in cellulose to generate CO and CO_2 gases. Moreover, the mineral oil produced more C_2H_2 than the soy-based biological insulating liquid. In addition, for the biological insulating liquid, the gas production content showed a growing trend with the increase of the micro water content, especially CO_2 and H_2.

The main characteristic gases in biological insulating liquid in case of partial discharge were CO_2, H_2, and CO, and the total hydrocarbon content was not high. C_2H_2 was produced when the insulating paper was immersed in biological insulating liquids. However, compared with the breakdown discharge, C_2H_2 content for partial discharge is very small. Under the same test conditions, the H_2 content of the biological liquid immersed paper is approximately two times higher compared to the pure liquid.

In case of a spark discharge fault, biological insulating liquids mainly produce CO_2, H_2, and CO.

6.9 Comparing the Dielectric Behavior of Different Insulating Liquids in Solid Insulation

Pratomosiwi performed investigations on homogenous solid samples which were impregnated with mineral oil Nynas Nytro 4000X (as reference), synthetic ester Midel I7131 and natural ester Envirotemp FR3 to evaluate the different specific dielectric properties [76]. Transformer insulation systems consist of oil-impregnated paper on the copper windings and several oil-impregnated pressboard barriers between the high and low voltage windings. Hence, the insulation capability of the transformer depends on the complex solid/fluid insulating system. For his investigation, Pratomosiwi used several materials. Their characteristics are listed in Table 6.15.

Table 6.15 List of tested solid sample

Manufacturer	Samples	Thickness (mm)	Density (g/cm^3)	BDV (kV/mm)		Comment
				In air	In oil	
Weidmann	B.3.1A (TIV)	2 and 4	1.19	n.a[a]	51	High density cellulose pressboard
Weidmann	B.4.1 (TIII)	2	0.9	n.a	37	Low density cellulose pressboard
DuPont	Nomex T-993	2	0.76	n.a	n.a	Aramid pressboard with excellent thermal stability and dielectric strength
Cottrell	CK 125-TU	0.076	1.15	10.2	48	Thermally upgraded extensible Kraft paper
Tullis russel	Rotherm HIHD	0.085	0.95	12	86.6	Thermally upgraded transformer paper
Nordic paper	Amortfors	0.08	0.8	7	n.a	Not thermally upgraded transformer paper

[a] No data are available

6.9.1 DC Conductivity of Solid Impregnated with Ester Fluids

Figure 6.54 presents typical long-time duration charging current of pressboard TIV impregnated with Nynas Nytro 4000X and Midel 7131. A charging voltage V_C of 1 kV for pressboard and 200 V for transformer papers was applied until the charging current has reach steady state value. Then DC conductivity value was calculated from the steady state current. Figure 6.55 presents the Arrhenius plot of logarithmic shift for solids impregnated with (a) Nynas Nytro 4000X, (b) Midel 7131 and (c) FR3. DC conductivity value is very sensitive to the rise of temperature. The increase in DC conductivity of solids impregnated with ester fluids follows Arrhenius relation (Eq. 6.23).

$$\sigma = A * e^{\frac{-E}{K*T}} \tag{6.23}$$

A Pre-exponential factor
E Activation energy (eV)
T Temperature (K)
K Boltzmann constant 8.6173324*10^{-5} (eV/K).

Figure 6.55 shows that DC conductivity of solids impregnated with Nynas Nytro 4000X is generally lower than of solids impregnated with ester liquids like Midel

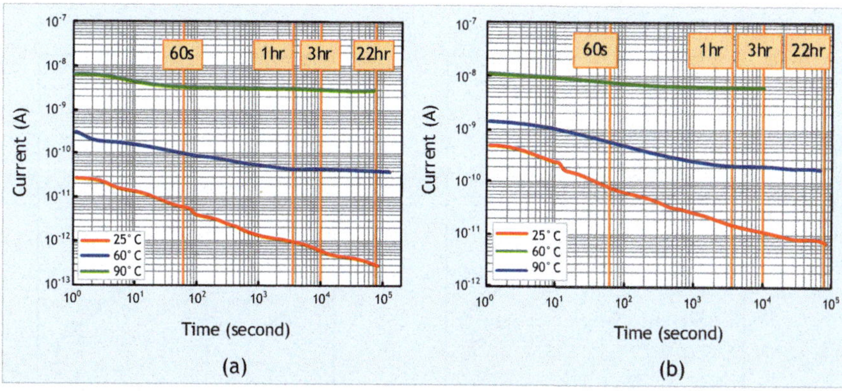

Fig. 6.54 Charging current of pressboard TIV impregnated with (a) Nynas Nytro 4000X and (b) Midel 7131 [76]

7131 and FR3. Ester liquids have higher conductivity and higher relative permittivity than mineral oil. High relative permittivity means that the insulating liquid contains more ionic impurities (higher ion concentration). Ionic impurities generate higher conductivity of ester liquids. Hence, solids impregnated with ester liquids have higher DC conductivity compared to solids impregnated with mineral oil. It indicates that the DC conductivity of solid impregnated material has a strong dependency on the conductivity of the insulating liquids. DC conductivity of transformer papers is generally higher compared to the pressboard. It might be due to transformer papers are very thin, therefore the fluid characteristic is more dominant than in pressboard. From all solid samples, Nomex has the lowest DC conductivity, and it also follows Arrhenius relation.

6.10 Retrofill of Mineral Oil Filled Equipment with Biological Insulating Liquids

Retrofilling is the process of removing the insulating liquid from an existing working transformer and replacing it with a new insulating liquid. There are some reasons why this is done and what new insulating liquid is chosen for what purpose. First retrofilling was applied for environmental reasons to replace polychlorinated biphenyl (PCB) contaminated liquids. Nowadays, mineral oil is replaced with either synthetic or natural ester. The reason for this retrofill is mostly the better fire property compared to mineral oil. The possibility of installing a transformer with a higher power or load with the same amount of space can also be a cause. One other reason is the environmentally friendly properties of the biological insulating liquids. Esters are readily biodegradable and natural esters have additionally a very low carbon footprint. Natural esters are miscible with mineral oil as well as other ester liquids in

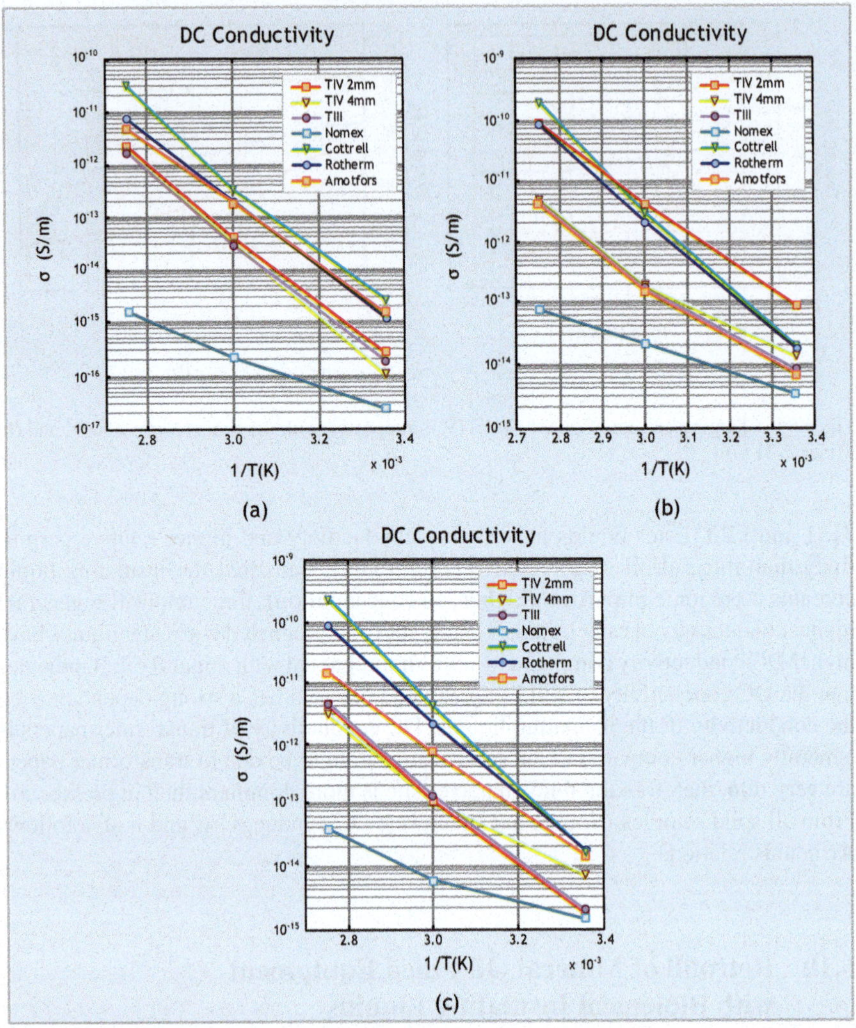

Fig. 6.55 DC conductivity of solid impregnated with **a** Nynas Nytro 4000X, **b** midel 7131 and **c** FR3 at temperature of 25, 60 and 90 °C [76]

any proportion. But synthetic and natural ester cannot be used to retrofill silicone oil filled units. Silicone oil mixed with ester liquids causes foaming.

Alternative ester liquids do not suffer from corrosive sulfur as some mineral oil. Replacing mineral oil in transformer where corrosive sulfur compounds have been detected in mineral oil but have not shown signs of copper corrosion would likely to prevent further damage. But it cannot reverse the effect of corrosion that already has occurred.

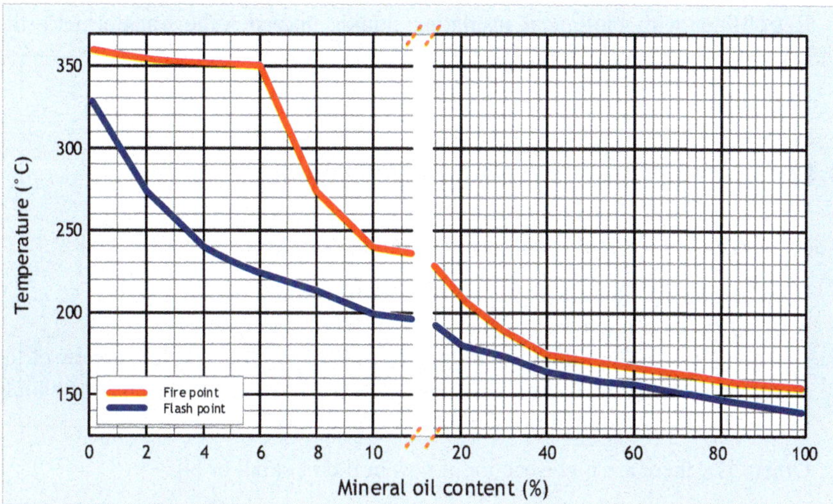

Fig. 6.56 Mixture of natural ester (FR3) with mineral oil (Nynas Nitro 4000X)—influence on flash- and fire point [76]

Further technical and nontechnical benefits related to the replacement of mineral with biological insulating liquids:

Biological insulating liquids will increase the transformer's life due to its ability to tolerate a higher concentration of water. Biological insulating liquids lead to a lower cost of insurance because the fire hazard is very low. Fire mitigation equipment can be removed from service, eliminating some maintenance expenses. Fire safety is greatly dependent on the quality of the liquid exchange treatment done (Fig. 6.56).

It should be ensured that the largest possible part of the mineral oil is removed from the system, as the mineral oil will adversely affect the properties of the newly filled liquid. Mineral oil left in the core will slowly diffuse into the new liquid and reverse. Up to about 10% of the mineral oil can be left in the cellulose of the core after retrofilling, sometimes also from the bottom of the tank as well as the walls of the tank. The ratio of the cellulose paper to insulating liquid might also affect the amount of mineral oil left after retrofilling. A rinsing procedure of the core and windings with heated biological insulating liquid would be helpful and should be used to have a more successful result. A properly retro filled transformer will contain a maximum of 2–3% of original insulating liquid.

Further costs can be saved because of

- additional 20 °C tolerance provided without accelerating the insulation system aging rate (achieved through additional loading).
- reduced risk of dielectric failure caused by bubble formation during overload.
- collateral damage to other equipment by not incurring dielectric pool fire [77].

Retrofilling with biological insulating liquids upgrades the transformer's fire safety and lowers the environmental risk of aged transformers.

6.10.1 Transformer Selection for Retrofilling

There are reasons for exclusion from retrofilling:

- It's a free breathing transformer and cannot be changed to a sealed one
- It's a silicone oil filled transformer
- Bad condition of the solid insulation. Assessment should include a review of the maintenance records. A critical parameter is the furan content. This one should be lower than 2 ppm in the old insulating liquid.

Otherwise, there are no restriction if new or old or small or big.

6.11 Advantages of Biological Insulating Liquids

Biological insulating liquids can help utilities providers to increase network efficiency and reduce overall expenditure and reducing the footprint of the substation. Compared to mineral oil, biological insulating filled transformers can be installed with no fire walls and the distances to surrounding buildings can be reduced up to ten times [53]. The result is a considerable reduction in both civil construction and land costs. In addition, ester transformers are increasingly being located underground and under buildings in urban areas, delivering significant cost savings and improved aesthetics [78]. Many electrical components holding dielectric liquids are situated near water or waterways. Leakage or spill of petroleum-based liquids can cause serious damage to water and marine life. Leaks or spills on land of petroleum-based liquids can threaten groundwater and contaminate soil. Biological insulating liquids can help to reduce overhead costs through the removal of ancillary equipment such as fire suppression systems, reduction in maintenance and allowing for shorter cable runs due a smaller construction footprint. This can ultimately help to reduce ongoing costs as well as capital expenditure. Also, less complexity in construction helps to bring substations online faster.

There are some technical aspects arguing for biological insulating liquids:

- Mineral oil can become saturated with small amounts of water, coming from outside or from the ageing of solid cellulose insulation, reducing the break down voltage and the effectiveness of the transformer. Because biological insulating liquids have a high moisture tolerance, they can absorb larger amount of water without compromising their dielectric properties. The uptake of water from aging paper by the biological insulating liquid and the protection of the cellulose by transesterification can also significantly extend the life of cellulose insulation.

- Replacing mineral oil will enhance fire safety, especially in populated or sensitive areas.
- Biological insulating liquids are classified as readily biodegradable and considered being environmentally friendly. Replacing mineral oil will greatly reduce environmental impact in case of spillage.
- Biological insulating liquids do not suffer from corrosive sulfur as in mineral oil. Replacing mineral oil in transformer where corrosive sulfur compounds have been detected in mineral oil but have not shown signs of copper corrosion would likely to prevent further damage. But it cannot reverse the effect of corrosion that has already occurred.
- There are many literatures that explain cellulose insulation last longer in ester liquids than in mineral oil. It is related to the ester's ability to retain moisture. Replacing mineral oil by biological insulating liquids will shift the moisture equilibrium between paper and liquids to the liquids. Biological insulating liquids will leach out the moisture from paper. Therefore, this process will delay the aging process of paper caused by hydrolysis.
- Biological insulating liquids are superior due to the higher operation temperature range of the liquid and reducing aging of the solid insulation. Hydrolysis loses importance when using a biological insulating liquid as the thermal capability of the transformer is primarily assigned by the solid insulation.

References

1. Nynas Transformer Oil Handbook. 2011. Stockholm Sweden, 10.2011.
2. IEC 62535. 2009. Insulating Liquids—Test Method for Detection of Potentially Corrosive Sulfur in Used and Unused Insulating Oil.
3. ASTM D130-19. 2019. Standard Test Method for Corrosiveness to Copper from Petroleum Products by Copper Strip Test.
4. Pagger, E. 2013. Alternative insulating liquids compared to the classic mineral oil, Doctoral thesis, Graz University of Technology.
5. Rafiq, M., et al. 2020. Sustainable, renewable and environmental-friendly insulation systems for high voltage applications, review, MDPI. *Molecules*. https://doi.org/10.3390/molecules25173901.
6. Vander T. 2009. DBDS & Corrosion Free, Program: Diagnosis and Countermeasures; EMPRESA ELÈTRICAS CHILENAS Y CIGRE.
7. Maina, R., et al. 2006. Dibenzyl disulfide (DBDS) as corrosive sulfur contaminant in used and unused mineral insulating oils, Sea Marconi Technologies, Collegno (TO)—Italy, Terna S.P.A, Venezia—Italy, University of Missouri Rolla—USA, Università degli Studi di Roma "La Sapienza" Roma, Italy.
8. Miller, K.W. 1992. Reductive Desulfurization of dibenzyl disulfide. *Biological Sciences Department, Illinois State University, Normal, Illinois* 61761: 2176–2179.
9. IEC 62697-1. 2012. Test Methods for Quantitative Determination of Corrosive Sulfur Compounds in Unused and Used Insulating Liquids—Part 1: Test Method for Quantitative Determination of Dibenzyl Disulfide (DBDS).
10. ASTM D129-18. 2018. Standard Test Method for Sulfur in Petroleum Products (General High Pressure Decomposition Device Method).

11. Hellebuyck, D. 2013. *Functional performance criteria for comparison of less flammable transformer oils with respect to fire and explosion risk, Report 5424*. Department of Fire and Safety Engineering and Systems Safety: Lund University.
12. Yan Z. W., et al. 2020. Measuring Low Molecular Weight Acids in Mineral and Ester Transformer Liquids. In *The 8th International Conference on Condition Monitoring and Diagnosis (CMD 2020)*, Thailand.
13. Transformer Insulation Life: Enviretemp® FR3™ Fluid and 80 °C Hot Spot Rise; Bulletin R900-20-4; Reference Information, December 2003.
14. Sumereder, C., and M. Muhr. 2006. Zuverlässigkeits- und Risikoabschätzung von elektrischen Betriebsmitteln; Institute of High Voltage Engineering and System Performance, Graz University of Technology, Symposium Energieinnovation—Graz.
15. Envirotemp™ FR3™ Fluid Guide for Storage, Installation, Commissioning and Maintenance of FR3 Fluid Filled Transformers G2300, April 2017.
16. Augusta, M., and G. Martins. 2010. Vegetable oils, an alternative to mineral oil for power transformers—Experimental study of paper aging in vegetable oil versus mineral oil. *IEEE Electrical Insulation Magazine* 26 (6).
17. Meissner, M. 2020. Alternative insulating liquids for high voltage application—Analyzes & interpretation. Doctoral thesis, University of Graz, Institute of Chemistry.
18. Lin, Y., et al. 2018. Effects of moisture on furfural distribution and aging estimation of transformer cellulose insulation. In *12th IEEE ICPADM*, Xi'an.
19. Pagger, E., and M. Scala. 2014. Diffusion and transport processes—"Post-Mortem Tests". In *5th Workshop, Oil/Paper- and Gas-insulated Systems in Components of Electrical Power Supply*, Graz.
20. IEC 62021-3. 2014. Insulating Liquids—Determination of Acidity—Part 3: Test Methods for Non-Mineral Insulating Oils.
21. COOPER Power Systems: Cellulose Interaction with Envirotemp® FR3™ Fluid; Bulletin R900-20-11; Reference Information; May 2004.
22. Ese, M. G., et al. 2013. Esterification of low molecular weight acids in cellulose. SINTEF Energy Research, Trondheim.
23. Münster, T., et al. 2021. Thermally accelerated aging of insulation paper for transformers with different insulating liquids. *Energies* 14: 3036.
24. Rapp, K. J., et al. 2005. Interaction mechanisms of natural ester dielectric fluid and Kraft paper. In *ICDL Conference*, Portugal.
25. Maneerot, S., N. Pattanadech. 2020. The comparative study of physical and chemical properties of palm oil and mineral oil used in a distribution transformer. In *The 8th International Conference on Condition Monitoring and Diagnosis (CMD 2020)*, Thailand.
26. Yuerekten, S. 2021. Transformer Components—Properties and Applications, Birikim Ofset Matbaacilik Ltd.
27. Du, Y., et al. 1999. Moisture equilibrium in transformer paper-oil systems. *Feature Article* 15 (1).
28. Koch, M., S. Tenbohlen. 2010. Signifikante Kenngrößen für die Alterung des Isoliersystems in Leistungstransformatoren, Stuttgarter Hochspannungssymposium.
29. Nomex® Pressboard. 2002. Technical Data Sheet, Nomex only by DuPont.
30. Nomex® Pressboard. 2002. Moisture Effects and Processing Information for Nomex® paper, Nomex only by DuPont.
31. Meyers Enzyklopädisches Lexikon; Bibliographisches Institut AG, Mannheim 1974, korrigierter Nachdruck 1981; Band 11; page 289.
32. Dai, J., et al. 2007. Investigation of the impregnation of cellulosic insulations by ester fluids. In *IEEE Annual Report Conference on Electrical Insulation and Dielectric Phenomena*.
33. Maneerot, S., et al. 2020. Polarization and depolarization current characteristics of natural ester impregnated pressboards with different periods of impregnation. In *The 8th International Conference on Condition Monitoring and Diagnosis (CMD 2020)*, Thailand.
34. IEC 60814. 1997. Insulating Liquids—Oil-Impregnated Paper and Pressboard—Determination of Water by Automatic Coulometric Karl Fischer Titration.

35. ASTM D1533–20. 2020. Standard Test Method for Water in Insulating Liquids by Coulometric Karl Fischer Titration.
36. ISO 12937. 2017. Petroleum Products—Determination of Water—Coulometric Karl Fischer Titration Method.
37. Sumereder, C., et al. 2009. Moisture determination in solid transformer insulation on the basis of capacitive oil sensors, CIGRE SC D1—Colloquium in Hungary Budapest.
38. Musai, F. 2009. Sensor based moisture determination in transformer insulating systems. Diploma thesis, Graz University of Technology, Graz.
39. CIGRE. 2018. WG D1.53, Ageing of Liquid Impregnated Cellulose for Power Transformers, Brochure N° 738.
40. Ghani Ab, S., et al. 2018. Methods for improving the workability of natural ester insulating oils in power transformer applications: A review. *Electrical Power Systems Research* 163. https://doi.org/10.1016/j.epsr.2017.10.008.
41. Zhong, Y., et al. 2013. Insulating properties and charge characteristics of natural ester fluid modified by TiO_2 semiconductive nanoparticles. *IEEE Transaction on Dielectrics and Electrical Insulation* 20. https://doi.org/10.1109/TDEL.2013.6451351.
42. Du, Y., et al. 2011. Effect of electron shallow trap on breakdown performance of transformer oil-based nanofluids. *Journal of Applied Physics*. https://doi.org/10.1063/1.3660783.
43. Totzauer, P., P. Trnka. 2019. Different ways to improve natural ester oils. In *13th International Scientific Conference on Sustainable, Modern and Safe Transport (TRANSCOM 2019)*. Published by Elsevier.
44. Maneerat, N., et al. 2020. AC breakdown and resistivity of natural ester based nanofluids. In *The 8th International Conference on Condition Monitoring and Diagnosis (CMD 2020)*, Thailand, October 2020.
45. ASTM 1816-12. 2019. Standard Test Method for Dielectric Breakdown Voltage of Insulating Liquids Using VDE Electrodes.
46. IEC 60247. 2004. Insulating Liquids—Measurement of Relative Permittivity, Dielectric Dissipation Factor (tan δ) and D.C. Resistivity.
47. Maneerot, S., et al. 2020. Effect of ZnO and $BaTiO_3$ nanoparticle on partial discharge characteristics of palm oil based nanofluids. In *The 8th International Conference on Condition Monitoring and Diagnosis (CMD 2020)*, Thailand.
48. Judendorfer, T. 2012. Oil-cellulose insulation systems for HVDC applications. Doctoral thesis, Graz University of Technology.
49. Fabian, J. 2012. Dielektrische Kennwerte von Isolierstoffen bei Gleichspannungsbeanspruchung, Leitfähigkeits- und Teilentladungsverhalten von mineralischem Isolieröl und Transfomer-Board, Doctoral Thesis, Graz University of Technology.
50. IEC 62631-3-1. 2016. Dielectric and Resistive Properties of Solid Insulating Materials—Part 3-1: Determination of Resistive Properties (DC Methods)—Volume Resistance and Volume Resistivity—General Method.
51. ASTM D257-14(2021)e1. 20121. Standard Test Methods for DC Resistance or Conductance of Insulating Materials.
52. ASTM D1169-19. 2019. Standard Test Method for Specific Resistance (Resistivity) of Electrical Insulating Liquids.
53. IEC 61936-1. 2015. Power Installations Exceeding 1 kV A.C.—Part 1: Common Rules.
54. Hera, U., et al. 2016. Planung von Elektromobilität im Großraum München "E-Plan München".
55. ee-news.ch. 2015. Energieforschung: Trafos Haben Noch Effizienzpotenzial.
56. WESTRAFO. 2017. Sustainable Peak Load Distribution Transformers, Vicenza.
57. Pagger, E., S. Bowers. 2017. Die Verwendung von Sojabohnenöl in der Hochspannungstechnik, 10. Internationale Energiewirtschaftstagung an der TU Wien.
58. Pagger, E., M. Thelen. 2017. Reduzierung der Umweltgefährdung durch den Einsatz natürlicher Ester, Oberlausitzer Energie-Symposium, Zittau.
59. Pagger, E., et al. 2018. New impacts on ecological transformer design. In *15th Symposium Energieinnovation*. Technische sUniversität Graz.

60. Pagger E. et al. 2018. Natural ester FR3 insulating liquid—Very paper friendly. In *12th IEEE ICPADM*, Xi'an.
61. Pattanadech, N., et al. 2018. Studies of electrical and thermal characteristics of natural ester immersed transformer compared with mineral oil immersed and palm oil immersed transformer. In *International Conference on Condition Monitoring and Diagnosis*, Perth 2018.
62. Borghardt, S. 2019. Transformatoren: Innovationen für die Umwelt, Natürliche Ester—eine Alternative zu Mineralölen, ZVEI, October 2019.
63. Thelen, M. 2018. Alternative Isolierflüssigkeiten—Natürliche Ester für einen flexiblen Energiebedarf, Life needs Power, Hannover.
64. Sitorus, H., et al. 2014. Comparison of streamers characteristics in Jatropha Curcas methyl ester oil and mineral oil under lightning impulse voltage. In *IEEE International Conference on Liquid*, Bled, Slovenia.
65. IEC 60567. 2011. Oil-Filled Electrical Equipment—SAMPLING of Gases and Analysis of Free and Dissolved Gases—Guidance.
66. Hamid, M.H.A., et al. 2017. Dissolved gas analysis (DGA) of natural ester oils under arcing. *Journal of Fundamental and Applied Sciences*. https://doi.org/10.4314/jfas.v9i3s.9.
67. Suksagoolpanya, S., et al. 2020. Dissolved gas analyzes of palm oil compared with mineral oil from different types of breakdown voltage. In *The 8th International Conference on Condition Monitoring and Diagnosis (CMD 2020)*, Thailand, October 2020.
68. IEC 60156. 2018. Insulating Liquids—Determination of the Breakdown Voltage at Power Frequency—Test Method.
69. IEC 60897. 1987. Methods for the Determination of the Lightning Breakdown Voltage of Insulating Liquids.
70. Azis, N., et al. 2016. Evaluation on the lightning breakdown voltages of palm oil and coconut oil under non-uniform field at small gap distances. *Journal of Electric Engineering and Technology*. https://doi.org/10.5370/JEET.2016.11.1.184.
71. Siriworachanyadee J., N. Pattanadech. 2020. AC and lightning impulse breakdown voltage of natural ester impregnated pressboards. In *The 8th International Conference on Condition Monitoring and Diagnosis (CMD 2020)*, Thailand, October 2020.
72. IEC 60243-1. 2013. Electric Strength of Insulating Materials—Test Methods—Part 1, Tests at Power Frequencies.
73. IEC 60243-3. 2013. Electric Strength of Insulating Materials—Test Methods—Part 3: Additional Requirements for 1,2/50 μs Impulse Tests.
74. Eberhardt, R. 2012. Elektrische Eigenschaften und Gasungsverhalten von biologischen und synthetischen Estern in Hochspannungs-Isolationssystemen unter Kälteeinfluss. Doctoral thesis, Graz University of Technology, Mai 2012.
75. Chen, C., et al. 2018. Research on the dissolved gas analyzes of electrical fault simulation of natural ester insulation oils. In *12th IEEE ICPADM*, Xi'an, 2018.
76. Pratomosiwi F. 2014. Dielectric properties of transformer boards and papers impregnated with alternative insulating fluids. Doctoral thesis, Graz University of Technology, March 2014.
77. DelFiacco, G. 2013. Retrofilling Transformers: A Financial Perspective, Version 1.0. Cargill, June 2013.
78. Lashbrook, M., A. Coker. 2020. The role of ester fluids in fire safe, greener, better performing power networks. *Transformer Technology* (9).

Chapter 7
Standardization

The calculation, design and operation of high voltage networks and high voltage systems is very complex and interdisciplinary. Therefore, a wide range of standards from international committees is used to test and check the quality and condition of the systems and equipment in order to avoid breakdowns. In addition, there are different standards for the same matter due to the non-uniform appearance of the large standardization institutes such as ASTM (American Society for Testing and Materials), IEEE (Institute of Electrical and Electronics Engineers), IEC (International Electronical Commission) and ISO (International Organization for Standardization). Regionality also plays a role here. ASTM and IEEE are headquartered in the United States, while IEC and ISO are headquartered in Europe. In this chapter, a summary of the most relevant standards for high voltage technology without any claim of completion is listed.

Table 7.1 shows the main standards in the field of insulating liquids.

Table 7.2 shows a summary of the two main standards for biological insulating liquids with the associated specific ones.

Table 7.3 shows the most commonly applied property tests of ASTM and provides guidance on how each test method can be applied to biological insulating liquids. For natural ester, the same standards [7] can be used as for mineral oil [1] but the value limits must be adapted as they can be significantly different from those established for mineral oil. Other standards relevant to the design of high voltage equipment and the use of dielectric insulating liquids are contained in [37–234].

E. P. Pagger et al., *Biological Insulating Liquids*,
https://doi.org/10.1007/978-3-031-22460-7_7

Table 7.1 International standards for mineral oil, synthetic ester, natural ester, and silicone oil

Liquid type	ASTM	IEC (unused liquid)	IEEE	IEC (used liquid)
Mineral oil	D3487 [1]	60296 [2]	C57.106 [3]	60422 [4]
Synthetic ester	–	61099 [5]	–	61203 [6]
Natural ester	D6871 [7]	62770 [8]	C57.147 [9]	62975 [10]
Silicone oil	D4652 [11]	60836 [12]	C57.111 [13]	60944 [14]

Table 7.2 Most relevant standards for biological insulating liquids

Properties outlined in	IEC 62770 [8]	ASTM D6871 [7]
Water content (ppm)	IEC 60814 [15]	ASTM D1533 [16]
Dissipation factor (%)	IEC 60247 [17]	ASTM D924 [18]
Dielectric breakdown voltage (kV)	IEC 60156 [19]	ASTM D877 [20]
Flash point (°C)	ISO 2719 [21]	ASTM D92 [22]
Fire point (°C)	ISO 2592 [23]	ASTM D92 [22]
Viscosity (mm^2/s)	ISO 3104 [24]	ASTM D445 [25]
Pour point (°C)	ISO 3016 [26]	ASTM D97 [27]
Relative density (g/cm^3)	ISO 3675 [28]	ASTM D1298 [29]
Corrosive sulfur	IEC 62697 [30]	ASTM D1275 [31]
Total acid content (mg KOH/g)	IEC 62021 [32]	ASTM D974 [33]
Visual aspect	IEC 16099 [5]	ASTM D1524 [34]
Color	ISO 2211 [35]	ASTM D1500 [36]

Table 7.3 ASTM methods and limits for dielectric insulating liquids

Property	ASTM D3487-16 [1] Mineral oil		ASTM D6871-17 [7] Natural ester	
	Specification	Method	Specification	Method
Water content (ppm)	≤ 35	D1533 [16]	≤ 200	D1533 [16]
Dissipation factor (%)	≤ 0.3 (100 °C)	D924 [18]	≤ 0.5 (90 °C)	D924 [18]
Dielectric breakdown voltage (kV)	≥ 30	D877 [20]	≥ 35	D877 [20]
Flash point (°C)	≥ 145	D92 [22]	> 250	D92 [22]
Fire point (°C)	–	D92 [22]	> 300	D92 [22]
Viscosity at 40 °C (mm^2/s)	≤ 12	D445 [25]	≤ 50	D445 [25]
Viscosity at 100 °C (mm^2/s)	≤ 3	D445 [25]	≤ 15	D445 [25]
Pour point (°C)	$\leq$ -40	D97 [27]	≤ -10	D97 [27]
Relative density (g/cm^3)	≤ 0.91	D1298 [29]	< 1.0	D1298 [29]
Corrosive sulfur	Not corrosive	D1275 [31]	Not corrosive	D1275 [31]
Total acid content (mg KOH/g)	≤ 0.03	D974 [33]	≤ 0.06	D974 [33]

References

1. ASTM D3487-16. 2016. Standard Specification for Mineral Insulating Oil Used in Electrical Apparatus.
2. IEC 60296. 2020. Fluids for Electrotechnical Applications—Mineral Insulating Oils for Electrical Equipment.
3. IEEE C57.106. 2007. IEEE Guide for Acceptance and Maintenance of Insulating Oil in Equipment.
4. IEC 60422. 2013. Mineral Insulating Oils in Electrical Equipment—Supervision and Maintenance Guidance.
5. IEC 61099. 2011. Insulating Liquids—Specification for Unused Synthetic Organic Esters for Electrical Purposes.
6. IEC 61203. 1992. Synthetic Organic Esters for Electrical Purposes—Guide for Maintenance of Transformer Esters in Equipment.
7. ASTM D6871-17. 2017. Standard Specification for Natural (Vegetable Oil) Ester Fluids Used in Electrical Apparatus.
8. IEC 62770. 2013. 1st Edition, Fluids for Electrotechnical Applications—Unused Natural Esters for Transformers and Similar Electrical Equipment.
9. IEEE C57.147. 2018. IEEE Guide for Acceptance and Maintenance of Natural Ester Liquids in Transformers.
10. IEC 62975. 2021. Natural Esters—Guidelines for Maintenance and Use in Electrical Equipment.
11. ASTM D4652-20. 2020. Standard Specification for Silicone Liquid Used for Electrical Insulation.
12. IEC 60836. 2015. Specifications for Unused Silicone Insulating Liquids for Electrotechnical Purposes.
13. IEEE C57.111. 1989. IEEE Guide for Acceptance of Silicone Insulating Fluid and its Maintenance in Transformers.
14. IEC 60944. 1988. Guide for the Maintenance of Silicone Transformer Liquids.
15. IEC 60814. 1997. Insulating Liquids—Oil-impregnated Paper and Pressboard—Determination of Water by Automatic Coulometric Karl Fischer Titration.
16. ASTM D1533-20. 2020. Standard Test Method for Water in Insulating Liquids by Coulometric Karl Fischer Titration.
17. IEC 60247. 2004. Insulating Liquids—Measurement of Relative Permittivity, Dielectric Dissipation Factor (tan δ) and D.C. Resistivity.
18. ASTM D924-15. 2015. Standard Test Method for Dissipation Factor (or Power Factor) and Relative Permittivity (Dielectric Constant) of Electrical Insulating Liquids.
19. IEC 60156. 2018. Insulating Liquids—Determination of the Breakdown Voltage at Power Frequency—Test method.
20. ASTM D877/D877M-19. 2019. Standard Test Method for Dielectric Breakdown Voltage of Insulating Liquids Using Disk Electrodes.
21. ISO 2719. 2016. Determination of Flash Point—Pensky-Martens Closed Cup Method.
22. ASTM D92-18. 2018. Standard Test Method for Flash and Fire Points by Cleveland Open Cup Tester.
23. ISO 2592. 2017. Petroleum and Related Products—Determination of Flash and Fire Points—Cleveland Open Cup Method.
24. ISO 3104. 2020. 3rd Edition, Petroleum Products—Transparent and Opaque Liquids—Determination of Kinematic Viscosity and Calculation of Dynamic Viscosity.
25. ASTM D445-21. 2021. Standard Test Method for Kinematic Viscosity of Transparent and Opaque Liquids (and Calculation of Dynamic Viscosity).
26. ISO 3016:1994. 2017. Petroleum Products—Determination of Pour Point, (DIN ISO 3016).
27. ASTM D97-17b. 2017. Standard Test Method for Pour Point of Petroleum Products.
28. ISO 3675. 1998. Crude Petroleum and Liquid Petroleum Products—Laboratory Determination of Density—Hydrometer Method.

29. ASTM D1298-12b. 2017. Standard Test Method for Density, Relative Density (Specific Gravity), or API Gravity of Crude Petroleum and Liquid Petroleum Products by Hydrometer Method.
30. IEC 62697-2. 2018. Test Methods for Quantitative Determination of Corrosive Sulfur Compounds in Unused and Used Insulating Liquids - Part 2: Test method For Quantitative Determination of Total Corrosive Sulfur (TCS).
31. ASTM D1275-15. 2015. Standard Test Method for Corrosive Sulfur in Electrical Insulating Liquids.
32. IEC 62021-3. 2014. Insulating liquids—Determination of Acidity—Part 3: Test Methods for Non-Mineral Insulating Oils.
33. ASTM D974-14e2. 2014. Standard Test Method for Acid and Base Number by Color-Indicator Titration.
34. ASTM D1524-15. 2015. Standard Test Method for Visual Examination of Used Electrical Insulating Liquids in the Field.
35. ISO 2211. 1973. Liquid Chemical Products—Measurement of Color in Hazen Units (Platinum-Cobalt Scale).
36. ASTM D1500-12. 2017. Standard Test Method for ASTM Color of Petroleum Products (ASTM Color Scale).
37. IEC 60050-212:2010/AMD3. 2020. Amendment 3—International Electrotechnical Vocabulary (IEV)—Part 212: Electrical Insulating Solids, Liquids and Gases.
38. IEC 60050-421. 1990. International Electrotechnical Vocabulary (IEV)—Part 421: Power Transformers and Reactors.
39. IEC 60052. 2002. Voltage Measurement by Means of Standard Air Gaps.
40. IEC 60060-1. 2010. High-Voltage Test Techniques—Part 1: General Definitions and Test Requirements.
41. IEC 60060-2. 2010. High-Voltage Test Techniques—Part 2: Measuring Systems.
42. IEC 60060-3. 2006. High-Voltage Test Techniques—Part 3: Definitions and Requirements for On-Site Testing.
43. IEC 60071-1. 2019. Insulation Co-Ordination—Part 1: Definitions, Principles and Rules.
44. IEC 60071-2. 2018. Insulation Co-Ordination—Part 2: Application Guidelines.
45. IEC 60076-1. 2012. Power Transformers—Part 1: General.
46. IEC 60076-2. 2012. Temperature rise for Liquid-Immersed Transformers.
47. IEC 60076-3. 2013. Power Transformers—Part 3: Insulation Levels, Dielectric Tests and External Clearances in Air.
48. IEC 60076-4. 2002. Power Transformers—Part 4: Guide to the Lightning Impulse and Switching Impulse Testing—Power Transformers and Reactors.
49. IEC 60076-5. 2006. Power Transformers—Part 5: Ability to Withstand Short Circuit.
50. IEC 60076-7. 2018. Power Transformers—Part 7: Loading Guide for Oil-Immersed Power Transformers.
51. IEC 60076-10. 2016. Power Transformers—Part 10: Determination of Sound Level.
52. IEC 60076-14. 2013. Power Transformers—Part 14: Liquid-Immersed Power Transformers Using High-Temperature Insulation Materials.
53. IEC 60085. 2007. Electrical Insulation—Thermal Evaluation and Designation.
54. IEC 60214-1. 2014. Tap-changers—Part 1: Performance Requirements and Test Methods.
55. IEC/IEEE 60214-2. 2019. Tap-Changers—Part 2: Application Guidelines.
56. IEC 60216-1. 2013. Electrical Insulating Materials—Thermal Endurance Properties—Part 1: Ageing Procedures and Evaluation of Test Results.
57. IEC 60216-2. 2005. Electrical Insulating Materials—Thermal Endurance Properties—Part 2: Determination of Thermal Endurance Properties of Electrical Insulating Materials—Choice of Test Criteria.
58. IEC 60243-1. 2013. Electric Strength of Insulating Materials—Test Methods—Part 1, Tests at Power Frequencies.
59. IEC 60243-2. 2013. Electric Strength of Insulating Materials—Test Methods—Part 2: Additional Requirements for Tests Using Direct Voltage.

60. IEC 60243-3. 2013. Electric Strength of Insulating Materials—Test Methods—Part 3: Additional Requirements for 1,2/50 μs Impulse Tests.
61. IEC 60270. 2000. High-Voltage Test Techniques—Partial Discharge Measurements.
62. IEC 60450. 2004. Measurement of the Average Viscometric Degree of Polymerization of New and Aged Cellulosic Electrically Insulating Materials.
63. IEC 60475. 2011. Method of Sampling Insulating Liquids.
64. IEC 60554-1. 1977. Specification for Cellulosic Papers for Electrical Purposes. Part 1: Definitions and General Requirements.
65. IEC 60554-2. 2001. Cellulosic Papers for Electrical Purposes—Part 2: Methods of Test.
66. IEC 60554-3-3. 1980. Specification for Cellulosic Papers for Electrical Purposes. Part 3: Specifications for Individual Materials. Sheet 3: Crêpe paper.
67. IEC 60554-3-5. 1984. Specification for Cellulosic Papers for Electrical Purposes. Part 3: Specifications for Individual Materials. Sheet 5: Special Papers.
68. IEC 60567. 2011. Oil-Filled Electrical Equipment—Sampling of Gases and Analysis of Free and Dissolved Gases—Guidance.
69. IEC 60590. 1977. Determination of the Aromatic Hydrocarbon Content of New Mineral Insulating Oils.
70. IEC 60599. 2015. Mineral Oil-Filled Electrical Equipment in Service—Guidance on the Interpretation of Dissolved and Free Gases Analysis.
71. IEC 60628. 1985. Gassing of Insulating Liquids under Electrical Stress and Ionization.
72. IEC 60641-2. 2004. Pressboard and Presspaper for Electrical Purposes—Part 2: Methods of tests.
73. IEC 60641-3-1. 2008. Pressboard and Presspaper for Electrical Purposes—Part 3: Specifications for Individual Materials—Sheet 1: Requirements for Pressboard, Types B.0.1, B.0.3, B.2.1, B.2.3, B.3.1, B.3.3, B.4.1, B.4.3, B.5.1, B.5.3 and B.6.1.
74. IEC 60666. 2010. Detection and Determination of Specified Additives in Mineral Insulating Oils.
75. IEC 60672-3. 1997. Ceramic and Glass-Insulating Materials - Part 3: Specifications for Individual Materials.
76. IEC 60695-11-10. 2013. Fire hazard testing—Part 11–10: Test Flames—50 W Horizontal and Vertical Flame Test Methods.
77. IEC 60763-1. 2010. Laminated Pressboard for Electrical Purposes—Part 1: Definitions, Classification and General Requirements.
78. IEC 60763-2. 2007. Specification for Laminated Pressboard—Part 2: Methods of Test.
79. IEC 60763-3. 2010. Laminated Pressboard for Electrical Purposes—Part 3: Specifications for Individual Materials—Sheet 1: Requirements for Laminated Precompressed Pressboard, Types LB3.1A.1 and LB3.1A.2.
80. IEC 60819-3-3. 2011. Non-Cellulosic Papers for Electrical Purposes—Part 3: Specifications for Individual Materials—Sheet 3: Unfilled Aramid (Aromatic Polyamide) Papers.
81. IEC 60851-4. 2016. Winding Wires—Test Methods—Part 4: Chemical Properties.
82. IEC 60867. 1993. Insulating Liquids—Specifications for Unused Liquids Based on Synthetic Aromatic Hydrocarbons.
83. IEC 60893-3-2. 2003. Insulating Materials—Industrial Rigid Laminated Sheets Based On Thermosetting Resins for Electrical Purposes—Part 3–2: Specifications for Individual Materials—Requirements for Rigid Laminated Sheets Based On Epoxy Resins.
84. IEC 60897. 1987. Methods for the Determination of the Lightning Breakdown Voltage of Insulating Liquids.
85. IEC 60970. 2008. Insulating Liquids—Methods for Counting and Sizing Particles.
86. IEC 61039. 2008. Classification of Insulating Liquids.
87. IEC 61061–1. 2006. Non-Impregnated Densified Laminated Wood for Electrical Purposes—Part 1: Definitions, Designation and General Requirements.
88. IEC 61061-2. 1992. Specification for Non-Impregnated, Densified Laminated Wood for Electrical Purposes—Part 2: Methods of Test.

89. IEC 61061-3. 1998. Non-Impregnated Densified Laminated Wood for Electrical Purposes—Part 3: Specifications for Individual Materials—Sheet 1: Sheets Produced from Beech Veneer.
90. IEC 61125. 2018. Insulating Liquids—Test Methods for Oxidation Stability—Test Method for Evaluating the Oxidation Stability of Insulating Liquids in the Delivered State.
91. IEC61181:2007+AMD1. 2012. CSV Consolidated Version, Mineral Oil-Filled Electrical Equipment—Application of Dissolved Gas Analysis (DGA) to Factory Tests on Electrical Equipment.
92. IEC 61198. 1993. Mineral Insulating Oils—Methods for the Determination of 2-Furfural and Related Compounds.
93. IEC 61212-3-1. 2013. Insulating Materials—Industrial Rigid Round Laminated Tubes and Rods Based on Thermosetting Resins for Electrical Purposes—Part 3: Specifications for Individual Materials—Sheet 1: Round Laminated Rolled Tubes.
94. IEC TR 61294. 1993. Insulating Liquids—Determination of the Partial Discharge Inception Voltage (PDIV)—Test Procedure.
95. IEC 61619. 1997. Insulating Liquids—Contamination by Polychlorinated Biphenyls (PCBs)—Method of Determination by Capillary Column Gas Chromatography.
96. IEC 61620. 1998. Insulating Liquids—Determination of the Dielectric Dissipation Factor by Measurement of the Conductance and Capacitance—Test method.
97. IEC 61628-2. 2007. Corrugated Pressboard and Presspaper for Electrical Purposes—Part 2: Methods of Test.
98. IEC 61629-1. 1996. Aramid pressboard for electrical purposes Part 1: Definitions, designations and general requirements.
99. IEC 61868. 1998. Mineral Insulating Oils—Determination of Kinematic Viscosity at Very Low Temperatures.
100. IEC 61869-1. 2007. Instrument Transformers—Part 1: General Requirements.
101. IEC 61936-1. 2015. Power Installations Exceeding 1 kV a.c.—Part 1: Common rules.
102. IEC 62021-1. 2004. Insulating Liquids—Determination of Acidity—Part 1: Automatic Potentiometric Titration.
103. IEC 62021-2. 2008. Insulating liquids—Determination of Acidity—Part 2: Colourimetric Titration.
104. IEC 62321-10. 2020. Determination of Certain Substances in Electrotechnical Products—Part 10: Polycyclic Aromatic Hydrocarbons (PAHs) in Polymers and Electronics by gas Chromatography-Mass Spectrometry (GC-MS).
105. IEC TS 62478. 2016. High voltage test techniques—Measurement of partial discharges by electromagnetic and acoustic methods.
106. IEC 62535. 2009. Insulating Liquids—Test Method for Detection of Potentially Corrosive Sulphur in Used and Unused Insulating Oil.
107. IEC 62631-2-1. 2018. Dielectric and Resistive Properties of Solid Insulating Materials—Part 2–1: Relative Permittivity and Dissipation Factor—Technical Frequencies (0,1 Hz–10 MHz)—AC Methods.
108. IEC 62631-3-1. 2016. Dielectric and resistive properties of solid insulating materials—Part 3–1: Determination of resistive properties (DC methods)—Volume resistance and volume resistivity—General method.
109. IEC 62697-1. 2012. Test Methods for Quantitative Determination of Corrosive Sulfur Compounds in Unused and Used Insulating Liquids—Part 1: Test Method for Quantitative Determination of Dibenzyldisulfide (DBDS).
110. IEC TR 62697-3. 2018. Test Methods for Quantitative Determination of Corrosive Sulfur Compounds in Unused and Used Insulating Liquids—Part 3: Test Method for Quantitative Determination of Elemental Sulfur.
111. IEC 62961. 2018. Insulating Liquids—Test Methods for the Determination of Interfacial Tension of Insulating Liquids—Determination with the ring method.
112. IEC 63012. 2019. Insulating Liquids—Unused Modified or Blended Esters for Electrotechnical Applications.
113. IEEE 4-2013. 2013. IEEE Standard for High-Voltage Testing Techniques.

114. IEEE 637-1985. 2007. IEEE Guide for the Reclamation of Insulating Oil and Criteria for Its Use.
115. IEEE 980-2013. 2013. IEEE Guide for Containment and Control of Oil Spills in Substations.
116. IEEE 1276. 2020. IEEE Guide for the Application of High-Temperature Insulation Materials in Liquid-Immersed Distribution, Power, and Regulating Transformers.
117. IEEE C37.015-2017. 2017. IEEE Guide for the Application of Shunt Reactor Switching.
118. IEEE C37.109-2006. 2007. IEEE Guide for the Protection of Shunt Reactors.
119. IEEE C57.12.00. 2015. IEEE Standard for General Requirements for Liquid-Immersed Distribution, Power, and Regulating Transformers.
120. IEEE C57.12.90. 2015. IEEE Standard Test Code for Liquid-Immersed Distribution, Power, and Regulating Transformers.
121. IEEE C57.21. 2012. IEEE Standard Requirements, Terminology, and Test Code for Shunt Reactors Rated over 500 kVA.
122. IEEE C57.91. 2011. IEEE Guide for Loading Mineral-Oil-Immersed Transformers and Step-Voltage Regulators.
123. IEEE C57.93. 2019. IEEE Guide for Installation and Maintenance of Liquid-Immersed Power Transformers.
124. IEEE C57.94. 2015. IEEE Recommended Practice for Installation, Application, Operation, and Maintenance of Dry-Type Distribution and Power Transformers.
125. IEEE C57.100. 2011. IEEE Standard Test Procedure for Thermal Evaluation of Insulation Systems for Liquid-Immersed Distribution and Power Transformers.
126. IEEE C57.104. 2019. IEEE Guide for the Interpretation of Gases Generated in Mineral Oil-Immersed Transformers.
127. IEEE C57.117-1986. 1988. IEEE Guide for Reporting Failure Data for Power Transformers and Shunt Reactors on Electric Utility Power Systems.
128. IEEE C57.125-2015. 2015. IEEE Guide for Failure Investigation, Documentation, Analysis, and Reporting for Power Transformers and Shunt Reactors.
129. IEEE C57.146. 2011. IEEE Guide for Interpretation of Gasses Generated in Silicone-Immersed Transformers.
130. IEEE C57.152. 2013. IEEE Guide for Diagnostic Field Testing of Fluid-Filled Power Transformers, Regulators, and Reactors.
131. IEEE C57.154. 2012. IEEE Standard for the Design, Testing and Application of Liquid-Immersed Distribution, Power, and Regulating Transformers Using High-Temperature Insulation Systems and Operating at Elevated Temperatures.
132. IEEE C57.155. 2014. Guide for Interpretation of Gases Generated in Natural Ester and Synthetic Ester-Immersed Transformers.
133. IEEE C57.637. 2015. IEEE Guide for the Reclamation of Mineral Insulating Oil and Criteria for Its Use.
134. ASTM D93-20. 2020. Standard Test Methods for Flash Point by Pensky-Martens Closed Cup Tester.
135. ASTM D94-17. 2017. Standard Test Methods for Saponification Number of Petroleum Products.
136. ASTM D117-18. 2018. Standard Guide for Sampling, Test Methods, and Specifications for Electrical Insulating Liquids.
137. ASTM D129-18. 2018. Standard Test Method for Sulfur in Petroleum Products (General High Pressure Decomposition Device Method).
138. ASTM D130-19. 2019. Standard Test Method for Corrosiveness to Copper from Petroleum Products by Copper Strip Test.
139. ASTM D149-20. 2020. Standard Test Method for Dielectric Breakdown Voltage and Dielectric Strength of Solid Electrical Insulating Materials at Commercial Power Frequencies.
140. ASTM D235-02. 2010. Standard Specification for Mineral Spirits (Petroleum Spirits) (Hydrocarbon Dry Cleaning Solvent).
141. ASTM D240-19. 2019. Standard Test Method for Heat of Combustion of Liquid Hydrocarbon Fuels by Bomb Calorimeter.

142. ASTM D257-14(2021)e1. 2021. Standard Test Methods for DC Resistance or Conductance of Insulating Materials.
143. ASTM D287-12b(2019). 2019. Standard Test Method for API Gravity of Crude Petroleum and Petroleum Products (Hydrometer Method).
144. ASTM D611-12(2016). 2016. Standard Test Methods for Aniline Point and Mixed Aniline Point of Petroleum Products and Hydrocarbon Solvents.
145. ASTM D664-18e2. 2018. Standard Test Method for Acid Number of Petroleum Products by Potentiometric Titration.
146. ASTM D892-18. 2018. Standard Test Method for Foaming Characteristics of Lubricating Oils.
147. ASTM D923-15. 2015. Standard Practices for Sampling Electrical Insulating Liquids.
148. ASTM D943-20. 2020. Standard Test Method for Oxidation Characteristics of Inhibited Mineral Oils.
149. ASTM D971-20. 2020. Standard Test Method for Interfacial Tension of Insulating Liquids against Water by the Ring Method.
150. ASTM D1169-19. 2019. Standard Test Method for Specific Resistance (Resistivity) of Electrical Insulating Liquids.
151. ASTM D1217-20. 2020. Standard Test Method for Density and Relative Density (Specific Gravity) of Liquids by Bingham Pycnometer.
152. ASTM D1218-12. 2016. Standard Test Method for Refractive Index and Refractive Dispersion of Hydrocarbon Liquids.
153. ASTM D1305-16. 2016. Standard Specification for Electrical Insulating Paper and Paperboard—Sulfate (Kraft) Layer Type.
154. ASTM D1401-19. 2019. Standard Test Method for Water Separability of Petroleum Oils and Synthetic Fluids.
155. ASTM D1481-17. 2017. Standard Test Method for Density and Relative Density (Specific Gravity) of Viscous Materials by Lipkin Bicapillary Pycnometer.
156. ASTM D1534-95. 2017. Standard Test Method for Approximate Acidity in Electrical Insulating Liquids by Color-Indicator Titration.
157. ASTM D1544-04(2018). 2018. Standard Test Method for Color of Transparent Liquids (Gardner Color Scale).
158. ASTM D1816-12. 2019. Standard Test Method for Dielectric Breakdown Voltage of Insulating Liquids Using VDE Electrodes.
159. ASTM D1868-20. 2020. Standard Test Method for Detection and Measurement of Partial Discharge (Corona) Pulses in Evaluation of Insulation Systems.
160. ASTM D1903-08. 2017. Standard Practice for Determining the Coefficient of Thermal Expansion of Electrical Insulating Liquids of Petroleum Origin, and Askarels.
161. ASTM D1934-20. 2020. Standard Test Method for Oxidative Aging of Electrical Insulating Liquids by Open-Beaker Method.
162. ASTM D2007-19. 2019. Standard Test Method for Characteristic Groups in Rubber Extender and Processing Oils and Other Petroleum-Derived Oils by the Clay-Gel Absorption Chromatographic Method.
163. ASTM D2112-15. 2015. Standard Test Method for Oxidation Stability of Inhibited Mineral Insulating Oil by Pressure Vessel.
164. ASTM D2129-17. 2017. Standard Test Method for Color of Clear Electrical Insulating Liquids (Platinum-Cobalt Scale).
165. ASTM D2140-08. 2017. Standard Practice for Calculating Carbon-Type Composition of Insulating Oils of Petroleum Origin.
166. ASTM D2144-07. 2013. Standard Practices for Examination of Electrical Insulating Oils by Infrared Absorption.
167. ASTM D2225-20. 2020. Standard Test Methods for Silicone Liquids Used for Electrical Insulation.
168. ASTM D2300-08. 2017. Standard Test Method for Gassing of Electrical Insulating Liquids Under Electrical Stress and Ionization (Modified Pirelli Method).

169. ASTM D2440-13. 2013. Standard Test Method for Oxidation Stability of Mineral Insulating Oil.
170. ASTM D2502 -14(2019)e1. 2020. Standard Test Method for Estimation of Mean Relative Molecular Mass of Petroleum Oils from Viscosity Measurements.
171. ASTM D2503-92(2016). 2016. Standard Test Method for Relative Molecular Mass (Molecular Weight) of Hydrocarbons by Thermoelectric Measurement of Vapor Pressure.
172. ASTM D2549-02(2017). 2017. Standard Test Method for Separation of Representative Aromatics and Nonaromatics Fractions of High-Boiling Oils by Elution Chromatography.
173. ASTM D2614-21a. 2021. Standard Test Methods for Electrical Conductivity of Aviation and Distillate Fuels.
174. ASTM D2668-07. 2013. Standard Test Method for 2,6-di-tert-Butyl- p-Cresol and 2,6-di-tert-Butyl Phenol in Electrical Insulating Oil by Infrared Absorption.
175. ASTM D2779-92(2020). 2020. Standard Test Method for Estimation of Solubility of Gases in Petroleum Liquids.
176. ASTM D2786-91(2016). 2016. Standard Test Method for Hydrocarbon Types Analysis of Gas-Oil Saturates Fractions by High Ionizing Voltage Mass Spectrometry.
177. ASTM D3238a-17. 2017. Standard Test Method for Calculation of Carbon Distribution and Structural Group Analysis of Petroleum Oils by the n-d-M Method.
178. ASTM D3239-91(2016). 2016. Standard Test Method for Aromatic Types Analysis of Gas-Oil Aromatic Fractions by High Ionizing Voltage Mass Spectrometry.
179. ASTM D3284-05. 2019. Standard Practice for Combustible Gases in the Gas Space of Electrical Apparatus Using Portable Meters.
180. ASTM D3300-20. 2020. Standard Test Method for Dielectric Breakdown Voltage of Insulating Liquids Under Impulse Conditions.
181. ASTM D3426-19. 2019. Standard Test Method for Dielectric Breakdown Voltage and Dielectric Strength of Solid Electrical Insulating Materials Using Impulse Waves.
182. ASTM D3455-11. 2011. Standard Test Methods for Compatibility of Construction Material with Electrical Insulating Oil of Petroleum Origin.
183. ASTM D3612-02. 2017. Standard Test Method for Analysis of Gases Dissolved in Electrical Insulating Oil by Gas Chromatography.
184. ASTM D3635-13. 2016. Standard Test Method for Dissolved Copper in Electrical Insulating Oil by Atomic Absorption Spectrophotometry.
185. ASTM D3756-18. 2018. Standard Test Method for Evaluation of Resistance to Electrical Breakdown by Treeing in Solid Dielectric Materials Using Diverging Fields.
186. ASTM D4056-16. 2016. Standard Test Method for Estimation of Solubility of Water in Hydrocarbon and Aliphatic Ester Lubricants.
187. ASTM D4059-00. 2018. Standard Test Method for Analysis of Polychlorinated Biphenyls in Insulating Liquids by Gas Chromatography.
188. ASTM D4176-21. 2021. Standard Test Method for Free Water and Particulate Contamination in Distillate Fuels (Visual Inspection Procedures).
189. ASTM D4243-16. 2016. Standard Test Method for Measurement of Average Viscometric Degree of Polymerization of New and Aged Electrical Papers and Boards.
190. ASTM D4310-20a. 2020. Standard Test Method for Determination of Sludging and Corrosion Tendencies of Inhibited Mineral Oils.
191. ASTM D4768-11. 2019. Standard Test Method for Analysis of 2,6-Ditertiary-Butyl Para-Cresol and 2,6-Ditertiary-Butyl Phenol in Insulating Liquids by Gas Chromatography.
192. ASTM D5222-16. 2016. Standard Specification for High Fire-Point Mineral Electrical Insulating Oils.
193. ASTM D5837-15. 2015. Standard Test Method for Furanic Compounds in Electrical Insulating Liquids by High-Performance Liquid Chromatography (HPLC).
194. ASTM D5949-16. 2016. Standard Test Method for Pour Point of Petroleum Products (Automatic Pressure Pulsing Method).
195. ASTM D5950-14. 2020. Standard Test Method for Pour Point of Petroleum Products (Automatic Tilt Method).

196. ASTM D6786-15. 2015. Standard Test Method for Particle Count in Mineral Insulating Oil Using Automatic Optical Particle Counters.
197. ASTM D6922-13. 2018. Standard Test Method for Determination of Homogeneity and Miscibility in Automotive Engine Oils.
198. ASTM D7042-21. 2021. Standard Test Method for Dynamic Viscosity and Density of Liquids by Stabinger Viscometer (and the Calculation of Kinematic Viscosity).
199. ASTM D7150-13. 2020. Standard Test Method for the Determination of Gassing Characteristics of Insulating Liquids Under Thermal Stress.
200. ASTM D7155-20. 2020. Standard Practice for Evaluating Compatibility of Mixtures of Turbine Lubricating Oils.
201. ASTM D7647-10. 2018. Standard Test Method for Automatic Particle Counting of Lubricating and Hydraulic Fluids Using Dilution Techniques to Eliminate the Contribution of Water and Interfering Soft Particles by Light Extinction.
202. ASTM D7752-18. 2018. Standard Practice for Evaluating Compatibility of Mixtures of Hydraulic Fluids.
203. ASTM E1269-11. 2018. Standard Test Method for Determining Specific Heat Capacity by Differential Scanning Calorimetry.
204. ISO 2049. 2001. Petroleum Products—Determination of Color (ASTM Scale) (ISO 2049:1996).
205. ISO 2160. 1998. Petroleum Products—Corrosiveness to Copper—Copper Strip Test.
206. ISO 4406. 2021. Hydraulic Fluid Power—Fluids—Method for Coding the Level of Contamination by Solid Particles.
207. ISO 5508. 1990. Animal and Vegetable Fats and Oils—Analysis by Gas Chromatography of Methyl Esters of Fatty Acids.
208. ISO 5509. 2000. Animal and Vegetable Fats and Oils—Preparation of Methyl Esters of Fatty Acids.
209. ISO 5555:2001/AMD 1. 2014. Animal and Vegetable Fats and Oils—Sampling—Amendment 1: Flexitanks.
210. ISO 5660-1. 2020. Reaction-to-Fire Tests—Heat Release, Smoke Production and Mass Loss Rate—Part 1: Heat Release Rate (Cone Calorimeter Method) and Smoke Production Rate (Dynamic Measurement).
211. ISO 5661:1883. 2020. Petroleum Products—Hydrocarbon Liquids—Determination of Refractive Index.
212. ISO 6614. 1994. Petroleum Products - Determination of Water Separability of Petroleum Oils and Synthetic Fluids.
213. ISO 6885:2016. 2021. Animal and Vegetable Fats and Oils—Determination of Anisidine Value.
214. ISO 12185. 1996. Crude petroleum and petroleum products—Determination of density—Oscillating U-tube method.
215. ISO 12937. 2017. Petroleum products—Determination of Water—Coulometric Karl Fischer Titration Method.
216. ISO 14596. 2007. Petroleum Products—Determination of Sulfur Content—Wavelength-Dispersive X-Ray Fluorescence Spectrometry.
217. ISO 31000:2018(en). 2018. Risk Management—Guidelines.
218. OECD Sustainable Manufacturing Indicator. 2011. P3: Renewable Materials Content of Products.
219. OECD 201. 2011. Freshwater Alga and Cyanobacteria, Growth Inhibition Test.
220. OECD 202. 2004. OECD Guideline for Testing of Chemicals, Daphnia Sp., Acute Immobilization Test.
221. OECD 203. 2019. Fish, Acute Toxicity Test, OECD Guidelines for the Testing of Chemicals, Section 2, OECD Publishing, Paris.
222. OECD 301. 1992. OECD Guideline for Testing of Chemicals, Ready Biodegradability.
223. OECD 420. 2002. Acute Oral Toxicity—Fixed Dose Procedure, OECD Guidelines for the Testing of Chemicals, Section 4, OECD Publishing, Paris.

224. EPA. 2009. OPPTS 835.3100 Aerobic Aquatic Biodegradation, Fate, Transport, and Transformation Test Guidelines.
225. EPA. 1998. OPPTS 835.3110 Ready Biodegradability, Fate, Transport, and Transformation Test Guidelines.
226. EPA 9045C. 1995. Soil and Waste pH, Revision 3.
227. AOCS Official Method Cd 1–25. 2019. Iodine Value of Fats and Oils Wijs Method.
228. AOCS Official Method Cd 18–90. 2017. p-Anisidine Value.
229. IUPAC 2.432. 1991. Standard Methods for the Analyzes of Oils, Fats and Derivatives, 1st Supplement of 7th Edition.
230. FM Global DS 5-4. 2019. Transformers, Property Loss Prevention Data Sheet.
231. FM Global DS 5-20. 2020. Electrical Testing, Property Loss Prevention Data Sheet.
232. FM Global 3990. 2021. Examination Standard for Liquid-Insulated Transformers.
233. UL Standard 340. 2017. Standard for Tests for Comparative Flammability of Liquids.
234. NEC Article 110. 2017. Requirements for Electrical Installations.

Chapter 8
Conclusion

Despite their attractive environmental properties and high availability, biological insulating liquids are still in the evolving stage at the time for most practical applications in high voltage engineering. Market and governing pressure are rising to solve the issues and challenges associated with the application of renewable and environmentally friendly insulating liquids to avoid the environmental hazards connected with the application of mineral oil in high voltage equipment. Furthermore, there are huge requirements to enhance effectiveness and approve more eco-friendly opportunities in electric network. Hence, high voltage industry has been looking for superior alternatives and novel concepts.

Besides its technical benefits, biological insulating liquids can have a socio-economic role by promoting income from the sale of seed oils to rural families, contributing to the countryside development, and avoiding rural exodus. Vegetable oils, generally extracted from seeds, which have two primary constituents. The oil portion and the solid segment possessing protein usually named as meal portion. In the case of soya, the meal portion is the main product, and the oil is a byproduct. Biological insulating liquids from vegetable oils like jatropha curcas can be considered as an agriculture alternative to arid and semi-arid regions, being used to recover degraded areas. Another non-edible vegetable oil alternative stems from the pongamia pinnata fruit. Historically, this plant has been used in India and neighboring regions as a source of traditional medicines, animal fodder, green manure, timber, water-paint binder, pesticide, fish poison and fuel.

Insulating liquids are derived from the crude base oil through refining, bleaching and deodorization. Moreover, winterization operation is applied to eliminate readily subzero (°C) saturate fats. The selection of appropriate base material for biological insulating liquids production is a huge challenge and sensitive process which may impact its characteristics.

Electrical analyzes show that biological insulating liquids present higher dielectric breakdown compared to mineral oil liquids, providing better insulation characteristics. In addition, these insulating liquids are not corrosive to copper and increase the

E. P. Pagger et al., *Biological Insulating Liquids*,
https://doi.org/10.1007/978-3-031-22460-7_8

insulation life of cellulose, since they present the ability to absorb moisture of the paper and protect cellulose from thermal aging.

Biological insulating liquids oxidation is a complex drastic chain reaction when oxygen breaks down fatty acids. The degree of oxidation is influenced by light, heat, the fatty acid profile, and the concentration of antioxidants in the liquid. Biological insulating liquids tend to deteriorate when oxidation takes place because of intrinsic unsaturation. These liquids loaded with polyunsaturated fatty acid are higher susceptible to oxidation than organic liquids which are plentiful in monounsaturated fatty acids. Chemical transformations may include fractional hydrogenation of the vegetable oils and a shifting of their fatty acids. Complete hydrogenation of oil may result in solid derivates like margarine. Depending on the required liquidity and pour point of the biological liquid, optimal hydrogenation is specified. Oxidation instability of biological insulating liquids is one of the biggest research challenges that obstruct its practical applications.

Another challenge associated with biological insulating liquids is their higher pour point. The freezing temperature is different for every vegetable oil due to the dissimilar configuration of each oil. A great amount of saturated chemical compounds in oil may augment the freezing point. Unsaturated oils generally possess pour point between -10 and -20 °C. This challenge may be partly handled using chemical additives (pour point suppressants).

Biological insulating liquids have a higher viscosity, and this can be further increased by oxidation. Thus, it is significant to be careful when filling the insulating liquid in transformers to eliminate oxygen. A reduction in viscosity by applying various chemical modifications (transesterification) is possible. The question is whether one can still speak of biological insulating liquids when the chemical structure of the vegetable oil is changed in such a way through the massive use of chemicals and energy. In any case, the information on the viscosity of the biological insulating liquid is necessary to design the heat transfer equipment.

In terms of cost saving, even biological insulating liquids and transformers are more expensive, the removal of ancillary equipment such as fire extinguishers and space reduction can give big savings and very quickly offset the extra capital expense. In addition, there is evidence that Kraft paper will live much longer if immersed in biological insulating liquid compared to mineral oil, and this extra lifetime can significantly reduce overall cost of an installation, if considered the whole lifetime.

During the previous couple of decades, nanotechnology applications have displayed extraordinary development in numerous engineering and scientific areas with nanofluids. Nanofluids are deliberated as a substitute for the contemporary generation of liquid dielectric insulation for high voltage equipment. Nanotechnology has been deliberated to enhance numerous features of renewable and eco-friendly liquid insulation for high voltage equipment. Presently, nanofluids are one of the most explored research fields concerning nanotechnology applications in high voltage apparatus.

Enhancement in the cold performance and oxidation stability along with a decline in viscosity, are leading subjects concerning studies in biological insulating liquids in electrical equipment. Moreover, non-traditional seed oils have been explored to formulate biological insulating liquids containing edible and non-edible plants.

Alternatively, it is projected that a large quantity of service-aged transformers would be deliberated for retrofilling/substituting mineral oil with environmentally friendly insulating liquids because of multiple benefits, for example enhancing its fire security, its consistency, and its rest of useful life.

In the view of environmental hazards, fire safety, health vulnerability, call for carbon-footprint decline, insulating liquids based on vegetable oils are the new generation of insulating liquids that are going to replace mineral oil liquids. Biological insulating liquids score over mineral oil on ecological apprehensions with abundant biodegradability, non-poisonous and are viable, ecological, and sustainable origin with carbon–neutral traits. In addition to environmental favorability, biological insulating liquids are the insulating liquids that have characteristics crucial to high volt equipment.

8.1 Advantages When Using Biological Insulating Liquids (Summary)

- Biological insulating liquids are environmentally friendly and come from renewable resources
- Biological insulating liquids are mixable with mineral insulating liquids in a large range
- Less flammable (K class, FM approved)
- Biodegradable
- More compatible with transformer materials (cellulose)
- Better water absorption
- Less decrease in the degree of depolarization
- An increase in service life of the transformer
- Using at higher temperatures
- With higher temperature, the waste heat from transformers can be better used in urban areas
- Reduction of the insurance premium, FM Global reduced the annual premium when using ester liquids in transformers due to the low fire load
- FM Global and IEC 61936 allow shorter distances between transformers with ester liquids and the building. This results in short busbars between generator and transformer, which means lower busbar losses and less space required.

8.2 Disadvantages When Using Biological Insulating Liquids (Summary)

- The price of biological insulating liquids is generally dependent on the feedstock cost and higher compared to mineral oil
- Storage and management of biological insulating liquids are more problematic and sometimes challenging
- Homogeneity of biological insulating liquids is generally determined by the trader, starting material and fabrication procedures
- Separate equipment for storage and handling on the manufactures workshop is necessary
- The difference between the dielectric and thermal behavior of biological insulating liquids in comparison to traditional mineral oil necessitates very often design modification
- High voltage equipment filled with biological insulating liquids must be airtight to inhibit the entrance of humidity and air into the unit
- Antioxidants must be present in the sealed unit due to the potential inclusion of impurities throughout the lifespan of the unit
- Biological insulating liquids application in high voltage equipment subjected to freezing environments is of serious concern
- The pour point of natural esters in no way proceeds beyond −30 °C even when following the addition of pour point depressants
- Less experience in analytics (Ostwald Coefficient)
- Lack of long-term experience.

8.3 Comparison of Applicability Performance of Different Insulating Liquids

Muhammad Rafiq et al. give in the review "Sustainable, Renewable and Environment-Friendly Insulation Systems for High Voltages Applications" few noteworthy explanations regarding to the applicability of different insulating liquids (Table 8.1).

Eco-friendly liquids are different in physical, electrical, chemical, and thermal characteristics than mineral oil. Consequently, certain design, engineering and functional contemplations should be taken care of when they are implemented in power transformers. Thermal design modification should be mandatory because of the augmentation of top insulating liquid, winding, and core temperatures. One way to do this is by enhancing the size of liquid conduits in coils. Biological insulating liquids possess greater thermal conductivity and, to some extent, larger heat capacity than mineral oil that marginally compensates for the adverse effect of its higher viscosity.

Table 8.1 Workability observations for various insulating liquids

Characteristics	Mineral oil	Synthetic ester	Natural ester
Miscibility with mineral oil	–	Miscible in all percentages	Miscible in all percentages
Appropriateness for cold environments	Acceptable	Acceptable	Questionable
Soluble particles with aging	Yes	Yes (with bigger aging)	Yes (with bigger aging)
Colloidal particles with aging	Yes	No (with lower/moderate aging)	No (with lower/moderate aging)
Major gases created post aging	H_2 and C_2H_2	CO and CO_2	CO and CO_2
Oxidation stability	Acceptable	Acceptable	Questionable
Absorption capacity	Rises briskly with aging	Rises moderately	Initially high
Antioxidants	Required	Required	Strongly required
Gelling	No	No	Partially yes (for breathing units)

Lightening impulse break down voltages of biological insulating liquids may be lower than mineral oil at huge gaps. Thus, broader insulation obstructions are usually deliberated when eco-friendly insulating liquids are applied to high voltage transformers. Alternatively, different from mineral oil, ester-based insulating liquids have permittivity values closer to paper and pressboard insulation that show superior electrical stress dispersal in insulation arrangement (liquid–solid) in high voltage transformers. This electrical stress is inversely proportional to permittivity, so the configurations with greater permittivity carry lesser intensities of stress.

Biological insulating liquids have outstanding features useful for the extension of transformer lifespan, which is mainly due to extensively slower aging of cellulose-based insulating paper in natural esters than mineral oil under similar thermal stress. This performance is partial because of the greater ability of biological insulating liquids to captivate more humidity than mineral oil. Moreover, there are certain indications that free acids emerged by the hydrolysis process may chemically react with cellulose, evolving reinforced, and better substance.

Design novelties and process adaptation for biological insulating liquids are effectively applied by transformer producers to meet greater voltage scale, up to 420 kV. Additional use of biological insulating liquids is retrofilling of transformers service-aged mineral oil. For this application, it is highly significant to consider that retrofilled transformer would have a blend of liquids, biological insulating liquid with small loading of remaining mineral oil. If the mineral oil loading is too large, benefits like fire opposition and biodegradability may be reduced.

Therefore, the authors of this book have tried to explain the different types of biological insulating liquids, their properties and production, as well as their advantages and disadvantages by using these insulation liquids in comparison to mineral oil. The possible uses and influence of these biological insulating liquids on the insulating system of transformers, as well as the status in research and development, are also presented. The aim is to create a better understanding and knowledge of these biological insulating liquids and to provide support in making a decision about their application.

The manufacturer's authorised representative in the EU is Springer Nature Customer Service Centre GmbH, Europaplatz 3, 69115 Heidelberg, Germany. If you have any concerns regarding our products, please contact ProductSafety@springernature.com

Printed and bound by CPI Group (UK) Ltd, Croydon, CR0 4YY
28/09/2026
02224077-0007